运输类飞机英汉词典

An English-Chinese Dictionary
of Transport Airplane

欧阳绍修 李亚萍 主编

西北工业大学出版社

西安

【内容简介】 本词典是以运输类飞机词汇为基础编制而成的航空类专用技术词典,内容涵盖运输类飞机构造、机载设备、动力装置、装配制造、飞行技术及试验、适航及质量、使用维护等航空技术领域专业词汇、缩略语、章节译文及用语,全书共收录各类词条 30 000 余条,供翻译情报人员、设计人员、工艺技术人员及市场售前/后人员使用。

本词典也可供航空业界的工程技术人员、科技管理人员、科技情报研究人员和航空类高校师生等相关人员使用。

图书在版编目(CIP)数据

运输类飞机英汉词典/欧阳绍修,李亚萍主编.
—西安:西北工业大学出版社,2017.12
ISBN 978-7-5612-5642-8

Ⅰ.①运… Ⅱ.①欧… ②李… Ⅲ.①运输机—词典—英、汉 Ⅳ.①V271.2-61

中国版本图书馆 CIP 数据核字(2017)第 288045 号

策划编辑:付高明
责任编辑:张 笛

出版发行:西北工业大学出版社
通信地址:西安市友谊西路 127 号 邮编:710072
电 话:(029)88493844 88491757
网 址:www.nwpup.com
印 刷 者:陕西金德佳印务有限公司
开 本:787 mm×1 092 mm 1/16
印 张:38
字 数:920 千字
版 次:2017 年 12 月第 1 版 2017 年 12 月第 1 次印刷
定 价:218.00 元

《运输类飞机英汉词典》
编委会

主　任　欧阳绍修

副主任　芮书梅　常　畅

委　员　尹晓宏　步恒祚　宋梦宁　何勇波

　　　　李亚萍　赵承周　罗喜东　侯　旺

　　　　黄如刚　黄智勇　崔颢元　易艳彤

《运输类飞机英汉词典》
编写组

主　　　编　　欧阳绍修　李亚萍
主要编写人员　　黄　征　黄　芬　张和平　陈　鑫
　　　　　　　　杨碧娟　刘　欢　陈芳明　黄如刚
　　　　　　　　李建平　商鹿鸣　孟永良　祝雨晴
　　　　　　　　黄天婧　黄　蓉　徐　坚　邓小玲
顾　　　问　　步恒祚

序

历时三年的苦心编撰，《运输类飞机英汉词典》终于要面世了。此项工作的缘起，一方面是基于编者多年在运-8/运-9系列飞机研发工作中的深刻体会，另一方面是由于航空工业日新月异的发展对航空专业英汉词典所产生的迫切需要。

近年来，我国国防建设和航空运输业需求不断增加，运输类飞机的研发与生产取得了令世人瞩目的成就。目前，从事运输类飞机的教学、科研和开发人员不断增多，国际性的合作研究和学术交流也日趋活跃。随着航空业界国际合作中技术交流和技术引进的不断深入，对航空专业英汉工具书的分类需求也愈加趋于精细化、专业化。基于这种现状，为应对运输类飞机专业英汉词典相对匮乏，编撰了这部《运输类飞机英汉词典》。

《运输类飞机英汉词典》共收录各类词条 30 000 余条，内容全面涵盖运输类飞机随机技术资料中的基础词汇、缩略语、随机资料目录清单、章节译文及用语。同时，结合编者在大中型运输机研发工作中积累的经验，该词典还收录了近年来在前沿航空文献中大量出现的新词汇。该词典编撰的宗旨是将《运输类飞机英汉词典》编写成一本专业性强、内容形式新颖、材料翔实的运输类飞机技术工具书。

《运输类飞机英汉词典》主要服务于航空业界的翻译情报人员、设计人员、工艺技术人员及市场销售人员，也可供航空业界的工程技术人员、科技管理人员、科技情报研究人员和航空类高校师生等相关人员使用，还可普及到与运输类飞机相关领域的从业人员。该词典兼具实用意义和科学意义。实用意义在于它有益于广大读者；科学意义在于它确定了英语和汉语丰富的词汇之间相对的唯一对应性，从而保证恰当地将英文翻译成中文，尽量减小英、中两种语言之间存在类型上差别而造成歧义的可能性。同时，英语是事实上的国际通用语，汉语是使用人口最多的语言，两种语言在国际上的影响力不分伯仲。该词典所收录的丰富词汇、准确严谨的表达，在一定程度上可以促进人们对两种语言的理解和交流，也能提升两种文化的渗透与融合。

编撰《运输类飞机英汉词典》期间，曾得到多位业界专家、教授的良好建议和大力帮助，在此一并致谢。同时，由于科技不断发展、技术领域的不断扩大、相关学科的不断交叉，以及编者的水平等因素，本书不足之处，欢迎读者批评指正。

欧阳绍修

2017 年 6 月

前　言

随着全球航空产品进出口贸易的迅速发展，以运输类飞机及其外围产品为代表的对外技术合作和交流日益增多，相关专业的广大技术人员对英语的依赖程度越来越高。在日益频繁的对外航空技术交流活动中，无论情报翻译人员，还是航空产品设计、制造、维护工程师和市场销售经理，都深感专业航空英汉词典的重要性。为满足运输类飞机等航空产品技术研究的需要，同时规范随机用户技术资料翻译标准，我们编制、出版了《运输类飞机英汉词典》。

《运输类飞机英汉词典》是一本以运输类飞机航空技术装备专业词汇为主，并适当吸收航空理论等专业术语的综合性词典，共收录 30 000 余词条，内容全面涵盖运输类飞机相关专业词汇、缩略语、随机资料清单、章节译文及用语。

《运输类飞机英汉词典》的编撰原则是"注重词典的先进性、确定性和实用性"。为此，在词条选取上，以当前国内外运输类飞机及其配套航空技术装备为基础进行词条甄选，并参考国内运-8 型系列飞机，以及国外波音、空客、C-130 等国外同类型飞机使用的专业术语，力求全面。此外，为保证词典的先进性，对航空技术发展中出现的最新词汇进行吸收，具体涵盖飞机设计、制造、试验、使用、维护等专业领域。同时，为确保用词准确、规范，在英文词条的选择上，遵循国际上主流通用的用法，摒弃冷僻的词汇和用法；在中文词条的译法上，所选择的术语词汇与现行国家标准、国家军用标准规定及航空界通用的用法相一致，并对不明确的用法进行了规范。

在词典的编排上，为了方便读者使用，除飞机基础词汇部分外，还增加了缩略语、用户技术资料专业词汇等内容以及英文索引。

经过不懈的努力以及很多同行的建议和帮助，完成了《运输类飞机英汉词典》的编撰工作。鉴于词典编撰工作的复杂性，加之水平和时间有限，其中难免存在不妥、遗漏之处，在此恳请读者批评指正，以便后续修订。

编　者

2017 年 6 月

体 例 说 明

一、飞机基础词汇为词典正文，以 ATA100 规范的各章进行划分，各章按照英文字母的顺序独立排序。词条由英文、汉语译文两部分组成。其中，英文后括号里内容为缩略语。

二、缩略语部分，按照英文字母顺序排序。词条由缩略语、全称、汉语译文三部分组成。修饰词、关联词在缩写时省略。为避免歧义，根据习惯用法，有些缩略语取单词中一个以上字母。

三、随机资料清单和随机资料用语中包括 21 种随机资料清单、参考用语。

四、维修手册章节和维修手册用语为维修手册章节名称、参考用语。

五、缩略语至维修手册用语的全部内容由序号、中文、英文三部分组成。

六、英文索引部分由按字母顺序排序的英文、页码两部分组成。其中，页码之间的"/"表示不同的页码，即该英文出现在不同的页码中。

七、词典中各种符号按以下顺序排序：空格、"&"".""/"、阿拉伯数字等。符号"－"不参加排序，当两个词条的字母组合完全相同时，含有符号"－"的排在后。

八、词典中的"/"为两种不同的表达；合成词之间用"－"连接；词条之间的","表示并列的词义；文中圆括号内的字词表示：①必要的注释或说明；②可以省略的内容；③缩略语；④特定的国家、军种专用；⑤文中方括号内的词可替换方括号前面的词。

九、词典中英文部分参考简化英语的通用规则及使用惯例。

目 录

飞机基础词汇

TIME LIMITS/MAINTENANCE
时限/维护

abnormal condition　非正常条件
abnormity　异常现象
accessory　附件
accumulator　蓄电瓶
actuator　作动筒
aileron　副翼
airborne electrical equipment　机载电气设备
airborne equipment　机载设备
aircraft control system　飞机操纵系统
aircraft registered number　飞机注册号
airtightness　密封性
airtightness check　气密检查
alignment　对齐
antenna　天线
anti-collision light　防撞灯
approval　批准
aramid fiber　芳纶
attachment fitting　固定接头
axle　轮轴
bearing　轴承
bearing cover　轴承盖
bend　弯曲
bird flock　鸟群
bird impact　鸟撞
blade　桨叶
blister　气泡
bolt　螺栓
bonding jumper　搭接线
braking device　刹车装置
buffer　缓冲器
buffet　抖振
bulge　鼓包
bulkhead　隔框
bump　颠簸
calibration　校准
canopy skeleton　舱盖骨架

carbon fiber　碳纤维
cargo door　货舱门
center of gravity　重心
center wing　中央翼
check　检查
circuit breaker　断路器
clearance condition　净空条件
clearance value　间隙值
cockpit　驾驶舱
compass　罗盘
component　部件
composite material　复合材料
conductor　导体
connecting bolt　结合螺栓
connecting mechanism　连接机构
connecting part　连接件
connecting position　连接部位
connector　接头
control device　操纵装置
control steel cable　操纵钢索
control surface　操纵面/舵面
control surface lock　舵面锁
control wiring　控制线路
cord fabric layer　帘线层
cowling　整流罩
crack　裂纹
crease　褶皱
damage　损坏
damaged zone　受损区
decolorization　脱色
decoration　装饰
deformation　变形
delamination　剥层
dent　凹陷
design center of gravity　设计重心
disassemble　分解
discriminatory criterion　判别标准
door mechanism　舱门机构
door-closing system　舱门关闭系统

door-operating mechanism　舱门操纵机构

downlock　下位锁

drum wheel　鼓轮

edge strip　边条

bonding resistance　搭接电阻

electrical equipment　电气设备

elevator　升降舵

elongation　拉长

emergency exit　应急出口

emergency window　应急窗口

engine mounting　发动机架

engine nacelle　发动机短舱

entry door　登机门

entry point　进口点

original equipment manufacturer　设备制造厂家

excessive braking　过量刹车

excessive control　过度操纵

exit point　出口点

exterior antenna　机外天线

exterior damage　外部损伤

fairing　整流包皮

fastener　紧固件

fixed bracket　固定支架

flap　襟翼

flight document　飞行文件

frame　框

fuel　燃油

fuel control unit　燃油调节器

full load　满载

full load running speed　满载滑跑速度

fuselage　机身

fuselage longitudinal line　机身纵向轴线

gale　狂风

generator　发电机

generator control panel　发电机控制板

glass fiber reinforced plastic　玻璃钢

grounding wire/earthing wire　接地线

hail　冰雹

hard landing　粗暴着陆

high-energy braking　高能制动

hinge　铰链

horizontal stabilizer　水平安定面

horizontal wing tip　平尾翼尖

hose　软管

hub spinner/hub fairing　桨毂整流罩

hydraulic fluid　液压油

hydraulic pipeline　液压管路

impact　撞击

inboard wing　中外翼

inclined bracing　斜撑杆

inner structure　内部结构

inspection　检验

instant wind speed　瞬时风速

instrument　仪表

instrument equipment　仪表设备

intake　进气道

jack　千斤顶

jack-up　顶起

landing gear　起落架

landing gear structure　起落架结构

landing light　着陆灯

latch groove　锁销槽

leading edge　前缘

leakage　漏油

level measurement　水平测量

lightning accident　雷击事故

lightning current　雷击电流

lightning rod　避雷条

lightning stroke　雷击

load characteristic　载荷性质

locking mechanism　上锁机构

loose　松动/松开

lower section　下部

magnetic compass　磁罗盘

magnetic heading reference system　磁航向基准系统

magnetize　磁化

magnifying glass　放大镜

main landing gear　主起落架

maintenance　维护

maintenance check　维护检查

mandatory life limit　强制性寿命限制

maximum overload　最大过载

mid fuselage　机身中段

misplace　错位

motion mechanism　运动机构

mounting edge　安装边

nacelle end　短舱末端

navigation cabin　领航舱

navigation light　航行灯

navigation system　导航系统

needle bearing　滚针轴承

nose　机头

nose landing gear　前起落架

nose wheel steering mechanism　前轮转弯机构

oil　滑油

on condition check　视情检查

outboard wing　外翼

overhaul interval　翻修时限

overheat　过热

overload　超载

overload landing　超载着陆

overload strength　过载强度

panel　壁板/面板

parameter　参数

peel-off　剥落

periphery structure　周边结构

pit　凹点

power distribution system　配电系统

power feeder　电源馈线

power plant　动力装置

pressure charging　充压量

propeller　螺旋桨

propeller tip　螺旋桨桨尖

pulley　滑轮

pulley bracket　滑轮支架

radar antenna　雷达天线

radio communication　无线电通讯

radio communication equipment　无线电通讯设备

radio equipment　无线电设备

radome　雷达罩

raid of hail　冰雹袭击

rear fuselage　机身尾部

rear spar　后梁

recognition　认可

recommendation　推荐

record　记录

red line mark　红线标志

reinforced frame　加强框

replace　更换

retractable actuator　收放作动筒

retraction/extension test　收放试验

rib　肋

rivet　铆钉

rocker arm　摇臂

rotor　转子

rudder　方向舵

safety factor　安全系数

sandwich plate　夹层板

scheduled maintenance check　定期维护检查

severe towing　剧力牵引

shock absorbing connector　减震接头

shock absorbing device　减震装置

shock absorbing strut　减震支柱

skin　蒙皮

stagnation　卡滞

static discharger　静电放电器

static tube　静压管

static-pressure hole　静压孔

steel structure part　钢结构件

stipulation　规定

stringer　桁条

strong airflow　强气流

structural factor　结构系数

structure piece　结构碎片

strut　支柱

support bracing　支臂

support bracket　支架

suspension connector　悬挂接头

suspension mount　悬挂架

swing　扳动

takeoff abortion　起飞中断

tension　张力

terminal switch　终点开关

thermal plug　热熔塞

time limit　时限

torque arm　防扭臂

transport aircraft　运输机

transverse beam　横梁

trim tab　调整片

two-arm mandrel　双臂紧轴

tyre　轮胎

tyre blown-out　轮胎爆破

unscheduled maintenance check　不定期维护检查

uplock　上位锁

vertical stabilizer　垂直安定面

water-cooling　水冷

waveguide tube　波导管

weight　重量

welding position　焊接处

wheel　机轮

wheel hub　轮毂

wheel rim　轮缘

windshield　风挡

windshield glass　风挡玻璃

wing　机翼

wing root　机翼根部

wing tip　机翼翼尖

DIMENSIONS AND AREAS
尺寸和区域

AC generator changeover box　交流发电机转换盒

accelerometer　加速度表

access cover　（出入)口盖

access facility　接近装置

accumulator battery　蓄电瓶

acquisition signal light　捕获信号灯

actuator　舵机/作动筒

aileron　副翼

aileron position feedback sensor　副翼位置反馈传感器

air bleed switch　输气开关

air conditioning pipeline　空调管路

air data computer　大气计算机

air separator　空气分离器

air speed indicator　空速表

air start switch　空中起动开关

air-bleed valve　放气活门

aircraft component　飞机构件

air-inlet duct　进气道

air-inlet guider　进气导向器

airtight cabin　气密舱

airtight cable　气密电缆

airtight connector　气密接头

aisle　过道

alarm cut off button　警铃断开按钮

altimeter　高度表

altitude differential pressure gauge　高度压差表

ammeter　电流表

amplifier　放大器

amplifier control　放大控制盒

angle of attack　攻角

angular velocity　角速度

antenna　天线

anti-icing heating pipe　防冰加温导管

anti-icing valve　防冰开关

aspect ratio　展弦比

assembly　组件

atmosphere thermometer　大气温度表

audio control box　音频控制盒

auto-feathering interlocking device　自动顺
桨联锁装置

automatic fuel consumption controller　自动
耗油控制器

aviation clock　航空时钟

aviation watch　航空计时表

balancing valve　连通开关

ball compensator　球形补偿器

barometric solenoid valve　气压电磁开关

base　底座

battery　电瓶

bearing　轴承

bilateral hydraulic lock　双向液压锁

blanking cap　堵帽

blocking annunciator　堵塞信号器

brace　撑杆

brace joint　撑杆接头

bracket　支座

brake clearance　刹车间隙

brake plate　制动片

brake pressure valve　刹车压力阀

braking cable　制动钢索

brightness regulator　亮度调节器

buffer strut　缓冲支柱

bulge　鼓包

bus bar　汇流条

butt attachment frame　对接框

butt bolt　对接螺栓

butt comb part　对接梳状件

butt position　对接处

butt section　对接型材

butt socket　对接插座

button lock catch　按钮锁扣

cabin altitude pressure difference gauge　座

舱高度压差表

capacitor　电容

carbon pile regulator　炭片调压器

cargo cabin　货舱

cargo emergency hatch　货舱应急窗

casing　机匣

cathedral angle　下反角

ceiling panel　顶板

center instrument panel　中央仪表板

center wing　中央翼

center wing fairing　中央翼整流罩

centerline　中心线

centrifugal switch　离心开关

changeover switch　转换开关

changeover valve　转换活门

charging pressure gauge　充气压力表

charging switch　充气开关

check button　检查按钮

check switch　检查开关

check valve　单向活门

chord line　弦线

chord plane　弦平面

circuit breaker　断路器/自保开关

cloud lock　云形锁

coarse filter　粗油滤

cockpit　驾驶舱

comb plate　梳状板

combination valve　组合活门

combustion chamber　燃烧室

common dome light　公用顶灯

compensator　补偿器

component　元件

connection hose　连接软管

console　操纵台

constant pitch line　定距油路

consumption indicator　耗油表

contactor　接触器

control box　操纵盒/控制盒

control cable　操纵钢索

control display　控制显示器

control surface　舵面

control switch　操纵开关

control system　操纵系统

cover plate　盖板

cross valve　交叉活门

cutoff valve　断流活门

damper　减震器

deflection angle　偏角

deflection angle sensor　偏角传感器

delay valve　延迟活门

differential under-voltage relay　差动低限保护器

dihedral angle　上反角

dimming potentiometer　亮度调节电位器

display process unit　显示处理部件

DME indicator　DME 显示器

dome light　顶灯

dorsal fin　背鳍

drain valve　放油开关

driving mechanism　传动机构

driving rod　传动杆

drop control box　空投操纵盒

dryer　干燥器

dual check valve　双单向活门

dual-lug rocker arm　双耳摇臂

dynamic pressure annunciator　动压信号器

electric cable　电气电缆

electric heating mechanism　加热电动机构

electrical mechanism/electric mechanism　电动机构

electrical transfer pump　电动输油泵

electronic clock　电子时钟

elevator　升降舵

elevator trim tab deflection mechanism　升降舵调整片偏转机构

emergency access　应急窗口

emergency access cover　应急口盖

emergency power supply box　应急电源盒

emergency pressure release switch　应急释压开关

encoder　编码器

engine automatic shutdown device　发动机自动停车器

engine mounting　发动机架

engine nacelle　发动机短舱

engine shutdown switch　发动机停车开关

engine starting check relay　发动机起动检查继电器

engine terminal box　发动机接线盒

engine vibration G-load indicator　发动机振动过载指示器

entry door　登机门

even number　偶数

exhaust gas temperature sensor　排气温度传感器

exhaust pipe　排气管

exhaust valve　排气活门

exit hatch　出入门

fairing　整流罩

fan switch　风扇开关

fault light　故障灯

feathering button　顺桨按钮

feathering pump　顺桨泵

feathering ready indication light　顺桨准备好指示灯

feathering relay box　顺桨继电器盒

fill plug　注油塞

filler opening　注油口

fillet　整流包皮

filter　滤波器

fine adjustment box　精调盒

fine filter　细油滤

fine pitch line　小距油路

fire bottle　灭火瓶

fire pipe　防火导管/灭火导管

fire switch　防火开关

fire warning sensor　火警传感器

fire-extinguishing control 灭火控制

fire-extinguishing microswitch 灭火微动开关

fire-extinguishing relay box 灭火继电器盒

fitting 接头

flap 襟翼

flap position sensor 襟翼位置传感器

flap retraction/extension mechanism 襟翼收放机构

flight data recorder 飞行数据记录器

flight date 航班日期

float valve 浮子活门

flow distribution valve 流量分配活门

flow indicator 示流器

flow regulating valve 流量调节活门

fluorescent light 荧光灯

follow-up device 随动器

follow-up pull rod 随动拉杆

foot button 脚踏按钮

foot heating switch 脚加温开关

forced landing 迫降

formation light 编队灯

four-hole socket 四孔插座

frame location 框位

front fitting 前接头

front spar fitting 前梁接头

fuel filter 燃油滤

fuel pressure sensor 燃油压力信号器

fuel pump regulator 燃油泵调节器

fuel quantity meter 油量表

fuel quantity sensor 油量表传感器

fuel solenoid valve 燃油电磁活门

fuel supply pump 供油泵

fuse 熔断器

fuse seat 保险丝座

fuselage 机身

fuselage axial 机身轴线

fuselage geometry 机身几何形状

general distribution box 总配电盒

generator 发电机

generator current divider 发电机分流器

generator power-on switch 发电机接通开关

glass heating pipe 玻璃加温导管

GRD-AIR changeover switch 地面-机上转换开关

ground power supply 地面电源

grounding cone 接地锥

guider 导向器

gyro group 陀螺组

hand pump 手摇泵

heating switch 加温开关

heating warning control 加温报警控制(盒)

high altitude oxygen annunciator 高空用氧信号器

high frequency cable 高频电缆

horizontal tail 平尾

hub spinner/hub fairing 桨毂整流罩

hydraulic feathering switch 液压顺桨开关

hydraulic fluid quantity sensor 液压油量表传感器

hydraulic lock 液压锁

hydraulic motor 液压马达

hydraulic oil filling filter 液压加油滤

hydraulic oil return filter 液压回油滤

hydraulic pipe 液压导管

hydraulic reservoir 液压油箱

hydraulic reservoir pressurization 液压油箱增压

hydraulic switch 液压开关

icing annunciator 结冰信号器

identification area 识别区

identifying number 识别号

ignition coil 点火线圈

ignition plug 点火电嘴

illumination light 照明开关

inboard wing 中外翼

incandescent light 白炽灯

incidence angle 安装角

inclined beam　斜梁

indicating light　指示灯

indicator　指示器

inertia fuse　惯性熔断器

initial breakdown　设计分离面

injection switch　引射开关

inlet pipe　进气管

inlet port　进气口

inspection heating changeover switch　检查加温转换开关

intercom　机内通话器

intermediate rocker arm　中界摇臂

inverter　变流机

key　电键

landing gear　起落架

landing gear compressed state　起落架压缩状态

landing gear free state　起落架自由状态

landing gear well　起落架舱

leading edge　前缘

leading edge fairing　前缘整流罩

lighting controller　照明控制器

lighting rheostat　照明变阻器

link　连杆

load-bearing segment　承力段

lock base　锁底座

lock sleeve　锁套

low altitude vent electric valve　低空通风电动活门

lower access cover　下盖

lower emergency door　下应急门

lower fuselage　机身下半部

magnesium alloy　镁合金

magnetic heading　磁航向

magnetic induction sensor　磁感应传感器

main air-bleed pipe　引气总管

main landing gear　主起(落架)

main shaft　主轴

main wheel tread　主轮距

maintenance access　维护舱口

maintenance access cover　维护口盖

maintenance bulkhead　维护隔板

maintenance door　维护舱门

major area　主要区域

manual control　手操纵

master/standby changeover switch　备用转换开关

mean aerodynamic chord　平均空气动力弦长

medium zone　中等区域

microswitch　微动电门

mounting board　安装板

mounting bracket　安装支架

multimeter　三用表

navigation compartment　领航舱

navigator　领航员

navigator controller　领航员控制器

negative thrust　负拉力

nose fairing　机头罩

nose wheel　前轮

odd number　奇数

oil drain port　放油口

oil pressure annunciator　滑油压力信号器

oil pump　滑油泵

oil-mist separator　油雾分离器

operating light　工作灯

operating nozzle　工作喷嘴

ornament plate　装饰板

outboard wing　外翼

outlet port　排气口

over station signal light　过台信号灯

overall height　停机高度

overall length　飞机长度

overvoltage protector　过压保护器

oxygen consumption indicating light　用氧信号指示灯

oxygen filling pipe　充氧管

oxygen pressure gauge　氧气压力表

oxygen regulator　氧气调节器

oxygen source gauge　氧源压力表

oxygen supply check button　用氧检查按钮

oxygen switch　氧气开关

panel　面板

parking angle　停机迎角

parking compression　停机压缩

part　零件

phase coupler　相位耦合器

pilot instrument panel　左仪表板

pin sleeve　销套

pipe connector　导管接头

pitot-static pressure system　全静压系统

pitot-static tube　全静压管

placard　标牌

portable light　手提灯

power distribution board　电源配电盘

power limiter　功率限制器

power supply changeover contactor　电源转换接触器

power terminal box for tail anti-icing　尾翼防冰左电源接线盒

precipitator　沉淀器

pressure annunciator　压力信号器

pressure gauge　压力表

pressure measure selector switch　电压测量选择开关

pressure reducer　减压器

pressure refueling nozzle　压力加油接头

pressure regulating shutoff valve　压力调节关断活门

pressure sensor　压力传感器

pressurization pipe　增压导管

principal dimension　基本尺寸

probe　探头

propeller anti-icing heating　螺旋桨防冰加温

propeller braking distribution box　螺旋桨制动配电盒

propeller stop release mid-pitch valve　螺旋桨解除中距限动活门

protective cover　保护盖

pull rod　拉杆

pulley mount　滑轮(支)架

radiator　散热器

radio compass　无线电罗盘

radio set　电台

radome　雷达罩

ramp door　货桥大门

rear cargo door　货舱后大门

receiver　接收机

refueling controller　加油控制器

refueling power distribution box　加油配电盒

refueling pump　加油泵

refueling system　加油系统

refueling valve　加油开关

regulator　调节器

remote switch　遥控开关

reserved cabin　备用舱

residual fuel pipe　余油管

residual fuel tank　余油箱

resistance box　电阻盒

retractable light　伸缩灯

rheostat　变阻器

rib　翼肋

rib location　肋位

rocker arm　摇臂

rocker arm assembly　摇臂组

rocker arm bracket　摇臂支架

rubber pipe　橡胶管

rudder actuator valve　舵机活门

rudder control　舵操纵

rudder lock mechanism　方向舵锁机构

rudder spring pull rod　方向舵弹簧拉杆

safety valve　安全活门

sealing tape sleeve　密封带套管

secondary zone　次级区域

selector switch　选择开关

sheet code　板件代号

sheet drawing 板件图

shield sleeve 防波套

shield wire 防波导线

shock absorbing strut 减震支柱

shunt valve 分路活门

signal flare 信号弹

signal gun 信号枪

single arm access cover lock 单臂口盖锁

single-phase converter 单向变流机

sliding rail 滑轨

small fairing door 小护板

solenoid valve 电磁活门

solenoid valve of starting fuel supply 起动供油电磁活门

spar 梁

spoiler 扰流板/扰流片

spring trim tab 弹簧力补偿片

starter power-off switch 起动机断电开关

starter/generator 起动发电机

starting box 起动箱

starting indication light 起动指示灯

starting regulation box 起动调节盒

static tube 静压管

steel cable 钢索

steering radius 转弯半径

steering wheel 驾驶盘

stringer 长桁

structural check 结构检查

structural component 结构部件

support frame 支撑框

switch-on ground power supply 接通地面电源

tachometer 转速表

tachometer sensor 转速表传感器

tail 尾翼

tail cabin 尾舱

tail cowling 尾罩

tail section 尾端

tailpipe 尾喷管

takeoff weight 起飞重量

tapered cantilever high wing 张臂式梯形上单翼

temperature compensator 温度补偿器

temperature thermal meter 温度表温度计

terminal 端子

terminal board 接线板

terminal box 接线盒

terminal post 接线柱

terminal switch 终点开关

three-digit number 三位数

three-phase converter 三相变流机

throttle lever 油门杆

tightening part 锁紧件

timing device 定时机构

top view 俯视

torque auto-feathering sensor 扭矩自动顺桨传感器

torque meter 扭力表

trailing edge 后缘

transceiver antenna 收发天线

transformer 变压器

trim tab 调整片

turbine blowing cover 涡轮吹风罩

two-arm rocker 双臂摇臂

unfeathering solenoid valve 回桨电磁活门

upper access cover 上盖

upper fuselage 机身上半部

vacuum valve 真空活门

vent pipe 通气管

vent system 通气系统

vent tank 通气油箱

vent valve 通风活门

ventilation drain valve 通气放油开关

vertical accelerometer 垂直加速度计

vertical speed indicator 升降速度表

vertical stabilizer area 垂直安定面面积

vertical tail span 垂尾展长

vertical tail/fin 垂尾

vibration meter　振动表

vibrator　振动仪

voltage dropping resistor　降压电阻

voltage dropping resistor contactor　降压电阻接触器

voltage regulation rheostat　电压调节变阻器

voltage regulator　电压调节器

voltmeter　电压表

wave-guide　波导

wheel base　前主轮距

winch　绞车

windshield framework　风挡玻璃窗骨架

windshield wiper　风挡刷

wing box　翼箱

wing span　翼展

wing sweep angle　机翼后掠角

wing tip　翼尖

work range　工作范围

LIFTING AND SHORING
顶起和支撑

accessory　附件

actuator　作动筒

adjustable height　调整高度

adjustable thread rod　调节螺杆

air-bleed valve　放气阀门

airborne equipment　机载设备

airframe　机体

alignment　对准

auxiliary thread rod　辅助螺杆

aviation grease　航空润滑脂

aviation hydraulic fluid　航空液压油

axis　轴线

ball socket　球窝

ball-head support thread rod　球头支承螺杆

base　底座

bearing　轴承

bolt　螺栓

bonding jumper　搭地线

brace rod　撑杆

brake　刹车

case　外壳

caution　注意

center of gravity　重心

check valve　单向活门

clamp base　卡座

clamp cover　卡盖

clearance　间隙

clockwise　顺时针

cloth cover　布套

component　部件

compression bush　压紧衬套

console　操纵台

counter-clockwise　逆时针

damage　损伤

damageable part　易损零件

disassembly　分解

drain valve　放油阀门

exposed quantity of thread　螺纹外露量

failure　故障

filler opening　加油口

filter　过滤器

flat end　扁平端

foreign object　外物

frame　框

friction operating surface　摩擦工作表面

front bracing　前撑杆体

front cabin　前舱

front safety top bar　前保险顶杆

fuselage　机身

grooved end　开槽端

ground equipment kit　地面设备放置箱

hand pump　手摇泵

handle　手把

heater　加温器

horizontal condition　水平状态

hydraulic control box　液压操纵箱

hydraulic reservoir　液压油箱

hydraulic system　液压系统

ice block　冰块

inboard wing　中外翼

inclined force　倾斜力

inner cavity　内腔

inner cylinder of jack　千斤顶内筒

jack　千斤顶

kerosene　煤油

knuckle bearing　关节轴承

level measurement　水平测量

lifting　顶起

lifting height　顶起高度

lifting load　顶起载荷

lifting mechanism　升降机构

lifting stroke　升程

lifting switch　顶起开关

limit height　极限高度

link　连杆

load　载荷

load/unload　装卸

long-distance transport　长途运输

low temperature grease　低温润滑油

lubricate　润滑

lug（fork）　耳叉

lug seat　耳座

main landing gear　主起落架

main piston rod　主活塞杆

main technical data　主要技术数据

maintenance practice　维修实施

manual nut　手动螺母

maximum diameter　最大直径

maximum height　最大高度

member　构件

minimum height　最低高度

moisture　水分

movable knuckle　活动关节

movable support head　活动支承头

nominal pressure　公称压力

non-operating condition　非工作状态

nose　机头

nose jack　机头千斤顶

nose landing gear　前起落架

nose wheel fork　前轮轮叉

nose wheel support　前轮支架

nut　螺母

obstruction　障碍物

oil return valve　回油开关

operating performance　工作性能

operating pressure　工作压力

operating principle　工作原理

operating regulation　使用规则

outer cylinder　外筒

outer of nut　螺母外体

part　零件

performance　性能

pin　销子

pipeline　管路/油路

piston　活塞

piston cylinder　活塞缸

piston rod　活塞杆

placard　标牌

plate　板片

plumb　测锤

positioning　定位

precaution　注意事项

pressure release valve　卸压开关

pressurization　增压

pressurization valve　增压开关

pull rod　拉杆

rated lifting height　额定起升高度

record of revision　更改记录

regulating nut　升降螺母

retraction/extension mechanism　收放机构

rib　肋

rocker arm　摇臂

routine maintenance　日常维护

safety nut　保险螺母

safety valve　安全活门

schematic diagram　原理图

scratch　划伤

screw　螺钉

screw limiter　螺钉限动器

screw plug　螺塞

screw-lifting mechanism　螺旋升降机构

sealing　密封

sealing ring　密封圈

sediment　沉淀物

semi-circle cover plate　半圈盖板

service ladder　工作梯

shoring　支撑

shroud　套罩

single-direction thrust ball bearing　单向推力球轴承

sleeve　套筒

spheral head support　球头支承

spheral lug　带球耳片

stop nut　止动螺母

stop pin　止动销

stop shaft　止动轴

storage　存放

stroke　行程

structure　构造

support base　支撑底座

support bolt　支承螺栓

supporting disk　支撑盘

supporting load　支撑载荷

supporting point　支承点

tail　尾部

tail cover　尾盖

tail jack　尾部千斤顶

tail support　尾部支撑

tank　油箱

thread　螺纹

thread rod　螺杆

tightening screw　紧定螺钉

top cap　顶帽

towing　牵引

towing bar　牵引杆

triangular frame　三角形构架

trouble shooting　排除故障

tyre　轮胎

union　接管嘴

vent bolt　通气螺栓

vent screw　通气螺钉

vernier potentiometer　油标计

vertical condition　垂直状态

vibration　振动

volume　容积

volume of reservoir　贮油量

washing hole　洗涤孔

weight　重量

weld　焊接

wheel　机轮

wheel axle　轮轴

wheel chock　轮挡

wheel jack　机轮千斤顶

wheel retracting/extending actuator　轮子收放作动筒

wind speed　风速

wing　机翼

wing jack　机翼千斤顶

working fluid　工作油

working medium　工作介质

LEVELING AND WEIGHING
调平与称重

abnormal phenomenon　异常现象

acceleration sensor　加速度传感器

accessory　附件

actual takeoff weight　实际起飞重量

actual weight　实际重量

adjusting bolt　调整螺栓

aileron　副翼

airborne component　装机(状态)部件

airborne equipment　机上设备

aircraft attitude　飞机姿态

airdropping platform component　空投货台装机件

airfoil　翼面

airframe　机体

alignment　对准

angle of attack（AOA）sensor　攻角传感器

arc clamp　弓形夹

armed soldier transporting airborne component　运送武装士兵机载部件

arrangement　布置

assembly　零组件

attain　获得

auxiliary tool　辅助工具

avoiding　避免

balance　平衡

balance arm　平衡臂

balance tab　补偿片

balance weight　配重

basic definition　基本定义

basic operating weight　基本操作重量

bearing force　承受力

blanking cover/blanking cap　堵盖

boarding　登机

bogie　车架

bonding jumper　搭地线

buffer strut　缓冲支柱

bulk cargo　散装

calculation formula　计算公式

calibrate　校正

cancel　取消

captain　机长

cargo　货物

cargo cabin　货舱

cargo cabin/compartment door　货（舱）大门

cargo loading mark line　装货线标记

caution　注意

center of gravity　重心

chassis/base　底座

check　检查

chock　挡好（轮挡）

clean　清除

clearance　间隙

cockpit　驾驶舱

column　栏

common scale　专用称

communicator　通讯员

component　部件

compress　压缩

configuration　配置

constant　常数

contour　外形

control mechanism　操纵机构

coordinate　坐标/协调

correctness　正确性

cowling　整流罩

crane　吊车

crew member　机组

damage　损坏

data　数据

datum line/reference line　基准线

datum point/reference point　基准点

deflection angle　偏角

deflection of inclined strut　斜支柱挠度

deflection position　偏度位置

deliver　交付

description/operation　说明和工作

design　设计

dihedral angle　上反角

disengagement　（千斤顶）脱离

dispart　脱离

distance　距离

drainage　放掉

draining port　排泄口

drawing No.　图号

effectiveness　有效性

electrical control　电气操纵

electronic weighing scale　电子称

elevator 升降舵

empty (weight) aircraft 空机

empty weight 空机重量

engine 发动机

ensure 确定

equipment 设备

equipment list 设备表

exceeding 超出

exposed quantity 外露量

extended section 延伸部分

external point/outer point 外点

fabric cover 蒙布

field 外场

fill in 填写

finished product 成品

fixed weight 固定重量

fixing pin 固定销

flap 襟翼

flight engineer 机械师

flight test 试飞

foreign object 外物

format 格式

frame 框

front/rear spar 前/后梁

fuel 燃油

fuel consumption 燃油消耗

full fuel 满油

full loading 完全加载

fuselage 机身

geometric measurement 几何测量

ground crew 地勤(人员)

group zero tank 零组油箱

group zero tank without refueling 零组不加油

gyro 陀螺

hardness 硬度

heading 航向

height difference 高度差

hoist bucket 吊斗

hub spinner 桨帽

hydraulic fluid quantity 液压油量

hydraulic jack 液压千斤顶

hydraulic lifting ladder 液压升降梯

hydraulic reservoir 液压油箱

hydraulic weighing unit 液压称

in parallel 平行

inboard engine 内发

inboard wing 中外翼

incidence angle 安装角

incline 歪斜

index 指数

indication 读数

indoor 室内

initial value 初始值

installation 装机

instrument panel 仪表板

internal point/inner point 内点

jack 千斤顶

jack socket 千斤顶窝

jack-up 顶起(飞机)

jet nozzle 喷管

landing deflection 着陆偏角

landing gear 起落架

landing weight 着陆重量

lateral leveling 横向调平

left/right utility hydraulic system 左/右通用液压系统

level measuring apparatus (level meter) 水平测量仪

leveling measurement 水平测量

leveling measuring ruler 水平测量尺

limit 限制

linen thread 亚麻线

list 清单

load 载荷

load sheet 载荷表

loading balance control procedure 载重平衡控制程序

loading balance sheet　载重平衡图表

lock　锁

log book　履历本

longitudinal leveling　纵向调平

magnetic field　磁场

manual　手册

marking　标记

material　材料

maximum takeoff weight　最大起飞重量

mean aerodynamic chord　平均空气动力弦

mean value　平均值

measurement data　测量数据

measuring item　测量项目

measuring point　测量点

measuring tool　测量工具

mechanic weighing scale　机械称

modification quality　改装质量

moment　力矩

mount/dismount　安装/拆卸

movable control surface　活动操纵面

movable tab　活动片

multiple　倍数

nacelle　短舱

navigation cabin　领航舱

navigator　领航员

neutral position　中立位置

nondestructive testing manual　无损检测手册

nose　机头

nose landing gear well　前起落架舱

nose wheel　前轮

nose/main landing gear　前/主起起落架

note　注释

observe　观察

offset　偏移

oil　滑油

oil tank　滑油箱

open end spanner　开口扳手

operating item　使用项目

operating lamp　工作灯

operating weight　使用重量

operation and maintenance manual/service instruction book　使用维护说明书

operation procedure　使用程序

operation regulation　使用规定

outboard engine　外发

outdoor　室外

overload　过载

oxygen　氧气

panel　壁板

parachuting airborne component　空降装机状态部件

parallelism　平行度

personnel　人员

pilot　驾驶员

plain view drawing/plane chart　平面图

plummet　铅锤

position measuring instrument/position level instrument　物位测量仪

power plant　动力装置

precaution　注意事项

precision　精度

preparations　准备工作

preset level　预调量

pressure gauge　压力表

principle　原理

procedure　步骤(程序)

profile　(发动机)外形

prohibition　禁止

project　投影

propeller　螺旋桨

propeller hub　桨毂

protective cover　保护套

pull rod　拉杆

qualified　合格

raise/fall　升高/降低

ramp door　货桥大门

range　量程

rear door 后大门

recheck 复查

record 记录

reduced scale coefficient 缩比系数

redundant/absent equipment 多装/少装设备

reference surface 基准表面

refueling 加油

reinforced frame 加强框

reinforced groove 加强槽

removal 拆卸

report 报告

requirement 要求

residual fuel 余(燃)油

residual oil 剩余滑油

restore 恢复

retraction/extension 放/收(起落架)

rudder 方向舵

series number（S/N） 序号

safety nut 保险螺母

screw downwards 下旋

select 选择

service ladder 工作梯

signal flare 信号弹

site 现场

sketch/schematic drawing 示意图

special state 特定状态

specify 规定

specimen 实例

spoiler 扰流板/扰流片

spring 弹簧

station 站位

steel cable control 钢索操纵

steel tape 钢卷尺

sum 总和

support concave 顶窝

survival kit 救生物品

symmetry axis 对称轴

symmetry plane 对称面

symmetry 对称性

tail 尾翼

tail cabin 尾舱

tail support 尾撑

take off 取下

take out 取出

takeoff deflection 起飞偏角

technical document 技术文件

tolerance 公差

torsion difference 扭转差

trapped fuel 死油

trim tab 调整片

trouble/failure/fault/malfunction 故障

unit 单位

unload 卸载

usable fuel 可用燃油

valid period 有效期

verify 核定

vertical deflection 垂直偏斜

vibration source 振动源

view M M 向

weighing scale 称重仪

wheel 机轮

wheel axle 轮轴

whole aircraft 全机

wind speed 风速

wing 机翼

wing rib 翼肋

wing tank 机翼油箱

zero fuel 零油

zero position 零位

TOWING 牵引

(steel) cable 钢索

accumulator 蓄压瓶

aluminum casting 铸铝块

balance weight 配重

bearing 轴承

bogie 轮架

brake distance 刹车距离

buffer strut 缓冲支柱

build-in axle 包轴式

castor 回转轮

caution 注意

centering mechanism 纠偏机构

clamp 卡箍

cloth cover 蒙布

commander 指挥员

communication 通讯

compression of tyre 轮胎压缩量

control column 操纵杆

corrosion 锈蚀

cross section 横断面

damper 缓冲器

destructive tension 破坏拉力

diameter of cable 钢索直径

dirt 脏物

double-lug fitting 双耳接头

dust 尘土

empennage 尾翼

exposed quantity 外露量

fastening condition 紧固情况

fixed small wheel 固定小轮

fork 叉子

fork joint 叉形接头

fracture 断裂

friction 摩擦

frictional surface 摩擦面

front baffle 前挡板

graphited calcium-base grease 石墨钙润滑脂

ground crew 地勤人员

grounding brush 接地刷

grounding wire/earthing wire 接地线

hand wheel 手轮

hangar 机库

high-temperature grease 高温润滑脂

hook 挂钩

inner cylinder 内筒

leveling 调平

lift range 升降范围

lifting 顶起

lifting wheel bracket 升降轮架

lubricating grease 润滑脂

lubricating oil 机油

lug 耳环

main rod 主体杆

main technical data 主要技术数据

maintenance practice 维修实施

mandrel 心杆

mounting point 安装点

movable knuckle 活动关节

mud pit 泥坑

navigation 导航

nose landing gear 前起落架

nose wheel steering bracket 前轮操纵架

nose wheel steering guide device 前轮操纵导向装置

nut-cover 螺帽挡盖

observer 观察员

obstacle 障碍物

orange yellow enamel paint 橘黄色磁漆（瓷漆）

parking apron 停机坪

pin 销子

pit 洼坑

pull force 拉力

pull ring 拉环

pulley 滑轮

quick-release stop pin 快卸止动销

release braking 解除刹车

removable connection 可拆连接

removal 卸下

rubber block 橡胶块

rubber piece 橡胶片

safety bolt 保险螺栓

servicing　保养

shaft rod　轴杆

sharp turning　急转弯

shear force　抗剪力

shed　棚

shock absorbing rubber ring　减震橡胶圈

shock load　冲击载荷

shore　支撑

sleeve　套筒

spiral lifting mechanism　螺旋升降机构

stop block　限动块

stop limit position　限动极限位置

structure　构造

tail forward　机尾朝前

tail support　尾部支撑

taxiing　滑行

tensile safety bolt　受拉保险螺栓

tensile strength　抗拉强度

towing　牵引

towing cable　牵引钢索

towing device　牵引装置

towing ring　牵引环

towing vehicle　牵引车

turning clamp　转弯卡箍

turning safety bolt　转弯保险螺栓

weighing　称重

wheel　机轮

wheel axle　轮轴

wheel chock　轮挡

PARKING, STORAGE AND RETURN TO SERVICE
停放、保存和恢复使用

access cover　口盖

accessory　附件

accessory casing　附件机匣

accumulation　堆积

accumulator　蓄压瓶

actuator　作动筒

aileron　副翼

air conditioning　空调

air inlet　进气道(口)

air pressure　气压

air temperature　气温

airborne equipment　机载设备

aircraft surface　飞机表面

aircrew　机组/空勤人员

audio　(声)音频

automatic oil temperature regulator　（滑油）温度自动调节器

automatic release display diaphragm　自动释放显示薄膜

auxiliary pump　辅助燃油泵

baffle　挡板

base　底座

battery　电瓶

bearing　轴承

blade angle　桨叶角

blanking cover/blanking cap　堵盖

bolt　螺栓

bonding jumper　搭地线/搭接线

bracing　撑杆

brake　刹车

brake lever　刹车杆

buffer　缓冲

buffer strut　缓冲支柱

butt joint strip plate　对接带板

cabin door　舱门

cable drum　钢索鼓轮

cargo cabin　货舱

cargo door　货舱门

cargo ramp　货桥

center wing　中央翼

chamfer fairing　包角片

charging nozzle　充气嘴

check　检查

cloth covering　蒙布

cockpit door　驾驶舱门

collide　撞击

condensation product　冷凝物

connector　接头

contact　接触

control cable　操纵钢索

control handle　操纵手柄

control surface　舵面(控制面/操纵面)

corrosion　锈蚀

cover　罩

cover plate　盖板

crack　裂纹

crew seat　机组座椅

cut-off valve　断油活门

damage　损坏

deaerator　空气分离器

decoration panel　装饰板

depreservation　启封

description　说明

designated position　规定位置

designation　指定

dip stick　油尺

disposal　排除(处置、安排)

double-layer glass　双层玻璃

downlock　下位锁

drainage　排出

edge　边缘

ejection hole　喷射孔

electric power station　充电站

elevator　升降舵

elevator driving shaft　升降舵传动轴

elimination　排除

emergency door　应急门

emergency window　应急窗

engine　发动机

engine nacelle　发动机舱

entry door　登机门

exceeding　超过

exhaust tailpipe　排气尾管

exposure　暴露

exterior　外部

fairing　整流罩

feathering pump　顺桨泵

filler cap　加油盖

fire bottle　灭火瓶

fire wall　防火墙

fixing clamp　固定卡箍

flap　襟翼

flap sliding rail　襟翼滑轨

flight record　飞行记录

foldable bracing　可折撑杆

fraction　片段

front cabin　前舱

fuel　燃油

fuel baffle device　挡油装置

fuel consumption meter sensor　耗量表传感器

fuel drain cock　放油开关

fuel filter　燃油滤

fuel regulating pump　燃油调节泵

fuel solenoid valve　燃油电磁阀

fuel tank　燃油箱

fuselage　(飞机的)机身

fuselage tail section　机身尾段

gale　大风

glass　玻璃

ground temperature　地面温度

hand and foot control mechanism　手脚操纵机构

handle　手柄

honeycomb　蜂巢

horizontal tail　平尾

hub spinner/hub fairing　桨毂整流罩

hydraulic accessory　液压附件

hydraulic fluid　液压油

hydraulic hose　液压软管

hydraulic reservoir　液压油箱

icing　结冰

igniter　电嘴

impact　碰伤

impurity　杂质/杂物

inboard wing　中外翼

inlet guide vane　导向器叶片

installation　安装

instruction book　说明书

instrument　仪表

landing gear　起落架

landing gear well　起落架舱

lavatory　卫生间

leading edge　前缘

leakage　渗漏

limit　极限

limit block　限动块

liquid　液体

lock　锁

lock hook　锁钩

lock latch　锁扣

lock mechanism　锁定机构

lock pin　锁销

looseness　松动

lower bracing　下撑杆

lubricating grease　润滑脂

magnesium alloy　镁合金

main fuel pump　主燃油泵

maintenance　维护

mechanical damage　机械损伤

metal　金属

mid-rear bearing　中后轴承

mooring　系留

movable clearance　活动间隙

movable knuckle　活动关节

movable surface　活动面

movable wing surface　活动翼面

noise　噪音

nose landing gear　前起落架

nose wheel　前轮

oil　滑油

oil pump　滑油泵

oil radiator　滑油散热器

oil tank　滑油箱

oil vent pipe　滑油通气管

oil/fuel filter　油滤

oil/fuel vent tank　通气油箱

oil-mist separator　油雾分离器

operating direction　工作方向

operating nozzle　工作喷嘴

operating procedure　工作程序

outboard wing　外翼

outer cylinder　外筒

overhaul period　检修期

paraffin paper　石蜡纸

parking　停放

pipe　导管

piston　活塞

piston rod　活塞杆

pitot-static tube　动静压管

portable fire bottle　手提灭火瓶

position　位置

power-on check　通电检查

preservation　油封

pressure gauge　压力表

pressure value　压力值

pressurized cabin　增压舱

procedure　程序

prohibition　禁止

propeller　螺旋桨

propeller blade　螺旋桨桨叶

protection cloth　防护罩布

pull rod　拉杆

quantity　数量

radar　雷达

radio　无线电

radome　雷达罩

refueling　加油

regulating turnbuckle　调节螺套

regulation　规定

removable panel　可卸壁板

removal　拆下

residual oil pipe　余油管

retraction/extension actuator　收放作动筒

rivet　铆钉

rod　杆

rotation　转动

rotor　转子

route　线路

rudder　方向舵

safety　安全

scavenge filter　回油滤

scratch　擦伤

screen　网罩

screw　螺钉

sealing rubber　密封橡胶

sealing rubber pad　密封胶垫

sealing rubber pipe　封严胶管

sealing strip　密封胶带

seam　接缝

section　节

sediment　沉淀物

sediment trap　沉淀槽

separator joint　分离接头

shaft　轴

shock absorber strut　减震支柱

short time parking　短停

shutter　风门

signal sensor　信号传感器

skin　蒙皮

sliding rail　滑轨

speed governor　调速器

stagnation　卡滞现象

start　起动

starting igniter　起动点火器

starting valve　起动活门

static discharge　静电释放

static discharger　放电刷

stop pin　限动钉

storage　保存

strut　支柱

support rod　支撑杆

surface　表面

tail cabin　尾舱

tail fairing　尾整流罩

tail support　尾撑杆

tailpipe　尾喷管

talcum powder　滑石粉

three-stage turbine vane　三级涡轮叶片

throttling　节流

tie down　系好

torque arm　防扭臂

torque sensor　扭矩传感器

triangle inclined bracing　三角斜撑杆

turbine cooler　涡轮冷却器

turbine starter generator　涡轮启动发电机（装置）

tyre　轮胎

vacuum valve shroud　真空活门外罩

value　量/值

valve　活门

vapor　蒸汽

vent handle　通风手柄

vent port　通气口

vertical fin　垂尾

waste　污物

welding part　焊接件

welding seam　焊缝处

wheel axle　轮轴

wheel chock　轮挡

wheel hub　轮毂

window　窗

windshield wiper　风挡雨刷

wing　机翼

wing leading edge　机翼前缘

wing tip　翼尖

PLACARDS AND MARKINGS
标牌和标志

access cover　口盖

accumulator　蓄压瓶

airline　航空公司

blank page　空白页

bottom view drawing　仰视图

bracket　托架

brake accumulator　刹车蓄压瓶

cargo ramp　货桥

caution　注意

change　变动

color strip　彩带

containerized cargo placard　集装货运标牌

description　说明

earthing wire　搭地线

elevator　升降舵

engine Ⅰ　Ⅰ发

engine nacelle　发动机短舱

exterior　外部

flap　襟翼

frame　框

fuel filler　油枪

fuselage　机身

general information　一般常识

horizontal tail　平尾

hydraulic reservoir　液压油箱

inboard wing　中外翼

interior　内部

jack　千斤顶

loading placard　装货标牌

location　位置

main landing gear　主起落架

maintenance　维护

maintenance description　维护说明

manufacturer　制造商

marking　标记

matrix　字模

movement　移动

nose landing gear　前起落架

note　注释

operation　使用

operation precaution　操作注意事项

outboard wing　外翼

paint　喷涂

painting color　喷漆色

photographic method　照相法

propeller　螺旋桨

right view drawing　右侧视图

rivet　铆接

rudder suspension lug　方向舵吊耳

safety　安全

sales contract of aircraft　售机合同

schematic diagram　示意图

special note　特别注明

top view drawing　俯视图

tread　踩压

user's insignia/aircraft insignia　用户标志（机徽）

vertical fin　垂尾

warning　警告

wing　机翼

SERVICING　保养

abrasive　磨料

access cover　口盖

accessory　附件

accessory cavity　附件腔

accessorygear box　附件传动机匣

accidental striking　偶然撞击

accumulator　蓄压瓶

accuracy check　准确性检查

acetone　丙酮

acidity　酸值

actual measurement　实测

actuating mechanism　执行机构

actuator　作动筒

adapter cylinder　转接筒

adjustment　调整

aileron actuator　副翼舵机

aileron balance tab　副翼补偿片

aileron differential rocker arm　副翼差动
摇臂

aileron lock　副翼锁

aileron pull rod joint　副翼拉杆接头

aileron rocker arm　副翼摇臂

aileron trim tab　副翼调整片

air bleed valve　放气活门

air bottle　气瓶

air extractor　抽风装置

air inlet shutter　进气风门

air leakage　漏气

aircrew　空勤人员

aircrew oxygen system　空勤用氧系统

aircrew seat　空勤座椅

airtight joint　气密接头

airtight joint of frame　框气密接头

airtightness　气密性

ambient temperature　大气温度

antenna abutment　天线支座

anti-explosion medium　防爆介质

anti-explosion system　防爆系统

appearance　外观

arc plate　弧形板

auto fuel consumption　自动耗油

automatic pressure refueling　自动压力加油

automatic unloading valve　自动卸荷活门

auxiliary oil pump　辅助滑油泵

auxiliary power unit　（APU)辅助动力装置

aviation hydraulic fluid　航空液压油

aviation lubricating oil　航空润滑油

bag tank　软油箱

balance tab　补偿片

balance tab control system　补偿片操纵系统

basic performance　基本性能

bearing　轴承

bearing bush　轴瓦

bearing cover　轴承盖

blanking cover/blanking cap　堵盖

block　卡住/堵塞

blow away　吹掉

bottle body　灭火瓶体

box member　盒形件

bracing　撑杆

bracking rocker arm　制动摇臂

brake accumulator　刹车蓄压瓶

braking pull rod　制动拉杆

braking system　刹车系统

bromide　溴化物

brush　毛刷

bubble　泡沫

buffer accumulator　缓冲蓄压器

buffer type　缓冲器型别

bushing　衬套

cabin　座舱

cabin pressure　机舱压力

canvas　帆布

carbon dioxide　二氧化碳

carbon dioxide bottle　二氧化碳瓶

carbon dioxide gas　二氧化碳气体

cargo door　货舱大门

cargo ramp actuator　货桥作动筒

cargo ramp door　货桥大门

cargo ramp lock actuator　货桥锁作动筒

casing/case　壳体

certificate　合格证

chain wheel　链轮

changeover switch　转换开关

changeover valve　转换活门

charge　充填

charging (of fire bottle)　（灭火瓶的)充填

charging amount　充气量

charging connector　充气接头

charging date　充填日期

charging equipment　充气设备

charging fixture　充填夹具

charging hose　充填软管

charging material　填充物

charging medium　充气介质

charging nozzle　充气嘴

charging quantity　填充量

charging room　填充间

charging unit　充气装置

charging valve　充气活门

charging valve hole seat　充气活门孔座

charging weight　充填重量

check date　检验日期

check for operation reliability　可靠性检查

check time　检验时间

check unit　检查装置

chrominum-plated pull rod surface　镀铬拉杆表面

circle azimuth gear　圆形方位齿轮

circulation section　流通截面

clamp　夹钳

cleaning　清洗

cleaning agent solution　清洗剂溶液

cleanliness　清洁度

cockpit front emergency door　驾驶舱前应急门

collar　套圈

combat flight　战斗飞行

combined lock mechanism　联合锁机

communication system　通讯系统

communicator seat　通迅员座椅

compressed air　压缩空气

compression force　压紧力

compressor casing/compressor case　压气机机匣

condensation point　凝点

conical flare tube　锥形喇叭管

connecting bearing　连接轴承

connecting bolt　连接螺栓

connecting sheet　连接片

connections　连接工作

console　操纵台

control mechanism　操纵机构

control pull rod　操纵拉杆

control pulley mechanism　操纵滑轮机构

control steel cable　操纵钢索

control stick/control column　驾驶杆

control surface lock control system　舵面锁操纵

coordination valve　协调活门

copper cable antenna　铜索天线

cotton cloth　棉布

cotton overall　棉织工作服

counterforce　反作用力

cover　罩盖

crack　裂纹

cutoff valve　断流活门

cylinder sliding block　圆柱滑块

dead fuel　不可用燃油

deflection limiter　偏角限制器

deformation　变形

dehydrated carbon dioxide　脱水二氧化碳

deicing　除冰

deicing fluid　除冰液

deleterious gas　有害气体

density　密度

de-pressurize　失密

diameter　直径

differential pressure　压力差

dip stick　油尺

discharge tube　放出管

disconnecting mechanism　断开机构

dissolving　溶解

distribution of fuel tank　油箱分布图

door pull rod　舱门拉杆

drain hose　放油软管

drain threaded plug　放油螺塞

drain valve　放油开关

draw back　抽回

driving gear　传动齿轮

driving rod　传动杆

driving rod bracket　传动杆支架

dropping pipe　滴液管

drum wheel　鼓轮

dryer　干燥装置

dual-lug rocker arm　双耳摇臂

dust-proof cover　防尘盖

effective certificate　有效合格证件

ejector radiation electric mechanism　引射散热电动机构

elbow nipple　弯管嘴

elbow union　弯管头

electric heating water tank　电加温水箱

electric mechanism　电动机构

electric mechanism extension rod　电动机构伸出杆

electric mechanism pull rod　电动机构拉杆

electric winch　电动铰车

electrohydrualic actuator　电液舵机

elevator　升降舵

elevator lock　升降舵锁

elevator rocker arm　升降舵摇臂

elevator rocker tube　升降舵摇管

elevator suspension lug socket　升降舵悬挂耳座

elevator trim tab control system　升降舵调整片操纵系统

elevator trim tab deflection mechanism　升降舵调整片偏转机构

emergency brake　应急刹车

emergency oxygen supply　应急供氧

emergent descent　紧急下滑

empty bottle　空瓶

empty bottle weight　空瓶重量

engine air intake anti-icing valve　发动机进气道防冰活门

engine control　发动机操纵

engine fixing joint　发动机固定接头

engine internal fire extinguishing system　发动机内部灭火系统

engine nacelle　发动机短舱

engine nacelle pulley bracket　发动机滑轮支架

engine throttle control lever　发动机油门操纵杆

environment control system　环控系统

equipment/furnishing system　装备/设备系统

extraction/extension mechanism　收放机构

expanding space　膨胀空间

expansion　膨胀

exposed quantity　外露量

external cleaning　外部清洁

extinguishant　灭火剂

extinguisher　灭火器

fabric covering　蒙布

feathering pump　顺桨泵

feathering standby oil quantity　顺桨备用油量

fill　加注

filler plug　注油塞

filler port　加油口

filling　灌充

filter element　过滤元件

filter screen　滤网

fire　（失）火

fire bottle　灭火瓶

fire bottle inner cavity　灭火瓶内腔

fire extinguishing equipment　灭火设备

fire extinguishing system　灭火系统

fire source　火源

fire/explosion　火灾/爆炸

fire-proof performance　防火性能

firing pin　撞针

fix　固定

fixed oxygen system　固定式用氧系统

fixture　夹具

flag mark　旗标

flammable abluent　易燃清洗剂

flap control system　襟翼操纵系统

flap retraction/extension mechanism　襟翼
收放机构

flash light　手电筒

flash point　闪点

flight control system　飞行操纵系统

flight engineer seat　空中机械师座椅

float valve　浮子活门

flow distribution valve　流量分配活门

fluid　油液

flush　冲洗

foldable inclined strut　可折斜支柱

follow-up mechanism　随动机构

foot pedal　脚蹬板

fork rocker arm　三叉摇臂

forward cabin floor door　前舱地板门

forward cabin floor door lock actuator　前舱
地板门锁作动筒

freezen　冻结

freezing point　冰点

front emergency door　前应急舱门

front fairing　前整流包皮

fuel　燃油

fuel consumption level　耗油油面

fuel consumption sequence　耗油顺序

fuel cutoff valve　燃油切断阀

fuel density　燃油比重

fuel level　油面

fuel measuring assembly　燃油测量组合件

fuel pressure　油压

fuel quantity gauge　油量表

fuel regulating pump　燃油调节泵

fuel sediment　燃油沉淀物

fuel supply　供油

fuel supply pump　供油泵

fuel supply system　供油系统

fuel system　燃油系统

fuel tank group　油箱组

fuel truck　加油车

fuel-oil accessory　燃油-滑油附件

fuselage tank　机身油箱

garage equipment　机库设施

gasoline　汽油

glass fiber reinforced plastic oxygen bottle
玻璃钢氧气瓶

glove　手套

grease gun　滑脂枪

greasing nozzle　注油嘴

ground AC and DC power supply　地面交流
和直流电源

ground hydraulic connector　地面液压接头

ground oil truck　地面滑油车

ground oxygen charging cart　地面充氧车

ground refueling connector　地面加油接头

grounding wire/earthing wire　接地线

guaranteed period　保险期

hand-electric pump system　手摇-电动泵
系统

handle　手柄

hand wheel　手轮

head case　头部壳体

heat screen　隔热屏(装置)

heat source　热源

heater　加热器

heating　加温

heating time　加温时间

heating unit　取暖装置

height　高度

high frequency（HF）radio set　高频电台

high temperature lubricating grease　高温润
滑脂

high-pressure flusher　高压冲洗器

high-pressure nitrogen bottle　高压氮气瓶

high-pressure oxygen bottle　高压氧气瓶

high-pressure pneumatic bottle　高压冷气瓶

hinging joint　铰接接头

hose　软管

hose connector　软管接头

hot shower　热水淋浴

housing screw/compression screw　压紧螺钉

hydraulic actuator　液压舵机

hydraulic drain access cover　液压放油口盖

hydraulic filter　液压油滤

hydraulic fluid　液压油

hydraulic fluid charging gun　液压油枪

hydraulic fluid container　盛油容器

hydraulic fluid level height　油面高度

hydraulic fluid quantity　加油量

hydraulic fluid suction connector　吸油接头

hydraulic fluid supply container　补油容器

hydraulic jack　液压千斤顶

hydraulic reservoir　液压油箱

hydraulic system　液压系统

ice and rain protection system　防冰、防雨系统

immerse　浸湿

impurity　杂质

inclined abutment　斜支座

inclined strut　斜支柱

indicated pressure　指示压力

indication　示值

industrial alcohol　工业酒精

industry nitrogen　工业用氮气

inlet pipe　进油导管

inner cavity (of airtight joint)　（气密接头）内腔

inner cylinder　内筒

inspection department　检验部门

inspection stamp　检印

installation　安装

installation position　安装位置

instrument panel　仪表板

jack　千斤顶

jack-up　顶起

joint bearing bush　接头轴瓦

kerosene　煤油

kinematic viscosity　运动粘度

knuckle bearing　关节轴承

landing gear fairing　起落架整流罩

landing gear system　起落架系统

lead seal　铅封（印）

leak　渗出

leak pipe　漏油管

left fuel supply system　左供油系统

left system reservoir　左系统油箱

level sliding rail　水平滑轨

lifting cylinder　升降筒

lifting latch　升降插销

light alkaline soapsuds　轻碱性肥皂水

living facility cabinet　生活设备柜

locating pin　定位销

lock base　锁座

lock hook　锁钩

lock key shaft　锁键轴

lock mechanism　锁机构

lock mechanism part　锁机零件

lock mechanism reset claw　锁机构复位爪

lock wire　保险丝

locking valve　锁闭活门

long-handle broom　长柄扫帚

low pressure gauge　低压压力表

low temperature lubricating grease　低温润滑脂

low-altitude vent valve　低空通风活门

lower braking pull rod　下刹车拉杆

lower spring pull rod　下弹簧拉杆

lower torque arm　下防扭臂

lubricant (grease)　润滑剂（脂）

lubricant type　润滑剂类型

lubricating method　润滑方法

lubricating period　润滑周期

lubricating position　润滑部位

lubrication 润滑

lug base 耳座

main landing gear door 主起落架舱门

main landing gear shock strut 主起落架缓冲支柱

main landing gear suspending ring 主起吊环

main pulley 主滑轮

main shock strut 主缓冲支柱

manual fuel consumption 人工耗油

manual pressure refueling 人工压力加油

manual refueling quantity 人工加油量

mass 质量

measure 措施

measurement 测量

mechanical impurity 机械杂质

mechanical indication rod 机械指示杆

medical oxygen standard 医学用氧标准

metal chip 金属屑

methanol 甲醇

middle connecting bolt 中部连接螺栓

mid-rear bearing 中后轴承

mixture 混合油

mounting bracket bearing 安装支架轴承

movable hinge 活动铰链

movable oxygen bottle 活动氧气瓶

navigation system 导航系统

navigator seat 领航员座椅

neutral gas bottle 中性气体瓶

neutral gas system 中性气体系统

neutral position 中立位置

neutral soapsuds 中性肥皂水

nitrogen 氮气

nitrogen cavity 氮气腔

nitrogen charging pressure 充氮压力

nitrogen charging unit 充氮设备

nitrogen pressure 氮气压力

normal operation 正常工作

normal parking compression value 正常停机压缩量

normal pressure 正常压力

nose jack 机头千斤顶

nose landing gear suspension ring 前起吊环

nose landing gear wheel bearing 前起机轮轴承

nose landing gear wheel shaft 前起机轮轮轴

nose shock strut 前缓冲支柱

nose wheel 前轮

nose wheel steering actuator 前轮转弯作动筒

nose wheel steering clamp 前轮转弯卡箍

nose wheel steering control cable 前轮转弯操纵钢索

nose wheel steering follow-up mechanism 前轮转弯随动机构

occupant oxygen system 乘员用氧系统

oil 滑油

oil assembly 滑油组合件

oil filter 滑油滤

oil nozzle 滑油喷嘴

oil pump 滑油泵

oil quantity gauge 滑油油量表

oil quantity indicator 油量表指示器

oil radiator 滑油散热器

oil radiator shutter 滑油散热风门

oil scavenge filter 回油滤

oil scavenge stage 抽油级

oil system 滑油系统

oil tank 滑油箱

oil-air separator 油气分离器

oiler 注油器

open fire source 明火

open shelter 露天敞棚

open wooden lookum 露天木棚

open-face of guide clamping plate 导向夹板开口面

operating fluid 工作液

operating pressure 工作压力

operating union 工作接管嘴

operation power supply 工作电源

operation temperature 使用温度

operator 操作人员

outer casing nut 外套螺母

oxygen charging cover 充氧口盖

oxygen charging field 充氧现场

oxygen charging filler 充氧嘴

oxygen charging hose 充氧软管

oxygen charging instrument panel 充氧仪表板

oxygen charging operator 充氧人员

oxygen charging port wrench 充氧口扳手

oxygen charging pressure 充氧压力

oxygen charging valve 充氧开关

oxygen equipment 氧气设备

oxygen flow indicator 氧气示流器

oxygen hose 氧气软管

oxygen part 氧气部件

oxygen pressure reducer 氧气减压器

oxygen regulator 氧气调节器

oxygen source pressure gauge 氧源压力表

oxygen system 氧气系统

oxygen valve 氧气开关

packing box 包装箱

pad 垫子

paint stripping position 掉漆部位

parking apron 停机坪

parking area 停放区域

parking brake 停机刹车

part 机件

partition 隔板

penetration 渗入

performance test/check 性能测试/检查

periodic check 定检期限

petroleum basis refined mineral oil 石油基精制矿物油

physical and chemical performance 物理及化学性能

pilot seat 驾驶员座椅

pin 销子

pipe 导管

pipe shaft 管轴

piston 活塞

piston rod 活塞杆

placement 放置

pneumatic bottle 冷气瓶

pointer 指针

poison 中毒

polish 抛光

portable carbon dioxide fire bottle 手提式二氧化碳灭火瓶

portable explosion-proof light 防爆手提式照明灯

portable oxygen system 便携式氧气系统

power distribution board 配电板

power plant 动力装置

power supply switch 电源开关

precise instrument oil 精密仪表油

predetermined pressure value 预定压力

pressing rod 压杆

pressing valve 压紧活门

pressing-through state 压通状态

pressure annunciator 压力信号器

pressure dropping value 压力下降值

pressure gauge 压力表

pressure measuring unit 测压装置

pressure refueling 压力加油

pressure refueling cap 压力加油口盖

pressure refueling quantity 压力加油量

pressure refueling signal 压力加油信号

pressure releasing 释压

pressure supply system 供压系统

pressure-charging 充压

pressure-charging check switch 充压检查开关

pressure-charging unit 充压设备

pressurization hose 增压软管

pressurization pipe connector 增压导管接头

pressurization stage　增压级

pressurization system　增压系统

pressurized cabin　气密舱

propeller blade　螺旋桨桨叶

pulley　滑轮

pulley bracket assembly bearing　滑轮支架组件轴承

pully bearing　滑轮轴承

pump　油泵

pump pressure pulse　油泵压力脉动

push rod　顶杆

pyrophoric cap　引火帽

quick-release stop pin　快卸止动销

rated value　额定值

rated weight　额定重量

reaction force　作用力

rear cargo door　货舱后大门

rear cargo door actuator　后货舱大门作动筒

rear fuselage auxiliary fuel tank bay　机身后副油箱舱

recharge　补充

red copper gasket　紫铜垫片

red signal light　红色信号灯

reduce　减小

reflecting screen　反射屏

refuel　加油

refueling automatic control system　加油自动控制系统

refueling command　加油口令

refueling connector　加油接头

refueling distribution box　加油配电盒

refueling gun　加油枪

refueling personnel　加油人员

refueling pipe　加油管路

refueling port　注油口

refueling procedure　加油程序

refueling quantity　加油量

refueling selector switch　加油选择开关

refueling switch　加油开关

regulating sliding rail　调节滑轨

reliability　可靠性

removable rocker arm　可卸摇臂

removal　拆卸

replacement　更换

residual extinguishant　残留灭火液

residual fuel quantity　剩余(燃)油量

residual oil quantity　剩余(滑)油量

residual snow　积雪

residual water　积水

restricting orifice　限流孔

retraction/extension actuator　收放作动筒

retraction/extension mechanism　收放机构

right system accumulator　右系统蓄压瓶

right system cock　右系统开关

right system pressurization connector　右系统增压接头

rocker arm　摇臂

rocker lug　摇臂支耳

roller　滚轮

roller with bush　带衬套的滚轮

rotating connector　转动接头

rotating device　转动装置

rotating latch　旋转插销

rotating mechanism　转动机构

rotating shaft　转轴

rotating shaft base/axle base　转轴座

rotation cylinder　转筒

rubber gloves　橡胶手套

rubber overshoes　胶鞋

rubber part　橡胶部件

rudder　方向舵

safety altitude　安全高度

safety device　保险装置

safety diaphragm　保险薄膜

safety piece　保险片

safety rule　安全规则

safety switch　保险开关

sample　试样

scheduled and unscheduled servicing　定期及非定期保养工作

scheduled lubrication　定期润滑

screw　螺钉

screw rod　螺钉杆

sealing　密封

sealing ring　密封圈

seat base　座椅支座

sector part　扇形件

sector pitching gear　扇形俯仰齿轮

security　安全性

sediment　沉淀物

sensor　传感器

serious accident　严重事故

serious rust　严重锈蚀

service life　使用期限

servicing　保养

seven-position distribution valve　七位分配开关

shatt　轴

shaft head　轴头

shock strut　缓冲支柱

shock strut outer cylinder　缓冲支柱外筒

side pulley　侧滑轮

signal light　信号灯

signal union　信号接管嘴

signature　签字

silent chain　无声链条

silk cloth　绸布

slide seat back　滑动椅背

sliding bracket　滑动架

sliding rail　滑轨

sling　吊挂

small air nipple　小气门嘴

small hand wheel　小手轮

small shaft/small axle　小轴

small shaft/small axle of link　联杆小轴

soapsuds　肥皂水

soft brush　软刷

soft cloth　软布

special bracket　专用托架

special hand wheel　专用手轮

specified value　规定值

speed reducing casing/case　减速器机匣

spherical fire bottle　球形灭火瓶

spiral cock　螺旋开关

spline　花键

spoiler　扰流板

spoiler rocker arm　扰流板摇臂

sponge　海棉

spraying direction　喷射方向

spring　弹簧

spring pull rod　弹簧拉杆

squib　传爆管

stable buffer　稳定缓冲器

stable buffer outer cylinder　稳定缓冲器外筒

stable shock strut　稳定缓冲支柱

starting lever　起动杠杆

steel bottle　钢瓶

steel cable　钢索

steel cable control mechanism　钢索操纵机构

steel thread plug　钢螺塞

steering actuator bracket　转弯作动筒支架

steering mechanism　转弯机构

still pot　沉淀槽

stop pin　止动销

stop refueling command　停止加油口令

stove　火炉

strong acid　强酸

structure integral tank　结构整体油箱

suspension lug socket　悬挂耳座

support　存放架

support abutment　支座

support abutment bearing bush　支座轴瓦

supporting arm　支臂

supporting bearing　支撑轴承

supporting plate　支板

supporting rod bearing　支撑杆轴承

suspension bracket　悬挂支架

suspension joint　悬挂接头

synthesis hydrocarbon aviation lubricating oil　合成烃航空润滑油

tabby gauze　平纹纱布

tail jack　尾部千斤顶

tank　油箱

tank No.　油箱编号

tank volume　油箱容积

technical data　技术数据

technical fine filter　工艺细油滤

technical standard　技术标准

temperature regulating electric valve　调温电动活门

terminal switch　终点开关

thermal decomposing　热分解

thermal radiation　热幅射

thermal water bottle　保温水瓶

thermal water tank　保温水箱

thread groove　螺纹槽

threaded plug　螺塞

three-position distribution valve　三位分配开关

tighten　拧紧

tolerance　容差

transport　搬运

trapezoidal screw thread　梯形螺纹

travel　行程

trigger connector　触发接头

trim tab control rocker arm　调片操纵摇臂

trim tab pull rod　调整片拉杆

trolley　小车

turbine starter/generator　涡轮起动发电装置

turbine starter/generator cabin　涡轮发电装置舱

two-stage gear type　双级齿轮式

union thread　接管嘴螺纹

universal coupling bearing　万向联轴节轴承

universal joint　万向接头

unlocking actuator　开锁作动筒

unpacked state　未装箱状态

unscheduled servicing　不定期保养

unscrew　拧出

uplock　上位锁

uplock hook　上位锁锁钩

upper/lower torque arms　上/下防扭臂

upper braking pull rod　上刹车拉杆

upper cover assembly　上盖组件

upper torque arm　上扭臂

valve core　气门芯

valve nozzle　气门嘴

vent hole　通气口

vent pipe　通气管

vent tank　通气油箱

ventilation window sliding device　通风窗滑动装置

vertical shaft　立轴

vertical sliding rail　垂直滑轨

virose gas polluted area　有毒气体污染区

volume　体积

water based cleaning agent　水基清洗剂

water purifying system　净水系统

water supply/waste disposal equipment　供水/排污物设备

weather radar antenna　气象雷达天线

weighing　称重

weight　重量

wheel shaft and bearing　机轮轴及轴承

windshield wiper　刮雨器

wing anti-icing valve　机翼防冰活门

wing fuel tank bay　机翼油箱舱

wing jack　机翼千斤顶

wooden support　木架

work site　工作场地

working cycle　工作循环

wrench　扳手

zero position 零位

AIR CONDITIONING 环控系统

absolute pressure 绝对压力

absolute pressure valve/vacuum aneroid 绝压活门

access cover 口盖

accessory 附件

accuracy 精度

acquisition 采集

action rod 作用杆

actuating mechanism 执行机构

adjusting bonnet 调节阀帽

aging 老化

air bleed overpressure valve 引气超压开关

air bottle/pneumatic storage bottle 冷气瓶

air conditioning 环控系统

air distribution system 空气分配系统

air filter 空气气滤

air horn valve 气喇叭开关

air source 气源

air supply thermometer 供气温度表

airborne equipment 机载设备

air-delivery pipe 输气导管

airflow 气流

airflow restriction ring 限流环

airtight cabin 气密舱

airtightness 气密性

alignment 对准

aluminum alloy 铝合金

amplifier 放大器

analog signal 模拟信号

anti-icing inlet nipple 防冰入口接嘴

A-type screw-in connector A型拧入式接头

auxiliary power unit 辅助动力装置

auxiliary ventilation valve 辅助通风活门

backward valve 反向活门

balance 平衡

ball joint 球形接头

bearing 轴承

bearing disk 轴承圆盘

bimetallic thermometer/bimetal thermometer 双金属温度表

binding wire 捆扎用线

blanking cap 堵帽

block 堵塞

block plate 堵板

bonding jumper 搭接线

box cover 箱盖

bracket piece 支架片

brush 电刷

bushing 衬套

button-head bolt 扁圆头螺栓

cabin 座舱

cabin altitude differential pressure gauge 座舱高度压差表

cabin movable door 活动舱门

cargo cabin temperature control box 货舱温控盒

cargo cabin thermometer 货舱温度表

cargo cabin/cargo compartment 货舱

cast 铸造

cavity 腔

center wing 中央翼

centrifugal force 离心力

changeover switch 转换开关

check valve 单向活门

circuit breaker 自保开关

circuit breaker board 自保开关板

clamp 卡箍

clamp ring 卡圈

clockwise 顺时针

cockpit 驾驶舱

cockpit cabin thermometer 驾驶舱温度表

cockpit evaporation cycle system 驾驶舱蒸发循环系统

cockpit temperature control box 驾驶舱温

控盒

cockpit temperature sensor　驾驶舱温度传感器

communicator　通讯员

compensation　补偿

compensator　补偿器

compressing　压缩

compression system　压缩系统

compressor　压气机

compressor impeller　压气机叶轮

compressor volute chamber/casing　压气机壳

condense　凝结

condenser　冷凝器

connecting plate　连接板

connecting rod/link　连杆

connector　插头

console　操纵台

cool　冷却

cool path　冷路

cooling path outlet header　冷边出口封头

cooling pipe valve component　冷管活门部件

coordination　协调

counter-clockwise　逆时针

cowling/fairing　整流罩

crack　开裂

crossover valve　交叉活门

current intensity　电流强度

cylinder　筒体

damping pad　减震垫

data transmission　数据传输

decline　倾斜

delay　延时

descending　下滑

dial/scale　刻度盘

diaphragm　膜片

differential-pressure sensor　压差传感器

diffuser　扩压器

diffusion cover plate　扩压盖板

displacement　位移

distribution valve　分配活门

diverter valve　分路活门

dust cover　防尘罩

dynamic pressure　动压

elasticity　弹力

elbow pipe　弯管

electric dryer　电吹风

electrical bridge　电桥

electrical butterfly valve　电动式蝶形活门

electrical mechanism　电动机构

electrical plug　电插头

electrical pump　电动泵

electrical wire　导线

electromagnet　电磁铁

electromagnetic valve/solenoid valve　电磁活门

elevator　升降舵

elimination　清除

emergency pressure release unit　应急释压装置

engine nacelle　发动机短舱

environmental control system　环控系统

error　误差

excessive pressure　余压

excessive pressure limiter　余压限制器

excessive pressure limiting mechanism　余压限制机构

excessive pressure valve　余压活门

expansion　膨胀

fastening part　紧固件

filter element　滤芯

filter screen　滤网

fixing bolt　紧固螺栓

flange　法兰盘

flap control　襟翼操纵

flow　流量

flow back　倒流

flow distribution valve　流量分配活门

fluid resistance　流体阻力

forward fluid resistance　正向流阻

fracture　断裂

free water content of inlet air　进口空气游离水量

friction　摩擦

gauze　纱布

gear　齿轮

generator　发电机

grease smudge　油污

ground air conditioning vehicle/cart　地面空调车

guide blade　导向叶片

guide header　导流封头

guide post　导向柱

hard core　硬芯

header body　封头体

heat　加温

heat exchanger core　芯体

heat exchanger core assembly　芯体组件

heat insulation layer　隔热层

heat path　热路

heat transfer efficiency　换热效率

heating ejector system　加温引射系统

heating path inlet header　热边入口封头

heating path outlet header　热边出口封头

heating pipe valve component　热管活门部件

heating valve　加温开关

hinge bracket　铰链支架

hydraulic reservoir pressurization system　液压油箱增压系统

icing　结冰

inductance load　电感性负载

inlet moisture content　进口含湿量

instrument panel　仪表板

insulation　隔离

integrated test bed　综合试验台

interconnect　交联

interface　接口

joint/contact　接头

knob　旋钮

kraft paper　牛皮纸

lead sealing　铅封

leading edge　前缘

leak　漏(气)

limit position　极限位置

linear-variation　线性变化

loosening　松开

low-altitude ventilation system　低空通风系统

lower cover　下盖

low-level ventilation valve　低空通风活门

magnetic field/transmission　磁场

main technical specification　主要技术指标

maintenance　维修

mandril　心轴

manganin　锰

mechanism　机构

metering hole　定径孔

microswitch　微动开关

mixing chamber　混合室

monitor procedure　监控程序

motor　电动机

mounting panel　安装板

moving-iron current ratio meter　动铁式电流比计

nacelle　短舱

navigator　领航员

needle valve　针形活门

nozzle ring　喷嘴环

oil plug　油塞

open air-cycle system　开式空气循环系统

opening aneroid/opening diaphragm capsule　开口膜盒

operating current　工作电流

operating environment temperature　工作环境温度

operating medium　工作介质

operating principle　工作原理

operating temperature　工作温度

operating voltage　工作电压

panel　壁板

parameter　参数

partition bar　隔条

partition/partition plate　隔板

pipe joint/coupling　管接头

pipe temperature sensor　管路温度传感器

pipeline high-temperature limiting sensor　管路高温限制传感器

pipeline low-temperature limiting sensor　管路低温限制传感器

placard　标牌

plug　插头

pneumatic control valve　冷气操纵开关

pointer　指针

pointer travel　指针行程

power consumption　功耗

power distribution system　配电系统

power supply system　供电系统

precipitator　沉淀器

preservation　油封

pressing plate　压板

pressing ring　压圈

pressure changing rate knob　压力变化速率旋钮

pressure control/shutoff valve　压力调节/关断活门

pressure difference　压力差

pressure regulator　压力调节器

pressure transducer/sensor　压力传感器

pressure-sensing tank　感压箱

pressurization　增压

primary heat exchanger　初级散热器

protective coating　保护层

protective lid/cover　保护罩

push rod　顶杆

quick-release bolt　快卸螺栓

radial bearing　径向轴承

radiator　散热器

ram air　冲压空气

rated voltage　额定电压

recovery　恢复

refrigeration system　制冷系统

regulating cover　调节盖

reinforce　加强

releasing　解除

repair　修理

resistance load　电阻性负载

resistor　电阻

retard　收（油门）

reverser　换向器

rib　肋

rivet　铆钉

rocker arm　摇臂

rotating speed　转速

rotation　旋动/转动

rubber ring　胶圈

safety valve　安全活门

safety wire　保险丝

secondary heat exchanger　次级散热器

sensitive component/sensing element　敏感元件

service ladder　工作梯

shaft base/axle base　轴座

shield　屏蔽

shock mount　减震座

shutoff valve　断流活门

signal sampling　信号采样

skin　蒙皮

special blanking cover/blanking cap　专用堵盖

speed reducer/decelerator　减速器

split cotter/cotter pin　开口销

spring　弹簧

spring leaf/spring sheet　弹簧片

square flange　方法兰盘

stagnate 卡滞

stainless steel 不锈钢

standard part 标准件

static pressure 静压

steel wire pin 钢丝销

stop cam 限位凸轮

stop ring 止动圈

sub-subsystem 分-分系统

subsystem 分系统

supporting ring 支撑环

swirler 旋流器

syphon wick/oil wick 油芯

tapered pin 圆锥销

temperature control system 温度控制系统

temperature control/electrical valve 调温电动活门

temperature selector 温度选择器

thermal sensitive element 热敏元件

three-way switch 三通开关

transition joint 过渡接头

troubleshooting 排除故障

turbine cooler 涡轮冷却器

ultrasonic cleaner 超声波清洗机

union 接管嘴

unit/module 单元体

upper cover 上盖

vacuity 真空度

valve base/valve seat 活门座

valve leaf/flap 活门瓣

valve plate component 活门板部件

valve shaft 活门轴

valve switchover duration 活门转换时间

valve tube casing 活门管壳体

vent valve/exhaust valve 排气活门

ventilation 通风

ventilation valve 通风开关

vibration 振动

washer 垫圈

water drainer 放水嘴

water jet 喷水器

water separator 水分离器

weight 重量

windshield glass 风挡玻璃

wing anti-icing valve 机翼防冰活门

wing heating valve 机翼加温活门

AUTO FLIGHT 自动飞行

(flight) track control 航迹控制

actuating cylinder 作动筒

actuating mechanism 执行机构

actuator 舵机

actuator piston rod 舵机活塞杆

adapting 转接

adverse roll 反坡现象

aerodynamic force 空气动力

aero-washing gasoline 航空洗涤汽油

aileron actuator 副翼舵机

aileron degree 副翼度

aircraft attitude 飞机姿态

aircraft control surface 飞机舵面

aircraft deflection 飞机偏转

aircraft inertia 飞机惯性

aircraft lateral axis 飞机横轴轴线

aircraft motion parameter 飞机运动参数

alcohol 酒精

altitude compensation signal 高度补偿信号

altitude difference sensor 高度差传感器

altitude difference signal 高度差信号

ambient factor 外界因素

ambient temperature 环境温度

amplifier 放大器

angle sensor 角度传感器

angular displacement 角位移

angular rate sensor 角速率传感器

angular rate/angular speed 角速率/角速度

armature deflection 衔铁偏转

armrest 扶手

attenuation oscillation 衰减振荡

attitude heading reference system（AHRS） 航向姿态系统

auto control system 自动控制系统

auto-flight 自动飞行

automatism navigation signal 自动导航信号

autopilot 自动驾驶仪

autopilot actuator 驾驶仪舵机

autopilot integrated circuit 驾驶仪综合线路

auto-preparation stage of autopilot 自动驾驶仪自动准备阶段

auto-return-to-zero system/automatic zero return system 自动归零系统

aviation hydraulic fluid 航空液压油

bank angle 倾斜角

bank angle holding 倾斜角保持

bank attitude 倾斜

bank channel 倾斜通道

bank coordinated signal 倾斜协调信号

bank drawer 倾斜抽盒

bank measuring axis 倾斜测量轴

banking turn 盘旋

bank-turn flight/banked turn 盘旋飞行

base 底座

base assembly 底座组件

bearing holder 轴承座

block diagram 方块图

bracket/support 支架

bridge arm 桥臂

brush 电刷

brush assembly 电刷组件

butterfly bolt 蝶形螺栓

button assembly 按钮组件

cam 凸轮

capacitor 电容

casing/housing 壳体

ceiling 顶棚

center vertical gyro 中心垂直陀螺

center wing 中央翼

center-return spring 回中弹簧

centralizing spring 对中弹簧

centripetal force 向心力

check/detection hole 检测孔

chock 轮挡

circuit breaker cover plate 自动保护开关盖板

clamp 卡箍

clutch 离合器

cockpit 驾驶舱

coil winding 线圈绕组

compensation 补偿

conductor 导线

cone pin 锥销

connection indicating light 接通指示灯

console 操纵台

constant interference moment 常值干扰力矩

constant speed 恒速

consumption material 消耗器材

contact 触点

contact semi-ring 触片半环

contact tab assembly 触片组件

contact unit 触片装置

control box 控制盒

control component 操纵元件

control condition 操纵状态

control flux 控制磁通

control handle 操纵手柄

control knob 操纵旋钮

control mechanism 操纵机构

control stick 驾驶杆

control surface 操纵舵面

control surface angle protractor 舵面角量角器

control surface deflecting angle 舵面偏角

control surface lock 舵面锁

control surface moment 舵面操纵力矩

control switching circuit 控制转换电路

control velocity 操纵速度

control wheel 驾驶盘

control winding 控制绕组

convertible cable 转换电缆

coordinated turn 协调转弯

coordination 协调

cotter pin 开口销

course 航线

cover plate 盖板

crank linkage/link mechanism 曲柄连杆机构

crosslink component 交联部件

current linear integration 电流线性综合

current-limiting resistor 限流电阻

damping cylinder 阻尼筒

damping moment 阻尼力矩

decompression orifice plug 减压节流塞

deflection 偏转

delay circuit 延时线路

deviation angle 偏离角

diagonal line 对角线

differential mechanism 差动机构

dimension 因次

displacement signal 位移信号

double gear 双片齿轮

driving 拖动

driving lever pin 拨杆销

dry air 干燥冷气

electric brake state 电制动状态

electric signal 电信号

electrical bridge 电桥

electrical schematic diagram 电气原理图

electrical zero position 电气零位

electric-mechanical device 电气机械手柄

electrohydraulic autopilot 电液自动驾驶仪

electrohydraulic control actuator 电液舵机

electrohydraulic valve 电动液压活门

electromechanical converter 机电变换器

electromechanical tachometer 机电测速表

electronic flight instrument system (EFIS) 电子飞行仪表系统

elevator 升降舵

elevator actuator 升降舵机

emergency cutoff button 紧急断开按钮

end point 末端

end surface 端面

engagement 吸合

equivalent signal 等量信号

erection moment 扶正力矩

excitation coil 激磁活门线圈

excitation flux 激磁磁通

excitation voltage 激磁电压

external plug 外部插头

extreme tightness 过度拧紧

feedback potentiometer 回输电位计

filter housing 油滤壳体

filter screen 滤网

fine filter element 细滤芯

flight altitude 飞行高度

flight attitude 飞行姿态

flight control box 飞行控制盒

(flight) track controll 航迹控制

fluid reservoir 贮油器

fluorine plastic ring 氟塑料圈

foot pedal 脚蹬

fork joint 叉形接头

free travel 自由行程

friction disk 摩擦盘

fuselage middle section 机身中段

galob positioning system (GPS) 全球定位系统

gear rotating angle 齿转角

general wiring diagram 总线路图

gimbal axis 环轴

glide 下滑

gradient/slope 梯度

ground wire 地线

gust 阵风

gyratory angular velocity 回转角速度

gyro measuring axis 陀螺测量轴

gyro output gradient 陀螺输出梯度

gyro platform 陀螺平台

hairspring 游丝

hand vacuum pump 手摇真空泵

handle control speed 手柄操纵速度

hard center 硬中心

heading channel 航向通道

heading deviation 航向偏差

heading indicator 航向指示器

heading integrated signal 航向积分信号

heading linkage box 航向联系盒

heading maintaining flight status 航向保持飞行状态

heading preset signal 航向预选信号

heading setting knob 航向给定旋钮

heading signal 航向信号

heading stability signal 航向稳定信号

high-pressure fluid line 高压油路

horizon 地平仪

horizon indication 地平指示

horizontal correction system/principle 水平修正原理

hydraulic amplifier 液压放大器

hydraulic filter 液压油滤

hydraulic fluid 液压油

hydraulic power cart 液压油泵车

hydraulic solenoid valve 液压电磁阀

hydraulic valve winding 液压活门绕组

hydraulic-return orifice plug 回油节流塞

indicating light cap 指示灯帽

inertial shaft 绕惯性轴

initial angle 起始角

initial position 初始位置

inlet oil pressure 进口油压

inner ring/inner gimbal 内环

instrument insensitivity 仪表不灵敏度

instrument panel 仪表板

instrument rack 设备支架

instrument screwdriver 仪表螺刀

insulating area 绝缘区

integrated circuit 综合电路/线路

interconnect 交联

interlock 联锁

iron core 铁芯

jack 插孔

jamming/block 卡死阻滞

joint body 接头体

knob position 旋钮位置

lag 落后

laser strapdown inertial 激光捷联惯性

lateral component force 横向分力

lateral side 侧面

lead sealing 铅封

lead signal 前置信号

level bubble 水准气泡

level flight 平飞

level installation 水平安装

level position 地平位置

level velocity 改平速度

level-off 改平

level-off button 改平按钮

level-off from tilt 改出倾斜

lift vector 升力向量

lightening hole 减轻孔

linear amplification 线性放大

linear electromagnet 线性电磁铁

linear rotating transformer 线性螺旋变压器

line-pressing screw 压线螺钉

link rod/connecting rod 连杆

locking wire 保险丝

longitudinal axis 纵轴

long-range flight 长途飞行

magnetic conductive body 导磁体

magnetic damping structure 磁阻尼结构

magnetic field 磁场

magnetic line 磁力线

magnetic pole 磁极

main pressure relief valve　主减压阀

maintainace department　修理部门

maneuvering flight　机动飞行

maximum flow　最大流量

measured angle　测量角

measuring point　测量点

mechanical amplification　机械放大

mechanical zero position　机械零位

middle point　中点

middle tapping　中间抽头

misalignment angle　失调角

moment motor excitation winding　力矩马达激磁绕组

motor　电机

motor-tachometer　机电测速机

mounting bracket　安装架

mounting hole　安装孔

mounting shim/mounting spacer　安装垫片

movable part　活动部分

multi-meter/tri-function meter　三用表

navigation computer　导航计算机

navigation selector indication　导航选择指示

navigation system relay box　导航系统继电器盒

navigator　领航员

navigator connection button　领航员接通按钮

navigator control knob　领航员操纵旋钮

navigator instrument panel　领航员仪表板

navigator turn knob　领航员转弯旋钮

neutral position　中间位置

neutral slot　中立卡槽

normal closed contact　常闭触点

nose-down pitching（steering）moment　下俯操纵力矩

nose-up signal　飞机抬头信号

null error/zero position error　零位误差

nut　螺母

oil delivery nipple　输油嘴

oil distributing piston　分油活塞

oil filter　油滤

oil inlet　进油

oil passing blind nut　通油螺盖

oil return　回油

oil-penetrating valve　透油活门

operating fluid　工作液

operating principle　工作原理

operating travel　工作行程

orthogonal excision network　正交切除网络

orthogonal excision stage　正交切除级

oscillating divergent curve　振荡发散型曲线

output cable　输出电缆

output force　输出力

output synchronizer　输出同位器

override force/forced control force　强迫操纵力

override the aircraft　强力操纵飞机

paired centripetal ball　双列向心球体

parameter adjustment potentiometer　调参电位计

phase discrimination　鉴相

phase-sensitive amplifier tube　相敏级的放大管

phase-sensitive stage　相敏级

pilot handle　飞行员手柄

pilot handle contact mechanism　飞行员手柄触片机构

pin　插针

pin number　插针号

pin rod of the depression throttling valve　减压节流阀针杆

pin rod slide valve　针杆滑阀

pipe joint　管接头

piston rod end　活塞杆末端

pitch angle　俯仰角

pitch channel　俯仰通道

pitch contact　俯仰触片

pitch drawer　俯仰抽盒

pitch potentiometer　俯仰电位计

pitch scale　俯仰刻度

plug number　插销号

plug-in board　插板件

plug-in unit　内插件

position feedback sensor　位置反馈传感器

position feedback sensor rod　位置反馈传感器连杆

potentiometer　电位计

potentiometer shaft　电位计轴

power unit　动力装置

preset course　预选航迹

preset heading　预选航向

preset pitch angle　给定的俯仰角

prestage　前置级

primary channel amplifier　主通道放大器

primary winding　初级绕组

push-pull amplification　推挽放大线路

quick correction　快速扶正

quick-release plug　快卸插销

rate gyro　速率陀螺

rate gyroscope casing/rate gyro box　速率陀螺盒

reflective mirror　反光镜

relay　继电器

resistor　电阻

restricting orifice/restricted orifice　限流孔

return spring　恢复弹簧

return-to-zero limit　回零极限

reverse heading　逆航向

reverse voltage　反方向电压

rocker arm　摇臂

rotary table　转台

rotating angle　转角

rotating shaft　旋转轴

rotor spindle　转子轴

rudder actuator　方向舵机

rudder deflection　偏舵量

satellite navigation system　卫星导航系统

scheduled flight altitude　预定飞行高度

scheduled heading　预定航向

schematic diagram　原理图

scissor　剪刀

screwed plug　螺塞

sealing ring　密封圈

sealing ring mounting slot　密封圈安放槽

secondary winding　次级绕组

sector gear　扇形齿轮

seepage　渗油

self-lock　自锁

self-oscillation　自振

sensitive component　敏感元件

service ladder　工作梯

servo(driver)　伺服器

servo(driver) system/follow-up system　随动系统

shaft gear　轴齿轮

shroud　罩盖

sideslip angle　侧滑角

signal gun　信号枪

single gear　单片齿轮

sliding tab　滑动片

slope　坡度

solenoid clutch/electromagnetic clutch　电磁离合器

special rotation table　专用转台

speed handle　速度手柄

spherical cover　球形罩

spherical handle　球形手柄

spring scale　弹簧秤

stable airflow　气流平稳

stable state/stable condition　稳定状态

static base　不动基座

static fitting/static union　静压接管嘴

static heading error　航向静差

static hose　静压软管

stator　静子

steel wire retainer　钢丝挡圈

steering force 操纵力

step signal 阶跃信号

straight flight 直线飞行

straight level flight attitude 水平直线飞行状态

support lug 支耳

synchronizer rotor 同步转子

synchronizer rotor coil output 同步器转子线圈输出

synchronizer stator 同步器静子

tail section 尾段

tail valve 尾阀

technological cable 工艺电缆

technological cap 工艺堵帽

terminal mark 终极刻线

test equipment 试验设备

tilt channel 倾斜通道

tilt transition mechanism 倾斜过渡机构

time integrated signal/time integral signal 时间积分信号

torque motor 力矩马达

torsion rod 扭杆

track 航迹

transistor 晶体管

transistor amplifier 晶体管放大器

transition mechanism 过渡机构

transmission ratio 传动比

trim tab 调整片

ultrasonic cleaner 超声波清洗机

unbalanced engine thrust 不平衡的发动机推力

universal joint 万向接头

vacuum diaphragm capsule 真空膜盒

vacuum tester 真空试验器

valve cover 活门盖

valve tab 阀片

vertical channel 升降通道

vertical gyro 垂直陀螺

vertical speed 升降速量

vertical speed indicator 升降速度表

vertical tracking speed 垂直跟踪速度

voltage difference 电压差

warning board 警告牌

washer 垫圈

weld spot 焊点

white silk cloth 白绸布

winding in series 串联绕组

wing root 翼根

workshop-adjusted position 内场调定位置

wrapping paper 包装纸

yaw angle 偏航角

yaw angle speed signal 偏航角速度信号

yawing angle signal 偏航角信号

zero adjusting screw 零点调节螺钉

zeroing state 回零状态

zero-returning extreme potentiometer 回零极限电位计

COMMUNICATIONS 通信系统

access cover 口盖

acoustic-electric signal 声-电信号

active noise-reducing sound insulation factor 主动降噪隔声量

adjustment 调整

airborne equipment 机载设备

aircraft formation 飞机编队

aircraft power self-calibration circuit 整机功率自校电路

aircrew 机组成员

airframe 机体

airport tower 机场塔台

amplify 放大

amplitude compatible modulation equivalent 兼容调幅

amplitude modulation 调幅

analog interface 模拟量接口

antenna download 天线引入线

antenna feeder 天线馈线

antenna switch unit 天线开关单元

antenna tuner 天线调谐器

antenna tuner standing wave 天调驻波

area 区域

audio amplification board 音频放大板

audio channel 音频通道

audio coding/decoding interface module 音频编/解码接口模块

audio identification signal 音频识别信号

audio interface unit 音频接口单元

audio monitor 音频监控器

audio power amplifier unit 音频功放单元

audio response 音频响应

audio signal 音频信号

audio system 音频系统

automatic calling 自动呼叫

automatic gain control signal 自动增益控制信号

axis 轴线

azimuth 方位

back power inspection 反向功率检测

band width 带宽

beacon 信标

bit 比特/位

blink 闪烁

bolt 螺柱

bonding jumper 搭接线

bonding limit 搭接限制

bonding resistance 搭接电阻

brightness 亮度

brightness control knob/brightness adjusting knob 照明亮度调节旋钮

brightness dimmer 亮度开关

buffer spring 缓冲弹簧

bulge 鼓包

bus 总线

bus signal interface module 总线信号接口模块

bushing 衬套

button set 按键组

cable 电缆

cable cover 钢索套

cable plug 电缆插头

calibration 校准

calling address 呼叫地址

calling setting state 呼叫设置状态

calling status 呼叫状态

cargo attendant/escort 押运员

center instrument panel 中央仪表板

centralize 集中

ceramic filter 陶瓷滤波器

channel crosstalk suppression 通道串音抑制

channel spacing 波道间隔

circuit protection status 线路保护状态

clamp 卡箍

cleat 线夹

clockwise 顺时针

clutter signal 杂波信号

cockpit 驾驶舱

cockpit audio signal 舱音信号

cockpit voice recorder 座舱音频记录器

cockpit voice recording system 座舱音频记录系统

code 代码

cold drawing steel-wire 冷拉钢丝

combination key 组合键

combined receiving device 组合接收设备

communication audio terminal 通信音频终端

communication protocol 通信协议

communication system 通信系统

communicator 通讯员

communicator circuit breaker board 通讯员自动开关板

communicator console 通讯员操纵台

communicator control panel 通讯员操纵板

communicator working desk 通讯员工作桌

compass 罗盘

component position/element position 元件位置

compress 压缩

conductive rubber plate 导电橡胶板

cone rod seat 圆锥杆底座

control box 控制盒

control box panel 控制盒面板

control mode 控制方式

copilot 右驾驶员

copilot control stick 右驾驶杆

copper-cable antenna 铜索天线

corrosion 腐蚀

counter-clockwise 逆时针

cover plate 盖板

crash survivability 坠毁幸存性

critical AOA warning 临界攻角告警

crosslink 交联

crystal filter 晶体滤波器

cursor 光标

data link 数据线

data link protection calling 链路保护呼叫

data processing module 数据处理模块

DC power 直流电源

deform 变形

demodulate 解调

digital process 数字化处理

digital volume potentiometer 数字音量电位器

digitron 数码管

dimming potentiometer 调光电位器

discharge needle 放电针

discharger 放电刷

discharging pulse 放电脉冲

display window 显示窗

disruptive field intensity 击穿场强

distortion 失真度

dorsal fin 背鳍

double sideband (DSB) 双边带

download 下载

driver stage 驱动级

driving circuit 驱动电路

dropping personnel 投放员

dust particle 尘埃粒子

duty cycle 工作周期

ear enclosure 耳罩

effective value 有效值

electric field intensity 电场强度

electronic-to-sound transform 电声转换

electrostatic charge 静电荷

emergency maintenance 应急维修

emitter follower 随射器

environment sound 环境声音

epoxy resin adhesive 环氧树脂胶

equipment rack 设备架

external communication selection knob 外通选择旋钮

failure analysis 故障分析

failure signal light 故障信号灯

fairing 整流罩

fastener 紧固件

fastening nut 紧固螺母

field 外场

field failure diagnosis and troubleshooting 外场故障诊断和排除

filter 滤除

filter unit 滤波器单元

final (output) stage 末级

fixed frequency 定频

fixed screw 固定螺钉

flight data field check processor 飞行参数外场检测处理机

flight data recorder 飞行参数记录器

flight data recording system 飞行参数记录系统

flight data recording system field tester 飞行参数外场检测仪

flight data system 飞行参数系统

flight data workshop check processor 飞行参数内场检测处理机

flight engineer 机上机械师

fork fixed head 叉形固定头

forked member 叉形件

forward power 正向功率

fracture 断裂

frequency error 频率误差

frequency integrated module 频综模块

frequency mixer 混频器

frequency modulation 调频

frequency range 频率范围

frequency response 频率响应

frequency synthesizer unit 频率合成器单元

full frequency range/full frequency band 全频段

functional check 功能检查

funnel 漏斗

general calling address 全呼地址

grease 润滑脂

ground brush/grounding brush 接地刷

ground check switch 地面检查开关

ground compass navigation station 地面罗盘导航台

ground data processing station 地面数据处理站

ground navigation station 地面导航台

ground proximity warning system 近地告警系统

grounding cone 接地锤

grounding wire/earthing wire 接地线

harmonic distortion 谐波失真

headset 头戴送受话器

headset/earphone 耳机

high frequency (HF) radio set 短波电台

high frequency (HF) signal 高频信号

high frequency (HF) transmission signal box 短波发射信号盒

high-pass filter 高通滤波器

holding bracket 固定支架

hook 挂钩

horizontal tail 水平尾翼

hot key 热键

housing 壳体

impedance 阻抗

inboard wing 中外翼

indicator 指示器

initialization 初始化

injection interface 注入接口

input stage 输入级

installation limit 安装限制

installation position 安装位置

instrument landing system (ILS) 仪表着陆系统

instrument panel 仪表板

insulator 绝缘子

intelligibility rate 可懂度

interchangeability 互换性

intercom 机内通话器

intercom control 机通控制盒

intercom expanding equipment 内通扩展设备

interface circuit 接口电路

interference input method 干扰输入法

interference source 干扰源

intermediate frequency (IF) signal 中频信号

intermediate level 中继级

internal bus 内部总线

isolation 隔离

jam nut 防松动螺母

key 密匙

keying signal/button-controlled signal 键控信号

lacquer coating 漆层

landing gear well 起落架舱

late entry (insertion) function 迟入(插入)功能

liaison　联络

life-saving condition　救生状态

line-of-sight short-distance communication
视距近距离通信

local oscillator (LO) signal　本振信号

locating pin　定位销

lock catch　锁扣

low frequency　低频

low-drag and noise-proof　低阻抗噪

lower sideband (LSB)　下边带

low-pass filter　低通滤波器

lubrication　润滑

main computer module/host computer
module　主计算机模块

main controller unit　主控制器单元

main display format　主显示画面

main technicalperformance　主要技术性能

main technical specification　主要技术指标

mainframe　主机

manual calling　人工呼叫

maximum output power　最大输出功率

mean ear enclosure sound insulation factor
耳罩平均隔声量

mean-time-to-repair　平均维修时间

medical staff　医务员

medium and long range communication　中
远距离通信

memory　存储器

memory channel/stored channel　存贮信道

metal wire　金属丝

micro-controller　微控制器

microhmmeter　微欧计

microphone（MIC）　话筒

modem module　调制解调模块

modulate　调制

mother board　母板

mounting base　安装底座

mounting bracket　安装架

mounting screw　安装螺钉

mounting surface　安装面

movable part　活动部件

narrowband　窄带

navigator　领航员

net calling address　网呼地址

noise　噪音

noise reduction function　降噪功能

non-memory channel/non-stored channel　非
存贮信道

normalized maintenance　规格化维修

nut　螺母

open-circuit/short circuit protection　开/短
路保护

operating mode　工作状态

operating principle　工作原理

operating voltage　工作电压

operation mode　工作方式

operation mode switch　工作模式开关

outboard wing　外翼

out-of-band interference　带外干扰

output voltage　输出电压

overall dimension　外形尺寸

overhead console　顶部操纵台

oxide layer　氧化层

oxygen mask microphone　氧气面罩送话器

page down　下翻

parameter　参数

parametric down-converter　参数向下换频器

parametric up-converter　参数上升变换器

password　密码

pedal button　脚踏按钮

pilot control stick　左驾驶杆

pin　插销

plug cover　塞盖

plug-pull structure　插拔式结构

pointer　指针

polarity　极性

polarization mode　极化方式

porous section　多孔型材

potential strength 电位强度

power amplification unit 功率放大单元

power amplifier module 功放模块

power capacity 功率容量

power circuit breaker 电源自保开关

power consumption 电源功耗

power indicating light 功率指示灯

power level 电平

power module 电源模块

power supply 电源

power switch 电源开关

power-on check 通电检查

power-on self-test 开机自检

pre/post selector 预/后选器

pre-amplifier 前置放大器

pre-amplifying circuit 前置放大电路

precaution 注意事项

preset 预置

programming condition 编程状态

pulley 滑轮

pulse 脉冲

quick-release screw 快卸螺钉

quick-release self-lock nut 快卸锁紧螺母

radio communication system 无线电通信系统

radio compass 无线电罗盘

radio frequency amplifier unit 射频放大器单元

radio set protector 电台保护器

real-time monitor 实时监测

receiver 接收机

receiving channel 收信道

receiving check 收信检查

receiving condition 接收状态

receiving indicating light 接收指示灯

receiving module 接收模块

receptacle 插头座

record maintenance 履历维护

recording module 记录模块

relay 继电器

relay box 继电器盒

removal 拆卸

replacement 更换

reverse process 逆过程

rib 肋

rust-proof oil 防锈油

sample 采样

scan 扫描

scanning channel group 扫描波道组

schematic diagram 原理框图

screwdriver 螺丝刀

second-line tester 二线检测仪

sediment particulate 沉降物微粒

selection knob 选择旋钮

self-adaptability 自适应

self-adaptive operation check 自适应工作检查

self-listening tone 自听音

self-lock nut 锁紧螺母

self-restoring locking wire 自恢复保险丝

self-test/built-in test 自检

semi-conductor memory 半导体存储器

sensitivity 灵敏度

service ladder 工作梯

shielded cable/shielded wire 屏蔽导线

shielded twisted pair/dual-entangled shielded wire 双纽屏蔽导线

shock absorber 防震架

shortcut key 快捷键

signal processor unit 信号处理器单元

signal source 信号源

signal-to-noise ratio 信噪比

simplified control box 简易控制盒

single set address 单台地址

single sideband（SSB） 单边带

slotted screwdriver 一字槽螺丝刀

socket 插座

sound pickup 拾音器

spatial condensation　空间凝结物

speaking status　发话状态

specification　指标

split pin/cotter pin　开口销

spring　弹簧

spring cushion　弹簧垫

squelch function　静噪功能

standing-wave ratio　驻波比

static discharge　静电释放

status light　状态灯

steel cable　钢索

steel fork joint　钢叉接头

stringer　长桁

strut　支柱

sub-band filter　分段滤波器

sub-menu　子菜单

summation circuit　求和电路

support component　配套件

support equipment　保障设备

suppress　抑制

switch power unit　开关电源单元

syllable squelch circuit　音节式静噪电路

synthesize　合成

system crosslinking block　系统交联方框图

TACAN　塔康

telegraph key　电键

temperature control circuit　温度控制电路

tensiometer　张力计

tension　张力

terminal box　接线盒

test　试验

thread　螺纹

timeinterval　时间间隔

tone key　单音按键

total flow sampling　总流取样

transceiver　收发机

transceiver indicating light　收发机指示灯

transceiver switching circuit　收发机转换电路

trans-illuminated panel　导光板

transmission channel　发信道

transmission check　发信检查

transmission indicating light　发射指示灯

transmission output power indicating light　发射输出功率指示灯

transmitter　发射机

transmitting condition　发射状态

transmitting power　发射功率

transportation and storage　运输和贮存

tune　调谐

tuning precision　调谐精度

tuning time　调谐时间

turn buckle　松紧螺套

underwater locator beacon　水下定位信标

unnecessary signal　无用信号

upper sideband (USB)　上边带

upper sideband report　上边带报

vertical polarization　垂直极化

vertical tail　垂尾

very high frequency (VHF) radio set　超短波(通信)电台

very-low altitude　超低空

visual range　视距

volume　音量

volume adjusting knob　音量调节旋钮

volume regulator　音量调节器

VOR　伏尔

warning placard/warning board　警告牌

warning signal　告警信号

washer　垫圈

wave-carrier　载波

web　腹板

wire terminal　导线端子

work site　工作现场

work type　工种

workshop　内场

ELECTRICAL POWER　电源系统

abrasive paste　研磨膏

absorb shock　减震

accelerometer　加速度表

access cover　口盖

accessory　附件

accessory box　附件箱

accessory contact　辅助触点

activation　活化

additional magnetic pole　附加磁极

adjustable resistor　可调电阻(器)

adjusting screw　调节螺钉

air gap　空气隙

airdrop　空投

airframe　机体

airport power supply power-on contactor　机场电源接通接触器

alkaline battery　碱性蓄电池

all-digital commutating pole　全数换向极

altimeter　高度表

aluminum wire　铝导线

ammeter　电流表

angle of attack（AOA）　迎/攻角

anti-icing heating PDB　防冰加温配电盒

antiseptic plaster　防腐膏

anti-vibration stability　耐振稳定性

anti-vibration strength　抗振强度

appearance check　外观检查

approach lighting　着陆照明

armament　军械

armature　电枢/衔铁

armature winding　电枢绕组

asbestos tape　石棉带

attitude and heading system　航姿系统

attitude heading reference system（AHRS）航向姿态基准系统

attitude system　姿态系统

attraction voltage　吸合电压

attractive force　吸力

attractive pile assembly　吸片组件

aviation gasoline　航空汽油

axle　轴线

balancing circuit　平衡电路

balancing coil　平衡线圈

balancing resistor　平衡电阻

ball track　滚珠轨道

batch　批架次

bearing　轴承

bearing ring　轴承环

bimetallic strip　双金属片

binding thread　绑线

blade　桨叶

blanking cover/blanking cap　堵盖

block diagram　方框图

bonding jumper　搭接线

bonding resistor　搭接电阻

boost charging　快速充电

bottom chassis　底盘

break-up corner　崩角

brush handle　刷握

brush spring　刷簧

bulge　鼓包

burr　毛刺

bushing　衬套

bushing-type capacitor　穿心电容器

butt joint　对接接头

button　按钮

cable bundle　电缆束

cable end　电缆收头

cable interconnection　电缆对接

cable laying/cabling　电缆敷设

cable through　电缆槽

cam　凸轮

capacitor　电容器

carbon brush　碳刷

carbon brush length　碳刷高度

carbon brush shelf assembly 碳刷架组合件

carbon brush short braid 碳刷小辫

carbon deposit 积碳

carbon pile 碳柱

carbon powder 碳粉

carbon tetrachloride 四氯化碳

carbonate 碳酸盐

carbon-pile 炭柱

carbon-pile voltage regulator 炭片调压器

carbon-pile voltage regulator 碳片调压器

cargo cabin 货舱

ceiling 顶棚

cell 单体蓄电池

center power distribution device 中央配电装置

center instrument panel 中央仪表板

centrifugal switch 离心开关

chamois leather 麂皮

changeover box 转换盒

changeover contactor 转换接触器

changeover switch 转换开关

charging efficiency 充电效率

charging mode/charging system 充电制度

charging station 充电站

check light 检查灯

choke/choke coil 阻流圈

circle ring 火花环

circuit breaker 断路器/自保开关

circuit network 网路

clamp 卡箍

clamp semi-ring 半卡环

clapper 拍合

claw 凸爪

clip 座夹

clutch 离合器

coaxial line 同轴线

coil circuit voltage 线圈电压

coil winding 线圈匝

cold run/crank 冷转

cold state 冷态

common cable 公用电缆

common connecting point 公共接点

communication radio set 通讯电台

communicator circuit breaker board 通讯员自动开关板

communicator DC instrument panel 通讯员直流仪表板

communicator PDB 通讯员配电盒

commutating segment/commutator segment 换向片

commutating spark 换向火花

commutator 整流子

compensation coil 补偿线圈

compensation winding 补偿绕组

complementary charging 补充充电

component vibration damping 机件减震

component/element 机件

composite power supply PDB 复合供电配电盒

compound motor/compound-wound motor 复激电动机

compressed air 压缩空气

conducting check 导通检查

conducting meter 导通表

conductive tape 导通带

connecting terminal 接线端子

connection plug 连接插头

console 操纵台

console panel 操纵台面板

constant magnetic flux 定值磁通

constant speed unit 恒速装置

constantan 康铜

contact coupling 接触偶

contact pile 接触片

contact point/contact 触点

contact resistance 接触电阻

contact voltage drop 触点电压降

contactor 接触体

contactor panel　接触器板

control device　控制装置

control switch　控制开关

convex lug　凸耳

cooling air　冷却空气

cooling fan　冷却风扇

copilot control stand　右驾驶台

copper grains　铜粒

copper layer　铜层

core　铁心

corrosion　腐蚀

cotter pin　开口销

cotton thread　锦丝线

cotton yarn　棉纱

counter sunk screw　沉头螺钉

counter-time-delay/anti-delay property　反延时特性

coupling　联轴器

cover　表蒙/蒙盖

cover plate　盖板

crack　裂纹

craze crack　龟裂

crest factor　波峰系数

cross-section area　截面积

current consumption　消耗电流

current divider　分流器

current stabilizer　稳流器

current transformer　电流互感器

damage　损伤

damper/shock absorber　减震器

damping elasticity　减震弹性

DC motor　直流电机

defective part　故障件

deformation　变形

demagnetize　去磁

density　密度

dent　压痕

depreservation　启封

deterioration　变质

dial gauge　千分表

diaphragm　膜片

differential coil　差动线圈

differential control relay　差动操纵继电器

differential low-voltage relay　差动低限保护器

differential relay　差动继电器

diode　二极管

diphenyl ethers lacquer　二苯醚漆

disassembly check　分解检查

distilled water　蒸馏水

distortion factor　失真系数

drive mechanism　传动机构

drive shaft/transmission shaft　传动轴

driving end　传动端

driving gear　传动齿轮

driving part　传动部件

dual-circuit power supply contactor　双路供电接触器

dust-proof cover　防尘保护罩

dynamometer　测力计

edge angle　棱角

elasticity deformation　弹性变形

electric arc　拉弧

electric iron　电烙铁

electric winch　电绞车

electrical circuit　电气线路

electrical component　电气元件

electrical connector　电连接器

electrical contact　电接触

electrical contact point　电接触点

electrical hot knife　电热刀

electrical instrument　电气仪表

electrical network　电气网路/电网

electricity　电力

electrode　电极

electrolyte　电解液

electromagnet　电磁铁

electromotive force　电动势

electron motion　电子运动

emergency airdrop PDB　应急空投配电盒

emergency bus bar/emergency bus　应急汇流条

emergency cut-off dropping fuse　紧急断开投放熔断器

emergency dropping fuse　应急投放熔断器

emergency horizon/standby horizon/standby gyro horizon　应急地平仪

emergency network　应急网路

emergency power supply　应急电源

emergency power supply box　应急电源盒

emergency power supply switch　应急供电开关

engine coupling　发动机联轴节

engine ground running-up　发动机试车

engine nacelle　发动机舱/发房

epoxy fiber glass reinforced plastic　环氧玻璃钢

excitation winding　激磁绕组

exciting circuit/excitation circuit　激磁电路

exciting coil　激磁线圈

exciting current　激磁电流

exposed core　裸露线芯

extension shaft　外伸轴

external resistor　外接电阻

factor　因数

fairing　整流罩

fastener　紧固件

fastening washer　紧固垫圈

faulty indicating light/failure indicating light　故障指示灯

faulty signal light　故障信号灯

faying surface　贴合表面

feathering system　顺桨系统

feed system　馈电系统

feedback coil　反馈线圈

feeder　馈(电)线

feeder diagram/feeding diagram　馈电图

ferric oxide　氧化铁

filtration　滤波

fine adjustment box　精调盒

fire wall　防火墙

fire warning system　火警信号系统

fire-protection control　防火操纵

fixed screw　固定螺钉

fixing clamp　固定卡箍

fixing nut　固定螺母

flange　法兰盘

flap position indicator　襟翼位置指示器

flashing beacon　闪光信标

flat hexagon nut　扁六角螺母

flexible shaft　软轴

fluctuate　波动

fluorescent mark　荧光标识

formation signal　编队信号

freezing zone　结冰区

frequency meter　频率表

front cabin　前舱

frontal airflow　迎面气流

fuel pump　燃油泵

fuse cap　熔断器帽

fuse pipe cap　熔断丝管帽

fuselage skin　机身蒙皮

gas turbine engine　燃气涡轮发动机

gasket　垫板

gasoline　汽油

gear box　传动机匣

gear shaft　齿轮轴

general power distribution board　总配电盒

glass tape　玻璃布带

graduation value　分度值

grease　润滑脂

grinding surface　磨合面

groove　压槽

ground power supply　机场电源

ground power unit (GPU)　地面电源车

grounding wire/earthing wire　接地线

half-round head screw 半圆头螺钉

hand clamp 手动钳

harmonic content 谐波含量

heat conduction direction 热传导方向

heat preservation measure 保温措施

heat-shrinkable tube 热缩管

hexagon head bolt 六方螺栓

high frequency communication system 高频通信系统

high-altitude oxygen supply signal 高空用氧信号

holding bracket/mounting bracket 固定支架

hollow shaft 空心轴

hot state 热状态

housing/casing 壳体

hub spinner 桨帽

humid heat test 湿热性试验

hydraulic control 液压操纵

identification mark 识别标志

igniter 点火器

illuminating light 照明灯

impurity 杂质

inductive load 电感性负载

injector 注射器

inner cavity 内腔

inspection light 检验灯

installation position 安装位置

instrument landing system (ILS) 仪表着陆系统

instrument panel 仪表板

insulating and dielectric strength 绝缘抗电强度

insulating layer 保温层

insulating resistance 绝缘电阻

insulating trough 绝缘槽

insulation layer 绝缘层

insulation tape 绝缘带

interlock relay 联锁继电器

interlocking contactor 联锁接触器

internal resistance 内阻

inversion output 逆变输出

inversion unit 逆变单元

inverter 变流机

inverter bridge 逆变桥

iron core sleeve 铁心筒

iron yoke 铁轭

junction position 接合处

key part/critical part 关键件

knob 旋钮

lead wire port 引线口

light cover 灯罩

linear acceleration 线加速度

load contactor 负载接触器

load current 负载电流

load power-on contactor 负载接通接触器

load torque 负荷力矩

lock latch 锁扣

lock nut 锁紧螺帽

locking wire 锁紧丝

log book 履历本

long-distance navigation system 远距导航系统

low-temperature grease 低温润滑脂

lubricating oil 润滑油

lug 接线片

Mach meter 马赫数表

magnetic amplifier 磁放大器

magnetic flux 磁通

magnetize 充磁

magnifier 放大镜

main cable 主干线

mainlanding gear well 主起落架舱

maintenance technician 机务人员

manufacture technology 制造技术

manufacturer 承制厂

mark sleeve 标识套管

measuring instrument 测量仪表

mechanical damage　机械损伤

mechanical pitting　麻点

mechanical self-lock system　机械自锁系统

metal brush　金属刷

metal luster　金属光泽

metal powder　金属末

metallic scrap/metal chip　金属屑

mica capacitor　云母电容器

microhmmeter　微欧表

microswitch　微动开关

modulating stage　调制级

monitor　监控器

motor　电机

mounting bolt　安装螺栓

mounting bracket　安装架

mounting panel　安装面板

movable contact　活动触头

movable member　活动构件

multimeter　三用表/万用表

navigation compartment/navigation cabin
领航舱

navigation signal　航行信号

needle file/handle file　什锦锉

negative pole　负极

negative wire　负线

neutral gas　中性气体

neutral position　中立位置

nickel cadmium battery　镉镍蓄电池

nickel-cadmium alkaline battery　铬镍碱性蓄
电池组

no-load voltage/zero load voltage　空载电压

nomenclature　名称

nominal section area　标称截面积

non-shielding wire　非防波导线

non-twisting wire　非扭绞导线

normally closed contact　常闭触点

normally opened contact　常开触点

off-set plier　偏口钳

oil department　油料部门

oil proof device　防油装置

oil stain　油迹

onboard circuit　机上电路

open bearing　开式轴承

open circuit　开路

operation indicating light　工作指示灯

operation state　工作状态

operation surface　工作面

outline dimension　外廓尺寸

output circuit　输出电路

output frequency　输出频率

output power　输出功率

output shaft　输出轴

output switch　输出开关

output voltage　输出电压

oven　烘箱

over charging　过充电

overhaul period　翻修期

overload　过载

over-voltage protector　过压保护器

over-voltage surge　过压浪涌

oxidate　氧化物

oxidation-reduction reaction　氧化还原反应

oxide layer　氧化层

oxidization　氧化

parachute cord　伞绳

parachuting signal　空降信号

parallel rating factor　并联修正系数

part　零件

partition　隔板

performance　性能

periodic chart　周期表

permanent magnetic field　永久磁场

permanent terminal　死接头

phase　相位

phase bus bar　相汇

pin　插针/销钉

pin connector　插销接头

placard　标志牌

planetary speed reducer　行星式减速器

plastic cloth　塑料布

pliers　压钳

plug　插销

plug/air lock　气塞

polarity relay　极性继电器

polyamide　聚酰胺

polyvinyl chloride（PVC）　聚氯乙烯

positive and negative sintering polar plate
正负烧结式极板

positive wire　正线

potential difference　电位差

potentiometer　电位计

power consumption equipment　用电设备

power distributionboard　电源配电盘

power distribution equipment　配电设备

power distribution panel　配电板

power distribution protection device　配电保护装置

power supply panel/distribution board　配电盘

power supply unit　电源装置

power-off reverse current　断开反流

power-on check　通电检查

precision instrument　精密仪表

preparation circuit　准备电路

preservation life　油封期

pressing stick　压杆

pressure gauge　压力表

printed circuit board assembly　印刷电路板组件

printing plate　印刷板

propeller　螺旋桨

propeller stop relay　螺旋桨限动继电器

protection circuit　保护电路

protective belt　防护带

protective cover　保护盖/保护罩

pulling force　拉力

pulsating button　脉动按钮

pulse signal　脉冲信号

purified water　净化水

push-pull circuit breaker　按拔式电路断路器

quick-release ring　快卸环

quick-release screw　快卸螺钉

radial swash value　径向跳动量

radiating system　散热系统

radiating tube　散热管

radio interference　无线电干扰

ramie thread　苎麻线

rated output power　额定输出功率

rated value　额定值

RC time delay circuit　阻容延时电路

rectifier　整流器

rectifying piece　整流片

reference point　基准点

refilling port/charging port　注液口

refueling distribution board　加油配电盘

relative humidity　相对湿度

relay protection device　继电保护装置

release voltage　释放电压

resistance compensation　电阻补偿

resistance load　电阻性负载

reverse current　反流

reverse force　换向力

reverse magnetic field　反磁场

reverser/commutator/converter　换向器

rheostat　变阻器

rheostat/variable resistor　可变电阻

root mean square current　均方根电流

rotor　转子

rubber band　皮筋

rust　锈蚀

safe hole　保险孔

safety paint　保险漆

safety wire/locking wire　保险丝

sand paper　砂纸

scale/dial　刻度盘

scratch　划伤

screw rod　螺杆

sealed bearing　密封轴承

sealed double-pole changeover switch　密封双极转换开关

sealing cover　密封盖

sealing partition　密封隔板

sealing place　封口处

sealing protection　密封保护

sealing putty　密封腻子

sealing sleeve　密封套管

sealing tape　密封带

sealing terminal post　密封接线柱

sealing washer　密封垫

section　截面/型材

selector switch　选择开关

self-lock nut　自锁螺母

sensitivity circuit　敏感回路

series coil　串激线圈

series-wound motor　串激式电机

service ladder　工作梯

shielding cable　防波电缆

shielding sleeve　防波套

shielding wire　防波导线

shock-absorbing pad　减震垫

short circuit　短路

sidewall　侧壁

silicon rubber　硅橡胶

silk cloth　绸布

simulation test　模拟试验

single line power supply　单线制供电

slip ring　滑环

socket　插头座

solder　焊料

solder/braze　钎焊

soldering point　焊接点

spanner/wrench　扳手

spar　大梁

spare cap　备用帽

spare wire　备用线

spark　火花

special tool　专用工具

specified level　规定级数

specified value　规定值

speed reducer　减速器

spiral washer　旋转式清洗器

spline　花键

spot　斑点

spring diaphragm　弹簧膜片

spring washer　弹簧垫圈

stagnation　紧滞/卡滞

standby power supply fuse　备用电源熔断器

start control system　起动控制系统

starter generator　起动发电机

starting box　起动箱

starting contactor　起动接触器

starting controller　起动控制器

starting ignition　起动点火

starting power supply　起动电源

static inverter　静止变流器

stator　定子/静子

steel cable clamp　钢丝箍

stop gear　止动齿

stop washer　止动垫圈

storage battery　蓄电池组

storage period　贮存期

strap　卡带/扣带

stringer　长桁

structural component　结构部件

surge voltage　浪涌电压

swaged connector　收压接头

swing bolt　摇摆螺栓

switching device　转换装置

switch-type sealing relay　开关式密封继电器

synchro generator　同步发电机

synchronous AC generator　同步交流发电机

tachometer　转速表

tail anti-icing　尾翼防冰

tail heating power distribution board　尾翼

加温配电板

talcum powder 滑石粉

technical data 技术数据

technicalservice 技术维护

temperature compensating resistor 温度补偿电阻

temperature control system 调温系统

temperature reducing measure 降温措施

tension bolt 拉紧螺栓

terminal bolt 接线螺栓

terminal box 接线盒

terminal post 接线柱/极柱

terminal switch 终点开关

test bench/test stand 试验台

thermometer 温度表

three-phase AC power circuit 三相交流电源电路

throttle lever 油门杆

throttle/air restrictor 节气门

tighten torque 拧紧力矩

time delay relay 延时继电器

time delay resistor 延时电阻

time-lag fuse 惯性熔断器

timing mechanism 定时机构

tin crimping 锡压接

tin soldering 锡钎焊

torque meter 扭矩表

torque wrench 定力扳手

traffic alert and collision avoidance system 空中交通告警防撞系统

transformer 变压器

transient voltage 瞬态电压

transistor 晶体

triode 三极管

turbine starter/generator 涡轮发电装置

turboprop engine 涡轮螺旋桨发动机

turn and slip indicator/turn coordinator 转弯仪

two-step charging 二阶段充电

ultrasonic washer 超声波清洗器

under-voltage protection 欠压保护

under-voltage surge 欠压浪涌

vent pipe 通风管

ventilator 通风器

vernier caliper 游标卡尺

very high frequency（VHF） 甚高频

VHF Omni-direction range（VOR） 伏尔

vibration indicator 振动指示器

visual check 目视检查

voltage collecting fuse 电压采集熔断器

voltage divider 分压器

voltage measuring circuit 电压测量线路

voltage modulation 电压调制

voltage oscillation 电压振荡

voltage reducing resistor 降压电阻

voltage regulating and switching equipment 调压转换设备

voltage regulating coil 调压线圈

voltage regulating device 调压装置

voltage regulating resistor 调压电阻

voltage regulator 调压器

voltage sealing relay 电压密封继电器

voltage stabilizer 稳压器

voltage stabilizing coil 稳压线圈

voltmeter 电压表

washer 垫圈

waterproof cloth 防雨布

wave filter 滤波器

waveform 波形

waxed ramie string 打蜡苎麻线

web 腹板

wedging state 楔紧状态

welding process 焊接工艺

welding seam 焊缝

wheel chock 轮挡

winding 绕组

windshield glass 风挡玻璃

wing root 翼根

wire bundle　线束

wire connector　导线接头

wire end　导线端头

wire gauge　线规

wire stripper　剥线钳

wiring board group　接线板组

wiring connection　导线转接

workshop　内场

worm　蜗杆

worm gear　蜗轮/螺旋齿轮

zero bus bar　零汇

EQUIPMENT/FURNISHINGS
设备/装备

abrasion-proof brake　防磨套制动器

acid　酸

activate　触发

adapter cylinder　转接筒

air charging device　充气装置

air transport　空运

airborne parachute horn　空降喇叭

airdrop　空投

aisle　通道

alkali　碱

aluminum foil　铝箔

antenna gain　天线增益

approval　批准

arc plate　弧形板

armrest with soft cushion　带软垫扶手

auxiliary ramp　辅助货桥

axial pulling off force　轴向拔脱力

bail　水杓

bakelite base　胶木座

battery capacity　电池容量

battery compound　电池组件

beacon transmitter　信标发射机

bench-type seat　平板式座椅

blowing nipple　吹气嘴

bonding jumper　搭接线

bonding resistance　搭接电阻

brake gear　制动齿

buckle　锁扣

buffer rope　缓冲绳

built-in antenna　内天线

buoyancy　浮力

buzz　蜂鸣声

caddice ends　毛线头

calculation　测算

cam　凸轮

canopy　船篷

capture　捕捉

carbon dioxide bottle　二氧化碳气瓶

cargo door control system　货舱门操纵系统

cargo loading device　装货设备

cargo transportation equipment　货运设备

cartridge chamber　弹舱

casting control system　投放控制系统

C-clamp　弓形卡箍

center instrument panel　中央仪表板

chest strap　胸带

combined harness-shoulder harness　组合式安全带-肩带

commutate　换向

convex block　凸块

cotton velvet-free absorbing cloth　无棉绒吸湿性布料

crack　断裂

crank　摇把

crash axe　应急斧

cross component force　侧向分力

crosshead pin joint　十字销接头

crosslink　交联

cushion net　垫网

cylindrical pin　圆柱销

deflection　挠度

detachable collar　可卸套圈

deterioration　变质

directional objective signal　定向目标信号

disengaging lock　脱离锁

dovetail plate　燕尾板

drift angle　偏流角

drifting deflection mechanism　偏流侧倾机构

drum wheel　鼓轮

electric percussion device　电燃撞击器

electric signal gun　电动信号枪

electric winch　电动绞车

electrical heating board　电热板

electrical oven　电烤箱

electromagnetic brake clutch　电磁制动离合器

emergency locator　应急定位仪

energy consumption brake　耗能制动

enforcing belt　加强带

escape procedure/departure process　离机过程

ethyl acetate　乙酸乙酯

evacuation　撤离

exciting manner　激磁方式

expiration　到期

external antenna　外天线

external manrope　外扶手绳

extraction parachute casting device　牵引伞投放装置

fastening pulley　拉紧滑轮

feed diagram　馈电图

field of view　视场

filler block　垫块

fixed lifting rope/fixed hanging cord　固定吊绳

flammable　易燃品

flat-head pin　平头销

flexible pin　软插销

flight engineer　机械师

float　浮筒

floating anchor　浮锚

foldable seat　折叠式座椅

footpedal　脚蹬

force component pad　分力垫

force-distributing chock　分力轮挡

fork rod　叉杆

formation airdropping and parachuting　编队空投空降

four-pole DC reversible automatic vent motor　四极直流可逆自动通风电机

gasket　垫片

gasoline　汽油

grease gun　注油枪

grease nipple　注油咀

ground reception station　地面接收站

hand transmission device　手摇传动装置

handbag　提包

handling pulley　装卸滑轮

hang/suspend　悬挂

high self-locking nut　高自锁螺母

high-pressure hose　高压软管

high-temperature engagement contactor　高温接通接触器

hinge　铰链

horizontal deflection angle　横偏角

horizontal sliding rail　水平滑轨

hull bottom　船底

identification　识别

imaging　成像

inflate　充气

interlining cloth　夹里布

intermittent sound　断续声响

iron stick　铁棍

jack-up　顶起

joint socket　接合座

level flight speed　水平飞行速度

life craft/life boat　救生船

life equipment cabinet　生活设备柜

life-saving HF radio　救生短波电台

life-saving parachute　救生伞

lifting control handle　升降操纵手柄

lifting cylinder　升降筒

lifting pin　升降插销

locating pin　定位销

lock ring body　锁环体

locking cone　锁锥

locking key　锁键

lubricating grease　润滑脂

magnesium alloy casting　镁合金铸件

main jacket band　主套带

main pulley　主滑轮

manual oar　手划桨

mayday signal　求救信号

mechanic's footboard　机械员脚踏板

medical staff　医务人员

middle orbit　中轨道

mildew　发霉

mission control center（MCC）　任务指控中心

modulation　调制

mooring equipment　系留设备

mooring fitting joint　系留接头

mooring net　系留网

mooring steel cable　系留钢索

mooring strap　系留带

mosquito repellent tissue　驱蚊纸

movable backrest　活动靠背

movable part　活动部位

multimeter　万用表

navigator cabin　领航舱

near-earth satellite　近地卫星

nicked teeth　刻齿

nitrogen　氮气

notch　槽口

observing angle　观测角

optical system　光学系统

overwater survival equipment　水上救生装备

oxygen chain　氧气锁链

oxygen flowmeter　氧气示流器

parachute　空降

parachute opener　开锁器

parachute steel cable　跳伞钢索

parachute support　伞托

parachute-cord box　伞绳盒

parachute-opening shock　开伞过载

parachuting signal device　空降信号装置

parachuting windshield plate　空降挡风板

paratrooper/parachutist　伞兵

patch　补丁

perforate/punch　穿孔

pillar　立柱

pilot chute　引导伞

pin socket　销座

pin support　插销支座

plateau oxygen supply apparatus　高原供氧器

polyamide belt　锦丝带

power distribution box　配电盒

precision instrument grease　精密仪表脂

press plate　压板

pulse current　脉冲电流

punctuate　刺穿

railing for parachute discharging operator　放伞员栏杆

ratchet mechanism　棘轮机构

reel　卷筒

remote electric control device　远距电控装置

request backup　求援

restraining chock　限动轮挡

retainer　挡件

roller device　滚棒装置

rollway　滚道

rotary connector　转动接头

rotation mechanism　转动机构

rotation pin　旋转插销

rough edge　毛边

saddle leather　马鞍具革

safety stopper　保险弹扣

salvage　打捞

sand paper　砂纸

scatter　分散

scrape　擦伤

seat surface/seat-top　椅面

sector piece　扇形件

semi-cross　半剖

series start coil　串激起动线圈

sew　缝

shaft pin　轴销

sharp nose pliers　尖嘴钳

shooting method　发射方式

shoulder strap　肩带

shroud　罩盒

side pulley　侧滑轮

sighting angle　瞄准角

sighting system　瞄准系统

signal flare　信号弹

signal gun chamber　信号枪舱

signal liaison　信号联络

single shot recoil　单发发射后座力

slant brace/diagonal strut　斜撑杆

sliding code　滑码

sliding seatback　滑动椅背

sliding wire　滑丝

slotted screwdriver　一字螺刀

smear/coat　涂(润滑脂)

soft brush　软毛刷

solid anodization　固态阳极化

solvent　溶剂

spacer　隔离片

spanner/wrench　扳手

spherical socket　球面窝

staggered arrangement　交错排列

staggering　错牙

stalling　失速

stationary tiedown ring/mooring ring　固定
式系留环

steel cable pad　钢索垫板

stop block　止动块

store room　库房

straighten　拉直

stretcher　担架

support　支撑

support cylinder　支撑筒

support roller　支撑滚轮

support tray　托盘

supporting stick　支撑辊

surveillance　侦察

survival kit　救生物品

suspension bracket　悬挂支架

tail support　尾撑

talc powder　滑石粉

terminal board　接线板

thread peeling-off　螺纹脱落

torque spanner　定力扳手

towing steel cable　牵引钢索

tracked equipment　履带式设备

trans-illuminated panel　导光板

turnbuckle　松紧螺套

unpowered wheeled device　无动力轮式设备

vertical landing speed　垂直着陆速度

vertical sliding bracket　垂直滑动架

vertical tail tip　垂尾翼尖

waist belt　腰带

warranty period　保证期

water bag　水囊

water boiler　开水器

wedge　楔子

wrench　扭伤

FIRE PROTECTION　防火

AC voltage　交流电压

acceleration　加速度

accessory gear box　附件传动机匣

acid liquid　酸性液体

adapter seat　转接座

adjusting nut　调整螺帽

altitude 海拔高度

aluminum foil membrane 铝箔膜

aluminum pipe 铝管

aluminum shell 铝外壳

ambient temperature 环境温度

analog sensor 模拟传感器

appearance 外形

armature(iron) 衔铁

asbestos cloth 石棉布

auto control fire bottle 自动控制灭火瓶

auto fire extinguishing 自动灭火

automatic air bleed port 自行放气信号口

automatic fire check box 自动灭火检查盒

automatic fire circuit 自动灭火电路

automatic return 自动返回

auxiliary travel 辅助行程

band switch 波段开关

bellow/corrugated tube 波纹管

belly 机腹

bipolar spring pressing mechanism 两极弹
簧按压机构

blaster 导火线

bonding jumper 搭铁线

bourdon tube 包端管

bracket 支架

breakdown 击穿

cable plug 电缆插头

cable socket 电缆插座

capacity 容量

carbon deposit 积炭

center wing 中央翼

centrifugal acceleration 离心加速度

certificate 证明书

changeover switch 转换开关

check relay 检查继电器

check valve 单向活门

chemical bond-breaking function 化学断键
作用

chemical chain reaction 化学链锁反应

chrome-nickel-bronze thermocouple 铬镍合
金-青铜热电偶

circuit breaker 自保开关

circuit of detonator 爆炸帽电路

clamp 卡箍

clamp ring 卡圈

clearance 间隙

cold end 冷端

combustion chamber/combustor 燃烧室

combustor casing 燃烧室壳体

compression nut 压紧螺母

conical horn tube/flared tube 锥形喇叭管

conical screw 锥形螺纹

connecting wires 连接导线

construction 构造

consumption 功耗

contact 接触点

control box 控制盒

control button 控制按钮

control panel 控制板

control signal 控制信号

converter 变换器

cover 罩子

crack 裂纹

critical temperature 临界温度

current leakage 漏电流

current strength holding 保持电流强度

cylindrical slide block 圆柱滑块

dent 压痕

detection 探测

detonator 爆炸帽

detonator circuit 传爆电路

dewatered kerosene 脱水煤油

differential amplifier 差动放大器

differential thermocouple 差动势热电偶

differential travel 差动行程

direct current voltage 直流电压

displacement 位移

driving component 传动元件

duration 持续时间

electric plug 电插销

electrical level 电平

electromagnet 电磁铁

emergency forced landing 应急迫降

end face 端面

engine air inlet fairing 发动机进气道整流罩

engine inner cavity 发动机内腔

exhaust exit of fire bottle 灭火瓶放气口

explosion protection 防爆

explosion starting device 爆炸起动装置

external fire detection system 外部火警探测系统

external fire extinguishing circuit 外部灭火电路

external screw thread/male screw 外螺纹

external warning circuit 外部告警电路

false signal 假信号

filter net 滤网

fire bottle 灭火瓶

fire bottle detonator indicating light 灭火瓶爆炸帽指示灯

fire circuit 灭火电路

fire detection 火警探测

fire detection system 火警探测系统

fire extinguishing 灭火

fire extinguishing agent 灭火剂

fire extinguishing button 灭火按钮

fire extinguishing check box 灭火检查盒

fire extinguishing check switch 灭火检查开关

fire extinguishing circuit 灭火电路

fire extinguishing control box 灭火控制盒

fire extinguishing device 灭火设备

fire extinguishing joint casing 灭火接头壳体

fire extinguishing pipeline 灭火输送管路

fire extinguishing selection switch 灭火选择开关

fire extinguishing signal 灭火信号

fire extinguishing test button 灭火检查按钮

fire extinguishing test changeover switch 灭火检查转换开关

fire protection system 防火系统

fire relay box 灭火继电器盒

fire signal 火警信号

fire source 火源

fire warning button indicating light 火警按钮指示灯

fire warning sensor 火警传感器

fire-warning signal light 火警信号灯

firing-pin 撞针

flight engineer 机上机械师

flow rate of the medium 介质流速

fluorescence powder 荧光粉

forced landing fire microswitch 迫降灭火微动开关

forced landing fire relay 迫降灭火继电器

free end 自由端

front casing area 前机匣区

gate 闸门

glass cloth 玻璃布

glass fiber 玻璃纤维

g-load factor 过载系数

glue 胶

glycerine 甘油

graduation value 分度值

grind 研磨

hairspring 游丝

hard paper plate 硬纸板

heating area 受热面积

heating muff 加温套

high stool 高凳

horn mouth 喇叭口

hot bath 热水淋浴

hot end 热端

hydraulic bed 液压床

hydraulic liquid 压力油液

hydraulic test 液压试验

hydrogen atom　氢原子

igniter　引火器

impact acceleration　冲击加速度

indication control system　指示控制系统

inertia　惯性

inlet hole　进气小孔

inlet pressure　进口压力

inlet union　进气管嘴

input end　输入端

insulation dielectric strength　绝缘抗电强度

insulation resistance　绝缘电阻

integrating circuit　积分电路

interchangeability　互换性

interlock relay　联锁继电器

isolation valve　隔断活门

lag error　迟滞误差

level-flight air-charging switch　平飞充气开关

light　灯光

liquid CO_2　液体二氧化碳

liquid fire extinguishing agent　液态灭火剂

litharge　铅黄

locating pin　定位销钉

location where easily catches fire　易起火部位

lock valve　锁闭活门

locking wire　保险

manual control fire bottle　人工控制灭火瓶

manual fire extinguishing　人工灭火

maximum operating pressure　最大工作压力

maximum scale value　最大刻度值

metal wire　金属丝/导线

microswitch　微动开关

minimum operative voltage　协动最小电压

multi-selection switch　多位选择开关

neutral gas bottle　中性气体瓶

nickel wire　镍丝

nipple　管嘴

non-operating end　非工作端

ohmmeter　欧姆表

oil chamber　滑油腔

onboard power source　机上电源

operating altitude　工作高度

operating end　工作端

operating pressure　工作压力

operating temperature　工作温度

operating travel　工作行程

operating voltage　工作电压

operation electrical level adjusting circuit　动作电平调整电路

outlet pressure　出口压力

outlet relay　输出继电器

outlet union　出气管嘴

output end　输出端

overload capacity　过负荷能力

overload current　过负荷电流

oxidizing　氧化

oxygen mask　氧气面罩

parchment　羊皮纸

perforator　穿孔器

pin　插钉

pinion　小齿轮

pipe holder　管夹座

pipeline heater　管路加温器

piston chamber　活塞室

pit　顶坑

plug　插销头

plumbous compound　铅化物

polytetrafluoroe thylene strip　聚四氟乙烯密封带

portable fire bottle　手提式灭火瓶

positive pole　正极

powder chamber　火药室

power amplifier　功率放大器

power stabilizer　电源稳压器

power switch of fire extinguishing system　灭火系统自动开关

power-on check　通电检查

pressure of driving component　传动元件的压力

pressure reducer　减压器

pressure releasing pipe　排压管

pressure solenoid valve　气压电磁阀

pressurization switch　增压开关

pressurized air　高压气体

pulse current　脉冲电流

rated load　额定负荷

rated weight　额定质量

rear fairing　后整流罩

red brass gasket　红铜垫片

red fire warning button light　火警按钮式红灯

relative humidity　相对湿度

resistance value　阻值

resistance wire　电阻丝

resistive current　电阻性电流

restricting diameter　限流直径

restricting heater　限流加温器

restricting nipple　限流嘴

restricting nozzle　限流喷嘴

restricting orifice　限流孔

reverse flow　逆向流通

reverse pressure　反向压力

roughness　粗糙度

round metal piece　圆金属片

rubber gloves　橡胶手套

safety diaphragm　保险膜片

sand hole　砂眼

scale/dial　刻度盘

scoop channel　凹槽

sector gear　扇形齿轮

sediment pool　沉淀池

self-test function　自检功能

sensor model　传感器型号

service condition　使用条件

service ladder　工作梯

shell　壳体

shielding cover　屏蔽套

short circuit　短路

signal light　信号灯

signal safety device　信号保险装置

signal union　信号接管嘴

siphon pipe　虹吸管

socket　插头座

solenoid valve　电磁活门

special condition　特殊情况

special cotton uniform　专用棉织工作服

spherical fire bottle　球形灭火瓶

spiral line　螺旋线

split-second response　瞬时动作

spot bolt　销钉

stainless steel　不锈钢

storage temperature　贮存温度

store room　库房

surface coating　表面镀层

surrounding medium temperature　周围介质温度

system operating inertia　系统动作惯性

system operating temperature　系统动作温度

system releasing temperature　系统释放温度

system warning stability　系统报警稳定性

temperature difference　温差

temperature relay　温度继电器

terminal pressure　终点压力

textolite gasket　夹布胶木垫片

thermal sensitive element　热敏元件

thermo couple assembly　热电偶组件

thermo electromotive force　温差电动势

thermo potential　温差电势

thermo potential value　热电势值

thermocouple　热电偶

thread die　扳牙

thread neck　螺纹颈

three-way pipe/three-port pipe　三通管

threshold mechanism　门限装置

touch down　接地

travel　行程

trigger joint　触发接头

turbine rotor area　涡轮转子区

turbine shaft cavity　涡轮轴腔

unipolar-converted　单极转换

utmost position　极端点位置

valve assembly　活门组件

valve base/valve seat　活门座

vibration frequency　振动频率

voice prompt　语音提示

voltage drop　电压降

warning cancelling inertia　消警惯性

warning electromotive force　报警电动势

warning inertia　报警惯性

warning temperature　报警温度

welding spot　焊点

wind speed　风速

wire　金属丝

wood stick　木棍

FLIGHT CONTROLS　飞行操纵

abrasive sand paper　研磨砂纸

actuating mechanism　执行机构

adapting sleeve　转接套

adjustable pull rod　可调拉杆

adjusting joint　调整接头

aileron actuator　副翼舵机

aileron control system　副翼操纵系统

aileron effect　副翼效应

aileron lock mechanism　副翼锁机构

aileron position feedback sensor　副翼位置反馈传感器

aileron servo tab　副翼随动补偿片

aileron trim tab　副翼调整片

aircraft control system　飞机操纵系统

airflow　气流

airfoil　机翼翼型

airtight box　气密箱

airtight connector　气密接头

airtight line　气密线

airtightness　气密性

airtightness test　气密实验

airtightness test bed　气密试验台

alignment　对正

anti-shedding shaft　止脱轴

autopilot　自动驾驶仪

autopilot emergency cutoff button　自动驾驶仪紧急断开按钮

auxiliary control system　辅助操纵系统

aviation cleaninggasoline　航空洗涤汽油

aviation grease　航空润滑脂

axis　轴线

balance　平衡

ball bearing　滚珠轴承

ball helical mechanism　滚珠螺旋机构

ball holder　夹珠圈

ball nut　滚珠螺母

ball rail　滚珠轨道

bending rigidity　抗弯刚度

bent nose pliers　斜口钳

bent shaft　弯头轴

bevel gear　伞齿轮/斜齿轮

binding-off/unbonding　收口

blanking cap　堵帽

boundary　边界

box spanner/box wrench　套筒扳手

bracing　撑杆

brake　制动器

brake accumulator　刹车蓄压瓶

brake handle　制动手柄

brake lock　制动锁

brake mechanism　制动机构

brake pin sleeve　制动销套

brake pressure-reduction valve　刹车减压活门

brake pull rod　制动拉杆

brake rocker　刹车摇臂

braking rocker/brake rocker　制动摇臂

braking steel cable　制动钢索

broken wire　断丝

bronze bushing　青铜衬套

brush　毛刷

bundle　捆

burr　毛刺

button seat　按钮座

casing assembly　壳体组件

catch pin　止脱销

ceiling　顶棚

center instrument panel　中央仪表板

chain plate　链板

chain sleeve　串动套罩

chain wheel　链轮

changeover switch　转换开关

changeover valve　转换活门

chrome cattlehide upper leather　铬制黄牛正鞋面革

chrome plate　镀铬

circuit breaker　保险电门/断路器

circuit breaker board　自动开关板

clamping plate　夹板

coating　包复层

combined bracket　组合支架

combined control　混合式操纵

combined flat nose pliers　组合平口钳

communicator circuit breaker board　通讯员保险电门板

conic bolt　锥形螺栓

connecting pipe　联接管

connector　接头

contact　触点

control force　操纵力

control hand wheel　操纵手轮

control handle　操纵手柄

control linkage　操纵线系

control mechanism　操纵机构

control surface lock system　舵面锁操纵系统

control switch　操纵开关

control wheel　驾驶盘

control wheel framework　驾驶盘骨架

coordination performance　协调性

copilot console　右驾驶员操纵台

copper nut　黄铜螺帽

cotton thread　棉线

cotton yarn　棉纱

crazing　龟裂

cross joint　十字接头

curvature　曲度

cutting pliers　克丝钳

cylinder wall　外筒壁

defect/flaw　缺陷

deflection　挠度

deflection angle limiting mechanism　偏角限制机构

deflection angle sensor　偏角传感器

deflection indicator　偏转指示器

deflection mechanism　偏转机构

delay/lag　迟滞

detergent oil　洗涤油

deterioration　变质

differential control mechanism　差动操纵机构

distributing column rocker　分配摇臂

distributor　分配器

double-lug rocker　双耳摇臂

driving　驱动

driving assembly　传动组件

driving mechanism　传动机构

driving pipe　传动管

driving rod　传动杆

driving shaft　传动轴

drum　鼓轮

dual-arm rocker　双臂摇臂

duralumin pipe　硬铝管

dynamometer　测力计

eccentric shaft　偏心轴

edge distance　边距

elastic deformation　弹性变形

electrical mechanism　电动机构

electrical plug　电插头

electrical signal　电讯号

electrical signal mechanism　电气讯号机构

electro-hydraulic control actuator　电液舵机

elevator control system　升降舵操纵系统

elevator trim tab　升降舵调整片

elimination　消除

ellipse degree　椭圆度

emergency extension system　应急放系统

emergency power-off button　应急断电按钮

emergency retracting/extending changeover switch　应急收放转换开关

end nut　端螺帽

end play　游隙/轴向间隙

end rib　端肋

engagement　啮合

extension travel　放行程

extension/retraction link rod　伸缩连杆

failure　失效

fastener　紧固件

feedback sensor　回输传感器

felt ring　毡圈

file　锉刀

filler　注油器

fillet　圆滑过渡

fine felt　细毛毡

fine rope　细绳

fine sand paper　细砂纸

fit　配合

fitting precision　配合精度

fixture　夹具

flap control circuit　襟翼操纵电路

flap control system　襟翼操纵系统

flap driving assembly　襟翼传动组件

flap driving rod　襟翼传动杆

flap hydraulic driving device　襟翼液压传动装置

flap hydraulic driving system　襟翼液压传动系统

flap position indicator　襟翼位置指示器

flap retraction/extension mechanism　襟翼收放机构

flap retraction/extension switch　襟翼收放开关

flap retraction/extension system　襟翼收放系统

flat wrench　平口扳手

flexible cable control　软式钢索操纵

flexible mechanical control system　软式机械操纵系统

flight accident recorder　飞行事故记录仪

flight control system　飞行操纵系统

fluorescent fluid　荧光液

foamed plastic　泡沫塑料

follow-up pull rod　随动拉杆

foot console　脚操纵台

foot pedal　脚蹬

fork joint　叉形接头

fork shoulder shaft　叉形肩轴

fork sleeve　叉形套筒

forked member　叉形件

forward inclination　前倾

friction　摩擦(阻)力

friction piece　摩擦片

front limit position　前极限位置

front-wheel turning　前轮转弯

function test　功能试验

fuse　保险丝(电)

gear shaft　齿轮轴

gears pair　齿轮副

grass green enamel lacquer　草绿色磁漆

grass green epoxy nitroenamel　草绿色环氧硝基磁漆

ground hydraulic testing system　地面液压

油车

guide plate　导向板

guide pulley　导向滑轮

guide pulley bracket　导向滑轮座

guide rod　导杆

guide seat　导向座

guider　导向器

guiding mechanism　导向装置

hand wheel control mechanism　手轮操纵机构

hand-electric pump system　手摇泵-电动泵系统

hangar　机库

heading　航向

helical actuator　螺旋作动筒

helical disk　螺旋盘

helical mechanism　螺旋机构

hexagonal head bolt　六角头螺栓

hexagonal nut　六角螺母

high speed taxiing　高速滑行

high-pressure fluid supply pipe　高压供油路

hinged cover plate　铰接盖板

hollow worm　空心蜗杆

hydraulic accumulator　液压蓄压瓶

hydraulic actuator　液压舵机

hydraulic fluid　液压油

hydraulic fluid line　液压系统油路

hydraulic fluid suction　吸油

hydraulic lock　液压锁

hydraulic motor　液压马达

identification color circle　识别色圈

idler rocker　中介摇臂

indicating mechanism　指示机构

industrial alcohol　工业酒精

initial power　始动力

inner ring　内圈

inner skin　内蒙皮

inspector　检验员

intercom button　机内通话按钮

interference　干扰

interlocking cable　连锁钢索

kerosene　煤油

landing　着陆

lateral control　横向操纵

lateral controllability　横向操纵性

lead out　引出

lead-out wire　引出导线

leather sleeve　皮革套

level flight　平飞

lift　升力

limit deflection angle　极限偏角

limit groove　限制槽

limit rod　限制杆

linear distance　直线距离

linen　亚麻绳

loading mechanism　加载机构

location hole　定位孔

location plate　定位板

lock nest　锁槽

lock plate　锁片

locking wire　保险

longitudinal clearance　纵向间隙

lower arm　下臂

lower limit position　下极限位置

lower panel　下壁板

lug　耳子

lug joint　耳形接头

machine oil　机油

magnesium alloy bracket　镁合金支架

magnetic testing　磁力探伤

main control system　主操纵系统

main driving rod　主传动杆

main wheel brake　主轮刹车

main wheel braking mechanism　主轮刹车机构

measuring personnel　测量人员

measuring point　测量点

mechanical driving system　机械传动系统

mechanical energy　机械能

metallic luster　金属光泽

mounting angle　装配夹角

movement signal　运动信号

moving angle　移动角度

natural vibration frequency　固有振动频率

neutral position　中立位置

nitrogen charging pressure　充氮压力

noiseless chain　无声链

non-adjustable pull rod　不可调拉杆

normal retraction/extension system　正常收放系统

nose wheel landing gear　前三点式

nylon　尼龙

oil tray　贮油盘

on-aircraft inspection　原位检查

open spanner/open wrench　开口扳手

operation angle　工作角度

outer fringe/outer rim　外沿

outer ring　外环

out-of-tolerance　超差

over-compensation　过补偿

overhaul　翻修

oxidant solution　氧化液

oxygen pipe　氧气管

paint remover　除漆剂

parking brake handle　停机刹车手柄

pedal control mechanism　脚操纵机构

pipe wall　管壁

plastic rubbing sleeve　塑料防磨套

platform　平台

plexiglass cover　有机玻璃罩

plug pliers　插头钳

pop rivet/hollow rivet　空心铆钉

positioning gasket　定位垫片

power-on check　通电检查

pressing plate　压板

pressure gauge　压力计

pressure lubricating nipple　压力注油嘴

pretension　预张力

primary driving part　主要传动件

process pull rod　工艺拉杆

profile clamp plate　外形卡板

propeller rotating zone　螺旋桨转动区域

pull rod　拉杆

pull rod linkage　拉杆线系

pulley shaft mechanism　滑轮轴机构

radial clearance　径向间隙

radio set button　电台按钮

ramie rope　苎麻绳

rear fuselage　机身尾段

reciprocating linear movement　往复直线运动

reducer　减速器

regulating　调节

repair tolerance　修理容差

repairing bushing　修理衬套

reset spring　回位弹簧

retainer　护圈

retract　退回

retraction travel　收行程

reversible hydraulic motor　可逆液压马达

rigid control　硬式操纵

rigid control linkage　刚性操纵线系

rigid driving part　硬式传动件

rigidpull rod　硬式拉杆

rocker bracket　摇臂支架

rocker group　摇臂组

rod body　杆体

roller　滚轮

root　根部

rotating control stick　转盘式驾驶杆

rotation circle　转动圈数

rotation movement　旋转运动

rubbing spring　防磨弹簧

rudder control system　方向舵操纵系统

rudder plate valve　舵板活门

rudder spring servo tab　方向舵弹簧补偿片

rudder trim tab 方向舵调整片

ruler 刻度尺

running distance 滑跑距离

safety cover 保险盖

safety measure 保险措施

safety nut 保险螺帽

scissors difference 剪刀差

screw joint 螺杆接头

screw rod 螺杆

screwdriver 螺丝刀

screwed sleeve 螺纹套筒

scrunch 嘎吱声

sealing adhesive putty 密封腻子

sector rocker arm 扇形摇臂

self-gravity 自重

separation disk 间隔盘

serrate spring washer 外齿形垫圈

seven-position distribution cock 七位分配开关

sew 缝合

shaft sleeve 轴套

sheath 包片

signal light 信号灯

sleeve 套筒

sliding nut 滑行螺母

socket 插头座

soft tank/flexible tank/bag tank 软油箱

solenoid valve 电磁阀

space movement 空间运动

spherical rubber core 球形橡胶芯

spline 花键

spline shaft 花键轴

spoiler 扰流板

spoiler control mechanism 扰流板操纵机构

spoke 辐条

spring pull rod 弹簧拉杆

spring washer 弹簧垫圈

square column 方形柱

stable climb 直线爬升

stain spot 斑点

stall warning system 失速警告系统

stand 托架

standard atmosphere pressure 标准气压

standard part 标准件

standard temperature 标准温度

standby control switch 备用开关

static friction 静摩擦力

steel ball 钢珠

steel ball tightener 钢珠收口器

steel cable 钢索

steel cylinder 钢筒

steel pipe 钢管

stop clamping plate 限动卡板

stop pin 限动钉

stop plate 限动片

stop travel 限死行程

stopwatch 秒表

straight deviation value 直线偏移值

strand 股

stratification 分层

sunk-head bolt 沉头螺栓

support rocker 支承摇臂

surface roughness 表面粗糙度

swing 摆动

tail rod 尾杆

takeoff 起飞

tangential line 切线

tangential plane 切面

tensiometer 张力表

terminal switch 终点电门

test bed 试验台

textolite 夹布胶木

thread 螺纹/丝扣

thread mechanism 螺纹机构

threaded connector 螺纹接头

three-arm rocker 三臂摇臂

three-position distribution cock 三位分配开关

three-position four-way solenoid valve 三位四通电磁阀

throttle lever/power lever 油门杆

throttle valve 节流活门

tightening area 收压部位

tolerance on fit 配合公差

torque 扭矩

trailing edge 后缘

transmission line 传动线路

travel 行程

trend 趋向

triangular rocker 三角摇臂

trim 配平

tube shaft 管轴

turnbuckle 松紧螺套

twist 扭转

ultraviolet ray illuminant 紫外线光源

unglued 脱胶

universal joint 万向接头

universal protractor 万能量角器

unthreaded section 光杆部分

upper arm 上臂

vernier 游标尺

vertical shaft 立轴

vibration zone 振动区域

volume 容积

wall thickness 壁厚

warning bell control mechanism 警铃控制机构

wheel edge 轮缘

wheel spoke 轮辐

wing chord plane 翼弦平面

wing tip 翼尖

wire 导线

wires bundle 导线束

workshop 内场

X-ray detection device X射线探伤装置

X-ray inspection X光检查

yellow-zinc primer 锌黄底漆

zero position 零位

FUEL 燃油系统

absorbent cotton cloth 脱脂棉布

accessory base rubber 附件底座胶

accessory central rubber ring 附件中间胶圈

active media 活性介质

adapter base 转接座

adapter pipe 转接管

aircraft wiring manual 飞机特设线路图册

amplitude 振幅

anti-vibration strength 抗振强度

arc bracket 弧形架

arrowhead 箭头

automatic fuel consumption control 自动控制耗油

automatic pressure refueling 自动压力加油

automatic refueling controller 自动加油控制器

auxiliary armature 辅衔铁

available fuel quantity 可用燃油量

aviation fuel 航空燃油

aviation kerosene 航空煤油

bag tank/flexible tank 软油箱

balance valve 连通开关

blind nut 盲孔螺母

bonding tape 绷带

bow clamp 弓形夹

box bracket 盒形架

bridge arm 桥臂

brush check access cover 电刷检查口盖

bulge 鼓包

bushing 衬套

cable plug connector 电缆插座插头

canvas 帆布

capacitance-type fuel quantity gauge 电容式油量表

carpet 地毯

cast aluminum 铸铝

cavitation corrosion/cavitation erosion 气蚀

centrifugal force 离心力

certificate 合格证

check valve 单向活门

clamp disk 卡盘

coarse filter 粗油滤

coaxial pole 同轴电极

combination valve/assembly valve 组合活门

compression disk 压紧盘

conductive plate 导电片

conical spring 锥体弹簧

consumption/refueling procedure 消耗/加油程序

contact tab 触片

continuous operating duration 连续工作时间

cooperation surface 配合面

cotton fabric product 棉织品

countersunk screw 埋头螺钉

cup 杯体

DC electric centrifugal fuel pump 直流电动离心式燃油泵

dead fuel 死油

delivery period 运输期

dent 压痕

deposit 沉淀物

diagonal line 对角线

diffuser 扩散器

disk 圆盘

division value 分度值

drain cock 放油开关

drain funnel 放油漏斗

drain pipe connector 漏油管接头

dry-reed tube 干簧管

duraluminum 硬铝

dust collector 吸尘器

dynamic pressure connector 动压接头

electric bonding seat 电搭接座

electricbridge 电桥

electric brush 电刷

electric socket 电器插座

electromagnetic relay 电磁继电器

empty tank 空油箱

excessive fuel tank/residual fuel tank 余油箱

explode 爆炸

explosion-proof light 防爆电灯

explosion-proof system 防爆系统

faying surface 贴合面

filer core/filer element 滤芯

filler 注油口

filler plug 注油塞

filler screen 滤网

filter screen cover 滤网罩

fine filter blocking annunciator 细油滤堵塞信号器

fire extinguishing equipment 灭火设备

fire-proof electromagnetic valve 防火电磁开关

fire-proof switch 防火开关

fixing bracket 固定架

flange 凸边

flashboard 插板

flexible chock 软挡

flexible connection 柔性连接

float assembly 浮子组合件

float valve 浮子活门

freezing point 冰点

fuel consumption meter sensor/fuel consumption gauge sensor 耗油量表传感器

fuel control unit 燃油调节器

fuel delivery pipe 输油导管

fuel drain system/defueling system 放油系统

fuel drip tray 接油盘

fuel full signal led wire 满油信号引出线

fuel gun 加油枪

fuel leakage quantity　漏油量

fuel level height　油面高度

fuel quantity balance switch　油量平衡开关

fuel quantity measuring control system　油量测量控制系统

fuel quantity measuring switch　油量测量开关

fuel rod　油柱

fuel supply system　供油系统

fuel tank　燃油箱

fuel tank fixing point　油箱固定点

fuel-full signal　满油信号

fuel-resisting layer　耐油层

fuel-throwing ring　甩油圈

full value error　满值误差

gear reducer　齿轮减速器

geometrical angle　几何角度

gland　压盖

gravity defuel　自流放油

gravity refueling　重力加油

grease　润滑脂

ground container　地面容器

ground power unit（GPU）　地面电源车

guide plate　导流板

guider　导向器

hydraulic jack　液压千斤顶

ice crystal　冰晶

impact of the power supply fluctuation　电源波动影响

impeller　叶轮

instantaneous pressure　瞬时压力

insulated pad　绝缘垫块

integral tank　整体结构油箱

interchangeability　互换性

inverse heading view　逆航向视图

inverter power box　逆变电源盒

isolated pore passage　隔离孔道

isolation gas mask　隔离防毒面具

jaw vice　虎钳

key slot　键槽

layout　分布图

lead sealing　铅封

lead sealing piece　铅封块

leaking port　漏油口

leather cup　皮碗

leather cup assembly　皮碗组件

limiter　限动器

lock base　锁座

locking piece　锁紧片

magnetic bar　磁棒

magnifier　放大镜

main fuel pump　主燃油泵

manual pressure refueling　人工压力加油

mechanical impurity　机械杂质

megger　兆欧表

metal pin　金属销

model/designation　牌号

moment　力矩

negative/positive pressure　正/负压力

net weight　净重

neutral gas interface　中性气体接口

oxide　氧化物

paint coat　漆皮

physical and chemical indexes　理化指标

piston ring　活塞环

pitting corrosion/spot corrosion　点蚀

potentiometer　电位器

pressing plate　压板

pressure annunciator　压力信号器

pressure defueling　压力放油

pressure gauge sensor　压力表传感器

pressure refueling adaptor　压力加油接头

primary electromagnet　主电磁铁

push rod　顶杆

ram air　冲压空气

rated value　额定值

reed　簧片

refueling adapter　加油接头

refueling bonding jumper 加油搭接

refueling power distribution board 加油配电盘

refueling system 加油系统

refueling truck 加油车

reinforced splice 加强条

relative humidity 相对湿度

remote control switch 遥控开关

removable panel 可卸壁板

restricting nipple 限流嘴

rocker lever 摇臂杠杆

rubber connecting disk 橡胶结合盘

rubber dreg 胶渣

rubber plate 胶板

rubber sleeve 胶套

safety valve 安全活门

schematic diagram 原理图

scrub 擦洗

sealing gasket 密封垫片

sealing pad 封严垫

self-locking nut 自锁螺母

servo motor 伺服电机

sharp edge 锐边

single-jack plug 单孔插头

sleeve pole piece 套筒极片

socket wrench/socket spanner 套筒扳手

special bracket 专用托架

sponge 海绵

static pressure connector 静压接头

steel ball lock spring 钢球锁弹簧

stop washer 止动垫圈

storage life 保管期

suds 肥皂液

sulfurize 硫化

supplement capacitor 凑值电容

talc powder 滑石粉

tank fuel-resisting layer 油箱外保护层

technical standard 技术标准

technological opening 工艺口

test bed 试验台

thermal expansion valve 热胀活门

thread 螺纹

three-way pipe 三通导管

throttling valve 节流阀

top cover 顶盖

torque wrench 定力扳手

transformer 变压器

transmission rod 传动杆

union/transition nozzle 接管嘴

vacuum valve 真空活门

valve base/valve seat 活门座

vent connecting disk 通气连接盘

vent plug 通气堵塞

vent system 通气系统

warning signal 告警信号

waxed paper 蜡纸

winding resistor 绕线电阻

wrinkling 褶皱

zero-position error 零位误差

HYDRAULIC SYSTEM 液压系统

absolute pressure 绝对压力

AC two-wire pressure gauge 交流二线式压力表

accessory board 附件板

acid value 酸值

actuating mechanism 执行机构

adjustment shim 调整垫片

aileron actuator 副翼舵机

air charging union 充气嘴

air filter 气滤

air guider shroud 导风罩

air pressurization system 空气增压系统

aircrew 空勤（人员）

annular groove 环形槽

anti-rust aluminum alloy 防锈铝合金

appearance 外观

artificial leather　人造革

autopilot　自动驾驶仪

aviation cleaning gasoline　航空洗涤汽油

aviation hydraulic fluid　航空液压油

aviation standard hose　航标软管

bearing ring　轴承环

big fluid distributing piston　大分油活塞

brake accumulator　刹车蓄压瓶

brake assembly　制动器组合件

brake valve　刹车活门

buffer accumulator　缓冲蓄压瓶

bushing　衬筒

cargo door　货舱大门

charge　充气

clamp　夹紧

cleaning cart　清洗车

cockpit bottom emergency door actuator　前应急舱门作动筒

cockpit floor door actuator　前舱门锁作动筒

cold air safety valve　冷气安全活门

communicating pipe　连通管

communication valve　连通活门

condensation　凝点

connection arm of relay　继电器连接支臂

connection pipe　连接导管

contacting piece assembly　接触片组合件

control distribution accessory　控制分配附件

control switch　操纵开关

coordination valve　协调活门

coupling　连轴节

cover bushing　盖衬套

cutoff valve　断油活门

damp-proof sand　防潮砂

delay valve　延迟活门

density　密度

diaphragm　薄膜

diode　二极管

discard　报废

distribution valve　分配活门

distributor disk　分油盘

downlock actuator　下位锁作动筒

drain outlet　漏油出口

driven gear　从动齿轮

driving gear　主动齿轮

driving rod/driving lever　拨杆

dryer filter　干燥过滤器

elbow union　弯管嘴

electric floating-type quantity indicator　电动浮子式油量表

electric pump　电动泵

elevator actuator　升降舵机

emergency brake valve　紧急刹车活门

end cover　端盖

engage　啮合

engine emergency feathering shutdown　发动机应急顺桨停车

engine gear box　发动机传动机匣

evenness　均匀性

feathering cock　顺桨开关

felt pad　毡垫

felt ring　毡圈

filter element　过滤元件

filter screen　滤网

filtering accuracy　过滤精度

finished product　成品

flange stand　凸台

flap driving mechanism　襟翼传动机构

flash point　闪点

flexible connection part　柔性连接件

flight altitude　飞行高度

flight attitude　飞行姿态

flight condition　飞行状态

flow regulator　流量调节器

flow restricting union/flow restrictor　节流管嘴

fluid channeling accessory　串油附件

fluid container　贮油器

fluid contamination degree　油液污染度

fluid drain cup   漏油杯

fluid film plunger assembly   油膜柱塞组合件

fluid inlet nozzle   进油管嘴

fluid intake hose   进油软管

fluid outlet nozzle   出油管嘴

fluid resistance   流体阻力

fluid return changeover valve   回油转换活门

fluid return hose   回油软管

fluid suction connector   吸油接头

fluid suction hose   吸油软管

fluid suction valve   吸油活门

fluid supply chamber   供油室

fluid supply valve   供油活门

fluorine plastic ring   氟塑料圈

follow-up piston/slave piston   随动活塞

fuse   熔断器

gas cavity   总容积

gauze   纱布

gear peak   齿峰

gear pump   齿轮泵

gear valley   齿谷

graphite bearing   石墨轴承

ground charging valve   地面充气活门

ground crew   地勤人员

ground hydraulic equipment   地面液压设备

ground hydraulic test cart   地面液压试验车

ground pressure source   地面压力源

guard valve   挡板活门

guide seat   导向支座

guide sleeve   导套

half-ball valve   半球活门

half-ring lock   半卡圈锁

hand pump   手摇泵

heat energy   热量

high pressure cavity   高压腔

high pressure valve combination   高压活门组合

hose   软管

hydraulic fluid   液压油

hydraulic fluid cylinder   油缸

hydraulic fluid filter   油滤

hydraulic fluid outlet connector   压力油出口接头

hydraulic fluid pump   油泵

hydraulic handle valve   液压手动阀

hydraulic lock   液压锁

hydraulic motor   液压马达

hydraulicreservoir   油箱

hydraulic reservoir pressurization system   油箱增压系统

hydraulic servo actuator   液压舵机

hydraulic switch   液压电门

hydraulic test stand   液压试验台

idling pressure   空转压力

inclined disk bearing   斜盘轴承

installation base   安装座

installation plate   安装板

jaw vice   虎钳

kinematic viscosity   运动粘度

landing gear well door   起落架舱门

locating pin   定位针

locking plate   锁片

lug   耳环

mandrel   芯杆

measuring cup   量杯

mechanic impurity   机械杂质

metallic wire   金属丝

mounting hole   安装孔

needle   顶针

needle flow restrictor valve   针状限流活门

nitrogen bottle   氮气瓶

normal brake   正常刹车

nose wheel control mechanism   前轮操纵机构

nose wheel steering   前轮转弯

nut washer   螺母垫圈

oil return pipe   回油导管

oil return position 回油处

oil-return passage 回油路

open wrench 螺帽扳手

operating pipeline 工作油路

outer shroud 外套

piston push rod 活塞顶杆

plastic tube 塑料管

plunger pump 柱塞泵

pneumatic bottle/nitrogen bottle 冷气瓶

pointer travel 指针行程

power mechanism 动力机构

preservation paper 油封纸

pressure changeover device 压力转换装置

pressure changeover valve 压力转换活门

pressure reduction valve 减压阀

pressure-restricting mechanism 限压机构

pressurization 增压

pressurization connector 增压接头

pressurization valve 增压活门

pressurized air inlet 增压进气口

pressurized air/compressed air 增压空气

pull rod 拉杆

pump driving shaft 油泵传动轴

push rod 顶杆

putty tape 腻子带

quantity meter 定量器

radiate 散热

radiator 散热器

ramp lock actuator 货桥锁作动筒

refueling cart 加油车

release 释放

remote induction pressure gauge 远距感应式压力表

restricting grid 节流栅

restrictor valve 节流阀

retraction/extension 收/放

return filter 回油滤

roller bearing 滚棒轴承

root of flare tube 喇叭口根部

rubber cup 胶碗

rubber cushion 胶垫

rubber particle 橡胶颗粒

safety valve 安全活门

sampling container 采样容器

separation connector 分离接头

service life 使用期

seven-position distribution cock 七位分配开关

shaft assembly 轴杆组合件

shimmy damper 减摆器

silicone rubber piece 块状指示硅胶

sleeve coupling nut 外套螺帽

sliding block 滑块

slotted screwdriver 一字螺刀

small actuator 小作动筒

solenoid valve 电磁阀

solid grain 固体颗粒

special washer 特型垫圈

spiral cover 螺盖

spiral strip clamp 螺旋带卡箍

spline shaft 花键轴

splinted tail rod 套齿尾杆

spring bracket 弹簧支座

spring seat 弹簧座

spring tension 弹簧张力

stainless steel wire 不锈钢丝

stamp/press/punch 冲压

standard code 标准编号

steel ball seat 钢球座

steel wire braided hose 钢丝编织软管

stop screw 止动钉

supporting cap 支承帽

swing 摆动

switch on/off pump pressure 接通/断开油泵压力

three-hole straight plug 三孔直插头

three-point socket 三点插座

throttle valve 节流阀

tooling equipment 工装设备

top emergency exit hatch of cockpit 前舱应急舱门

total capacity 总容积

ultrasound 超声波

unload valve/unloading valve 卸荷活门

valve lever 活门杠杆

viscosity 黏性

vulcanized rubber cushion 硫化橡胶垫

warning placard 警告牌

wheel braking hose 机轮刹车软管

winding 缠绕

windshield wiper 风挡雨刷

windshield wiper valve 风挡雨刷活门

work site/working site 工作场地

ICE AND RAIN PROTECTION
除雨和防冰

absolute pressure 绝对压力

AC generator 交流发电机

AC power distribution board 交流配电盘

acceptance inspection 验收检验

access cover 口盖

accident 事故

adiabatic shroud 绝热套罩

adjusting screw 调整螺钉

adjustment procedure 调整程序

adjustment/test 调整/调试

aerodynamic characteristic 气动特性

aerodynamic configuration 气动外形

air bleed pressure valve 引气压力开关

air chamber 气室

air compressor 压气机

air conditioning system 空调系统

air horn 气喇叭

air inlet heating connection signal light 进气道加温接通信号灯

air inlet/air intake 进气口

air outlet tube 出气管

air pipe connector 空气接头

air pressure signal sensor 空气压力信号传感器

air restricting valve 节气活门

airborne tool 随机工具

aircrew 机组人员

air-delivery pipe 输气管

air-delivery switch 输气开关

airdrop-paradrop button 空投空降按钮

airspeed tube 空速管

airtightness 气密性

alcohol 酒精

alignment 对准

alternating current power 交流电

aluminum 铝

ambient condition 环境条件

ambient temperature 外界气温/环境温度

ammeter 电流表

amplify 放大

annular ejecting tube 环形引射管

anti-icing control circuit 防冰操纵电路

anti-icing heating element 防冰加温元件

anti-icing nozzle 防冰喷嘴

anti-icing power distribution box 防冰配电盒

anti-icing pressure annunciator 防冰压力信号器

anti-icing single-stage ejector 防冰单级引射器

anti-icing system 防冰系统

anti-icing valve 防冰活门

arch bracket 弓形架

area 面积/区域

assemble 装配

atmosphere 大气

auto-transformer 自偶变压器

aviation hydraulic fluid 航空液压油

axial line　轴线

ball joint　球形接头

ball nipple　球形管嘴

base　底座

bellow aneroid　膜盒

bellow compensator　波纹补偿器

belted heater　带状加温器

bilayer structure　双层结构

blade　桨叶

blanking cap　堵帽

bleed-air over-pressure switch　引气超压开关

bleed-air port　引气口

blinking frequency/glitter frequency　闪烁频率

block　堵/端头

block plug　堵塞

bonding jumper connector　搭接线接头

bracket　支架

brass bus　黄铜汇流条

built-in test button/BIT button　自检按钮

built-in test normal indicating light　自检正常指示灯

built-in test switch　自检开关

bulkhead　隔框

burn-out　（电机)烧坏

bus bar　汇流条

bushing　衬套

bushing assembly　衬套组件

butt　对接

butt connector/butt joint　对接接头

butt joint section　对接处

butterfly valve　蝶形活门

button light　按钮灯

cable plug　电缆插头

cam　凸轮

cam mechanism　凸轮机构

canopy　天窗

cap　帽盖

cap nut　螺盖

capacitor　电容器

carbon brush　碳刷

cargo cabin　货舱

casing/case　壳体

cause　故障原因

cavity　腔

ceiling　顶棚

center instrument panel　中央仪表板

certificate　合格证

chamois leather　麂皮

changeover mechanism　转换机构

charging　充电

check button　检查按钮

check procedure　检查步骤

check valve　单向活门

check/inspection　检查

circuit　电路

circuit breaker　断路器/自动保护开关

circuit fuse　电路熔断器

circular nut　圆顶螺母

circulation heating　循环加温

clamp　卡箍

clearness　清晰度

climb　爬升

clockwise　顺时针

coat　涂层

cockpit　驾驶舱

cockpit door　驾驶舱门

coil circuit　线圈电路

combustion chamber/combustor　燃烧室

communicator circuit breaker board　通讯员自动开关板

compensation　补偿

complex weather condition　复杂气象条件

component/element　元件

composited circuit　复合线路

composition　组成

compound power　复合电源

compressor 压气机

conductive film 导电薄膜

conductive ring 导电环

conductivity 导电性

cone needle 锥形油针

conical needle plunger 锥形针塞

connecting bolt 连接螺栓

connecting disk 结合盘

connection support point 连接支点

connector assembly 插头座组件

constant heating 恒加温

construction 构造

consumption 消耗

contact 接点

contactor 接触器

continuous operation 连续工作

control signal 控制信号

control switch 操纵开关

cool 冷却

copper ring 铜环

correctness/validity 正确性

corrugated plate 波纹板

cotter pin 开口销

counter-clockwise 逆时针

cover plate 盖板

cover/shroud 套罩

crack 裂纹

crash 碰撞

critical speed 临界速度

cross valve 交叉活门

current leakage 漏电流

current relay 电流继电器

current required 需要电流

current-limiting resistor 限流电阻

damage 破坏

DC power voltage 直流电源电压

de-icing 除冰

delay 延迟

delay circuit 延时电路

deviation 偏离

diameter 直径

differential amplification circuit 差动放大电路

differential amplifier 差分放大器

diode 二极管

direct current power 直流电

discharge 放电

displacement 位移

distribute 分配

divider valve/division valve 分路活门

driving disk 导动盘

driving gear 传动齿轮

driving shaft 传动轴

duration 持续时间

durite hose 夹布胶管

dynamic static pressure sensor heating 动静压受感器

ejecting device 引射装置

ejecting sleeve 引射套管

ejection 引射

ejection radiating switch 引射散热开关

ejector 引射器

ejector radiating tube 引射散热管

ejector switch 引射开关

elastic washer 弹性垫圈

elasticity 弹性

elbow 弯管

elbow assembly 弯管组件

electric bridge 电桥

electric control circuit 电气控制线路

electric heating anti-icing system 电热防冰系统

electric heating glass 电加温玻璃

electric mechanism 电动机构

electric plug 电气插头

electric principle 电气原理

electricity contact 电接触点

electro-thermal ice protection system 电加

温防冰系统

elimination 排除

emergency bus 应急汇流条

engagement 吸合

engine accessory casing 发动机附件机匣

engine curve test 发动机试车

engine inlet 发动机进气道

engine nacelle 发房

engine shutdown 发动机停车

(engine) run-up （发动机)开车

(engine) start （发动机)起动

enhancement 增强

entry door 登机门

equipment 设备

equipment rack 设备架

evenness/uniformity 均匀性

excitation return circuit 激磁回路

exhaust gas 废气

exhaust hole 排气孔

exponential curve 对数曲线

failure analysis 故障分析

false weld 虚焊

faulty warning circuit 故障告警回路

feather 顺桨

feature 特点

fire-extinguishing 熄灭

fixed screw 固定螺钉

fixing bracket 固定架

flange 法兰盘

flight altitude 飞行高度

flight safety 飞行安全

flight speed 飞行速度

flow 流量

flow divider 分流活门

flow rate/flow velocity 流速

flow restricting rod 节流杆

flow restrictor valve 限流活门

flow-restricting sleeve 限流套

fluid/fuel/oil tray 贮油盘

follow-up unit 随动器

foot voltage 脚电压

fracture 断裂

frame 框

framework 骨架

frequency 频率

front canopy/forward cabin 前舱

front section 前段

fuel control unit（FCU） 燃油调节器

function 功能

fuselage 机身

gap/clearance 缝隙

gasket 衬垫

gauze 纱布

general-purpose tool 通用工具

glass heating circuit fuse 玻璃加温熔断器

glass heating transformer 玻璃加温变压器

glass insulating cloth 玻璃绝缘布

groove 槽

ground check 地面检查

ground power/ground electrical power source 地面电源

guide bushing 导套

guiding groove 导槽

hand wheel 手轮

hand wheel assembly 手轮组件

heading 航向

head-on airflow 迎面气流

heat energy 热量

heat exchange 热交换

heat knife 热刀

heat shield 隔热罩

heating annunciator 加温接通信号器

heating check button 检查加温按压开关

heating element 加温元件

heating power supply 加温电源的配给

heating resistance/heating resistor 加温电阻

heating signal 加热信号

heating signal control box 加温信号控制盒

heating switch 加温开关

heating type　加温类型

heating warning box　加温报警控制盒

heating wire　电热丝

heat-insulating material　绝热物

high altitude oxygen-using annunciator　高空用氧信号器

horizontal stabilizer　水平安定面

horizontal tail　水平尾翼

horn joint/flared join　喇叭形接头

hot air　热空气

hub spinner/hub fairing　桨毂整流罩

hydraulic driving mechanism　液压传动机构

ice and rain protection　除雨和防冰

ice-detector main warning light　结冰信号主告警灯

icing　结冰

icing annunciator　结冰信号器

icing area　结冰区域

icing detecting system　结冰探测系统

illuminate/blink/glitter　闪烁

illumination　燃亮

immerse　蘸

impact acceleration　冲击加速度

impulse load　冲击过载

impulse speed　冲击速度

impurity　脏物

inboard wing　中外翼

indicating light　指示灯

indicating signal　指示信号

inlet gas vane　进气道导向器

inletguide vane　导向器叶片

inlet guide vane heating pipe　进气导向器加温管

inner section　内段

inner skin　内蒙皮

input signal　输入信号

inspection heating changeover switch　检查加温转换开关

inspection light　检测灯

inspector　检验员

installation hole　安装孔

installation piece　装机件数

installation position　安装位置

instrument panel　仪表板

insulated resistance　绝缘电阻

insulating layer　绝缘层

insulating ring　绝缘圈

insulation connector　绝缘接头

interlock relay　联锁继电器

judge　判断

kinematic velocity/velocity of movement　运动速度

landing　着陆

landing gear well　起落架短舱

laterial element　横向元件

lead wire　引线

leading edge　前缘

leading edge ejector　前缘引射器

leak　漏气

length　长度

limit position　极限位置

linear acceleration　线加速度

liner/lining　内衬

link rod　连杆

load current　负载电流

locking link rod　锁紧连接杆

locking wire　保险丝

longitudinal element　纵向元件

lower section　下段

main technical data　主要技术指标

mainfold　总管

maintenance practice　维修实施

mal-function　失灵

mass　质量

material　材料

maximum oil consumption　最大耗油量

maximum oil return pressure　最大回油压力

maximum operating condition　最大工作

状态

mechanic energy　机械能

mechanic vibration frequency　机械振动频率

metal sealing ring　金属密封圈

micro-ejector nozzle　微引射管

microswitch　微动开关

middle section　中段

mixing chamber　混合室

mixture　混合

moisture　湿气

motor　电(动)机

mounting bracket　安装架

mounting sleeve　安装套

multimeter　万用表

multivibrator　多谐振荡器

navigator　领航员

needle plunger valve　针塞开关

negative control　负控

negative wire　负线

neutral position　中立位置

neutral soap water　中性肥皂水

nickel　镍

normal closed contact　常闭触点

normal operation　正常工作

normal temperature　常温

normally open contact　常开触点

nose radome/nose fairing　机头罩

nut　螺帽/螺母

observation　观察

oil distributing valve　分油活门

oil radiator　滑油散热器

oil return line　回油路

oil return nipple　回油管嘴

oil tank　滑油箱

operating cause　工作原因

operating cycle sequence　工作循环顺序

operating fluid　工作液

operating pressure　工作压力

operating principle　工作原理

operating specification　工作规范

operating state　工作状态

operating temperature　工作温度

operating voltage　工作电压

operation voltage range　工作电压范围

operation/maintenance　使用/维护

original position　原位

outboard wing　外翼

outer section　外段

outer skin　外蒙皮

out-of- control　失控

output signal　输出信号

output signal channel　输出信号通道

output swing angle　输出回转角度

overhead panel　顶部操纵台

overload　过载

oxide skin　氧化皮

pad　垫板

panel　面板

parallel connection　并联

parking　停放

partition　隔板

penetrate　穿入

pilot/copilot　左/右驾驶员

pipe connector　管接头

pipe end　导管端头

piston　活塞

piston rod　活塞杆

pitot-static pressure heating system　全静压加温系统

pitot-static pressure sensor　全静压受感器

plate nut　托板螺母

pliers　钳子

polish　抛光

polishing paper　抛光纸

position/location　位置

positive control　正控

positive electricity　正电

potentiometer　电位器

power consumption　消耗功率

power plant　动力装置

power supply　电源

power supply system　电源系统

power supply voltage　电源电压

power-on check　通电检查

power-on switch　接通开关

precaution　预防措施

prediction　预报

preparations　准备工作

press　按压

press force　压紧力

pressure energy　压力能

pressure gauge　压力表

pressurerefueling　压力加油

pressure regulating/cutoff valve　压力调节/关断活门

pressure sensor　压力传感器

pressure supply nipple　供压管嘴

pressure supply system　供压系统

pressurization　充压/加压

pressurized terminal　气密接线柱

printed circuit board　印制电路板

procedure　程序

propeller　螺旋桨

propeller brake　螺旋桨制动

property　性质

protection ring　保护圈

protective diode　保护二极管

pull rod　拉杆

pulse amplifier　脉冲放大器

pulse generator　脉冲发生器

pulse signal　脉冲信号

quantity　数量

quick-release clamp　快卸卡箍

rack piston　齿条活塞

radiation switch　散热开关

radioactive requirement　辐射要求

rated voltage　额定电压

reading　读数

receiver　接收机

reciprocating motion　往复运动

relative humidity　相对湿度

relay　继电器

reliability　可靠性

removal　拆下

replacement　更换

reset　复位

resistance　电阻

resistance signal　电阻信号

resistance wire　电阻丝

resistor disc　电阻片

restricting machine　限流机

restrictor valve　限流开关

rib　肋

rivet　铆接

riveting placard　铆接标牌

rocker arm　摇臂

rotating shaft　转轴

rotation　旋转

rubber washer　橡胶垫

rust　锈蚀

sampling bleeder circuit　取样分压电路

sandwich　夹层

schematic diagram　原理图

scrap　刮伤

screw　螺钉

screw restrictor　螺钉限制器

screwdriver　螺丝刀

sealing device/unit　密封装置

sealing ring　密封圈

seat　座椅

seat plate　座板

semi-annular ejector　半环形引射器

serial connection　串联

serial number（S/N）　序号

service　保养

service ladder　工作梯

shaft sleeve 轴套

short circuit 短路

shutoff valve/cutoff valve 断流活门

signal 信号

signal source 信号源

signal weight 单台重量

skin 蒙皮

sleeve 套筒

slide 滑动

slow-blow fuse 难熔熔断器

small shaft/small axle 小轴

socket 插座

spanner/wrench 扳手

spark quenching circuit 灭火花电路

speed reducer converter 减速转向器

spherical compensator 球形补偿器

spiral plug 螺塞

spline joint 套齿

spline shaft 花键轴

spline washer 内齿垫圈

spring 弹簧

spring balance 弹簧秤

spring leaf 弹簧片

spring washer 弹簧垫圈

sprinkle 喷(水)

stabilivolt 稳压器

stabilized voltage source 稳压电源

stagnate 卡滞

stainless steel strip 不锈钢带

stall 失速

starting distribution box 起动配电盒

static pressure cavity 静压腔

steel bushing 钢制衬筒

stop screw 限动螺钉

stop watch 秒表

storage temperature 储存温度

straight-line motion 直线运动

stringer 长桁

strobe 频闪

stroboscope 频闪仪

stroke 冲程

strong/weak heating 强/弱加温

support base 支座

surface 表面

swing angle 摆角

symptom 故障现象

tail cabin 尾舱

tail wing/empennage 尾翼

takeoff 起飞

technical requirement 技术要求

technical specification/data 技术性能/数据

technological adapter 工艺转接头

temperature compensator 温度补偿器

temperature sensing probe 感温棒

temperature sensor 温度传感器

terminal handle 接线手柄

terminal screw/binding screw 接线螺钉

terminal switch 终点开关

terminal voltage 端电压

test 试验

tester 试验器

thermal relay 热自动开关

thermal switch 过热保护开关

thermistor 热敏电阻

thread 螺纹

thread hole 螺纹孔

thrust 拉力

tightness 拧紧程度

timing mechanism 定时机构

transceiver 收发机

transformer 互感器

transistor 晶体管

transistor base electrode 晶体管基极

transmitter 发送机

transparency 透明度

transponder 应答机

travel 行程

triode 三极管

troubleshooting　排除方法

turbine generating device　涡轮发电装置

type No.　型号

unit/device　装置

upper section　上段

utilization　利用

valve bushing　活门衬套

valve changeover time　活门转换时间

valve plate　阀板

ventilation condition　通风条件

vertical stabilizer　垂直安定面

vibration　振动

vibration acceleration　振动加速度

view A　A向

view/vision　视界

visibility　能见度

void/pore　孔隙/穴

voltage signal　电压信号

voltage-regulator diode　稳压二极管

warning　警告

washer　垫圈

washer/gasket/spacer　垫片

water steam　水蒸气

waveform　波形

web　腹板

weight　重量

weld　焊

welding point　焊点

wheel chock　轮挡

window and windshield system　窗及风挡系统

windshield glass　风挡玻璃

windshield wiper　风挡雨刷

wing　机翼

wing anti-icing ejector　机翼防冰引射器

wing anti-icing electrical switch　机翼防冰电动开关

wing heating changeover switch　机翼加温转换开关

wing surface　翼面

wing tip　翼尖

wiper rubber　刮水胶皮

wiring diagram　电气电路图册

INDICATING/RECORDING SYSTEMS
显示/记录系统

absolute pressure altitude　绝对气压高度

AC amperemeter　交流电流表

AC generator　交流发电机

AC generator control switch　交流发电机控制开关

AC generator fault indicating light　交流发电机故障指示灯

AC power distributionboard　交流配电盘

AC synchronizer signal interface module　交流同步器信号接口模块

AC voltmeter selector switch　交流电压表选择开关

accelerometer　加速度表

accident button　事件按钮

accident signal　事件信号

accumulator　蓄压瓶

accumulator pressure indicator　蓄压瓶压力表

ADF control box　无线电罗盘控制盒

ADF control panel　无线电罗盘控制板

AHRS control box　航姿控制盒

AHRS power availability　航姿电源有效性

aileron actuator deflection angle sensor　副翼舵偏角传感器

aileron deflection angle　副翼偏角

aiming sight　瞄准具

air data computer　大气数据计算机

air data system　大气数据系统

air delivered by left/right engine　由左/右发动机输气

air inlet icing signal switch　进气道结冰开关

air supply cock　供气开关

air supply thermometer　供气温度表

air temperature indicator　大气温度指示器

air temperature meter　大气温度表

airborne crashworthy recording system　机载抗坠毁的记录系统

air-charging valve　充气活门

aircraft horizontal datum line　飞机水平基准线

airdrop signal　空投信号

airframe icing　机体结冰

airspeed high annunciator　速度大信号器

airspeed high warning light　速度大告警灯

airspeed indicator　空速表

allowable erasable times　可擦写次数

allowable swapping times　可插拔次数

altimeter brightness adjusting knob　高度指示器亮度调节旋钮

altitude　高度

altitude alert light　危险高度指示灯

altitude annunciator　高度信号器

altitude indicator　高度指示器

altitude setting value　高度订数值

altitude signal　高度信号

amperemeter　电流表

amplification/enlarge　放大

amplifier　放大器

analog distributor　模拟量分配器

analog quantity　模拟量

analog signal　模拟量信号

analog signal interface module　模拟量信号接口模块

angular displacement　角位移

angular displacement sensor　角位移传感器

annunciator　信号器

anti-collision light　防撞灯

anti-fire indicating light　防火指示灯

anti-fire switch　防火开关

AOA sensor　攻角传感器

AOA vane heating switch　攻角加热开关

ATC transponder control box　空管应答机控制盒

attitude　姿态

attitude director indicator（ADI）　地平指示器

attitude heading reference system（AHRS）　航向姿态系统

audio monitor　音频监控器

audio signal　音响信号

aural warning system　音响警告系统

auto and manual fuel consumption switch　自动人工耗油开关

auto feathering switch　自动顺桨开关

auto fuel consumption　自动耗油

auto fuel consumption switch　自动耗油开关

auto navigation signal selection button　自动导航信号选择按钮

auto navigation signal selection indicating light　自动导航信号选择指示灯

auto-feathering check switch　自动顺桨检查开关

autopilot　自动驾驶仪

autopilot console　自动驾驶仪操纵台

autopilot engagement　自动驾驶仪接通

autopilot solenoid valve　自动驾驶仪电磁活门

auxiliary instrument　辅助仪表

auxiliary vent valve　辅助通风活门

average transient error　平均瞬时误差

aviation clock　航空时钟

axial acceleration　轴向

azimuth　方位

balance position　平衡位置

band changeover selector switch　波段转换选择开关

band suppressor　波段抑制器

bank angle　倾斜角

barometric altimeter　气压高度表

base 底座

battery 电池

battery amperemeter 电瓶电流表

battery box 电池盒

battery leakage 电池漏液

battery overheat indicating light 电瓶超温指示灯

beacon light box（OM/MM/IM） 信标灯盒（远/中/近）

beacon sensitivity 信标灵敏度

beacon sensitivity switch 信标灵敏度开关

bleed air overpressure indicating light 引气超压指示灯

bleed air pressure indicator 引气压力表

block diagram 方框图

bottom plate 底板

brake pressure gauge indicator 刹车压力表指示器

brake release indicating light 解除刹车指示灯

brake release switch 解除刹车开关

brightness adjusting button 亮度调节按钮

brightness adjusting knob 亮度调节开关

brightness adjusting potentiometer 亮度调节电位器

built-in computer 嵌入式计算机

built-in test/self-test 自检

bus（bar） 总线

bus data 总线数据

butterfly bolt 蝶形螺栓

cabin altitude and pressure differential gauge 座舱高度压差表

cabin altitude annunciator 座舱高度信号器

cabin audio recording system 座舱音频记录系统

cabin door signal 舱门信号

cabin emergency pressure release switch 座舱应急释压开关

cabin pressure release switch 座舱释压开关

calibration 标定/校准

capacity sensor 电容传感器

capacity value 电容值

card slot 卡槽

cargo cabin altitude annunciator 货舱高度信号器

cargo cabin horn stop button 货舱喇叭断开按钮

cargo cabin temperature control 货舱温度控制

cargo door motion indicating light 货舱门运动指示灯

ceiling light switch 顶灯开关

center of gravity 重心

central warning system 中央警告系统

centrer instrument panel 中央仪表板

certification 证明

changeover switch 转换开关

changing value 变化量

check button 检查按钮

check in real time 实时检查

chip 芯片

chronometer 计时仪器

circuit breaker 自动保护开关

clamping device 夹紧装置

climbing decision 爬升决策

clock brightness adjusting knob 时钟亮度调节旋钮

cockpit ceiling light 前舱顶灯

cockpit light 驾驶舱照明灯

cockpit overhead console 驾驶舱顶部操纵台

cockpit oxygen supply annunciator 用氧信号器

cockpit temperature control 驾驶舱温度控制

cockpit temperature meter 驾驶舱温度表

cockpit temperature regulation selector switch 驾驶舱温度调节开关

collector 采集器

combined control 组合控制

communication agreement 通讯协议

communication control 通讯控制

communicator AC instrument panel 通讯员交流仪表板

communicator circuit breaker board 通讯员自动开关板

communicator control panel 通讯员操纵板

communicator DC instrument panel 通讯员直流仪表板

communicator radio instrument panel 通讯员无线电仪表板

compass fast slave button 罗盘快速协调按钮

compass lighting switch 罗盘照明开关

composite error 综合误差

conductive ring 导电环

conductive ring bush 导电环衬套

connector 连接器

console 操纵台

constant illuminating 常亮

contact 触点

control box 控制盒

control coil 控制线圈

control display unit（CDU） 控制显示器

control surface 舵面

control surface lock control handle 舵面锁操纵手柄

control switch 控制开关/操纵开关

converter 变流机

cover plate for power 用电盖板

crab angle 偏航角

crash worthiness techniques 抗坠毁技术

critical angle of attack 临界攻角

critical angle of attack annunciator 临界攻角信号器

critical angle of attack power 临界攻角电源接通指示灯

critical angle of attack sensor 临界攻角传

感器

critical angle of attack signal system 临界攻角信号系统

critical angle of attack voltage 临界攻角电压

critical angle of attack warning light 临界攻角告警灯

cross valve 交叉活门

crossfeed switch 连通开关

current signal 电流信号

cutoff button 切油按钮

cutoff valve 断流活门

cylinder screw 圆柱螺钉

dangerous（critical）angle of attack 危险（临界）攻角

data card 数据卡

data unloader 数据卸载器

DC amperemeter 直流电流表

DC consumption 耗电（直流）

DC converter DC 转换器

DC voltage signal 直流电压信号

DC voltmeter 直流电压表

dedicated protractor 专用量角器

deflection angle sensor 偏角传感器

descending decision 下降决策

detection circuit 检测电路

deviation 偏差

dialog box 对话框

dimension 外型尺寸

discrete signal 离散量信号

display 显示器

division valve 分路活门

downloading duration 下载时间

drift angle 偏流角

dual-port RAM 双口 RAM

duty pump 值班泵

dynamic pressure lead-in port 动压引入口

dynamic static pressure hose 动静压软管

easting acceleration 东向加速度

easting speed　东向速度

ejector radiation switch　引射散热开关

elastic contact　弹性接触点

electric accelerometer　电动加速度表

electricbell　电铃

electric bridge principle　电桥原理

electric mechanic instrument panel　电气员仪表板

electric zero alignment mark　电气零位对准标记

electric zero datum point　电气零位基准点

electrical plug　电气插头

elevator deflection angle　升降舵偏角

elevator illumination switch　升降舵照明开关

elevator trim tab hand wheel　升降舵调整片手轮

ELT control box　应急定位仪发射器控制盒

emergency airdrop indicating light　应急投放指示灯

emergency airdrop switch　应急投放开关

emergency bus bar　应急汇流条

emergency cargo airdrop　应急投货

emergency control circuit　应急操纵电路

emergency door indicating light　应急舱门指示灯

emergency door opening　应急开舱门

emergency horizon　应急地平仪

emergency horizon brightness adjusting knob　应急地平仪亮度调节旋钮

emergency horizon power switch　应急地平仪电源开关

emergency light　应急照明灯

emergency locator　应急定位仪

emergency locator transmitter（ELT）　应急定位仪发射器

emergency power supply control switch　应急电源控制开关

emergency power supply switch　应急供电开关

emergency start cutoff button　紧急断开起动按钮

encoding　译码处理

engine air inlet heating indicating light　发动机进气道加温指示灯

engine air inlet icing indicating light　发动机进气道结冰指示灯

engine auto-shutdown indicating light　发动机自动停车指示灯

engine emergency feathering shutdown knob　发动机紧急顺桨停车旋钮

engine exhaust gas temperature indicator　发动机排气温度指示器

engine fault indicating light　发动机故障指示灯

engine feathering　发动机顺桨

engine fire-extinguish device　发动机灭火设备

engine inlet guide heating switch　发动机进气导向器加温开关

engine inner cavity fire-extinguishing button　发动机内腔灭火按钮

engine instrument　发动机仪表

engine lighting switch　发动机照明开关

engine over-vibration indicating light　发动机振动过载指示灯

engine over-vibration indicator　发动机振动过载指示器

engine revolution indicator　发动机转速指示器

engine shutdown switch　发动机停车开关

engine start button　发动机起动按钮

engine start selection switch　发动机起动选择开关

engine start voltmeter　发动机起动电压表

engine vibration instrument　发动机振动仪表

entry door warning indicating light　登机门

告警指示灯

error of AOA   攻角误差

exhaust temperature indicator   排气温度指示器

exit   （安全)出口

explosion cap   爆炸帽

failure discrimination   故障判别

failure indication   故障指示

failure report   故障报告

fan switch   风扇开关

fault code   故障代码

fault light   故障灯

faulty component   故障件

feathering check cutoff switch   顺桨检查断开开关

feathering control panel   顺桨操纵板

feathering indicating light   顺桨指示灯

feathering signal light   顺桨信号灯

feedback system   反馈系统

field test   外场检测

filtration   滤波

fine filter blocking signal switch   细油滤堵塞信号开关

fine filter blocking indicating light   细油滤堵塞指示灯

finished plug   成品插头

fire extinguishing button   灭火按钮

fire extinguishing check switch   灭火检查开关

fire extinguishing selector switch   灭火选择开关

fixed mount   固定架

flange   法兰盘

flap changeover switch   襟翼转换开关

flap control hydraulic motor   襟翼操纵液压马达

flap control switch   襟翼操纵开关

flap electrical mechanism   襟翼电动机构

flap position indicator   襟翼位置指示器

flap signal   襟翼信号

flight archives   飞行档案

flight data recorder   飞行参数记录器

flight data recording system   飞行数据记录系统

flight engineer fan switch   机上机械师电风扇开关

flight instrument   飞行仪表

flight landing signal light box   航行着陆信号灯盒

flight parameter   飞行参数

flight parameter field check processor   飞参外场检测处理机

flight statistical data   飞行统计数据

flight time   航时

flight time driving   航时传动

flight time handle   航时手柄

flight training   飞行训练

flow distribution valve   流量分配活门

flow regulation switch   流量调节开关

fluorescent light   荧光灯

fluorescent light switch   荧光灯开关

foot console   脚操纵台

forced-landing fire extinguishing   迫降灭火

formation light   编队灯

forward cabin oxygen supply check button   用氧检查按钮

forward cabin oxygen supply signal   用氧信号

frequency division   分频

frequency meter   频率表

fuel consumption indicator   燃油耗量表

fuel consumption indicator switch   燃油耗量表开关

fuel consumption sequence   耗油顺序

fuel cross-feed valve   连通开关

fuel pressure annunciator   燃油压力信号器

fuel pump operating light   油泵工作指示灯

fuel quantity gauge changeover switch   油量

表转换开关

fuel quantity gauge power switch　油量表电源开关

fuel quantity indicator　燃油油量指示器

fuel quantity measuring switch　油量测量开关

generator　发电机

generator voltmeter　发电机电压表

glass heating　玻璃加温

glass heating switch　玻璃加温开关

glide slope　下滑道

gliding deviation　下滑偏差

gliding deviation warning signal　下滑偏差告警信号

ground data processing station　地面数据处理站

ground proximity warning system　近地告警系统

ground speed　地速

ground station　地面站

harness　绑带

heading angle　航向角

heading deviation　航向偏差

heading deviation warning signal　航向偏差告警信号

heating check indicating light　加温检查指示灯

HF radio set control box　短波电台控制盒

high frequency quartz resonator　高频石英谐振器

high-altitude and oxygen device instrument　高空及氧气设备仪表

high-altitude oxygen supply signal　高空用氧信号

horizon selection switch　地平仪选择开关

horizontal situation indicator　航向位置指示器

horn release button　解除喇叭按钮

hose union　软管接嘴

hour hand　时针

hydraulic fluid quantity gauge indicator　液压油油量表指示器

hydraulic fluid quantity gauge switch　液压油量表开关

hydraulic pump switch　液压泵开关

icing annunciator　结冰信号器

icing indicating light　结冰指示灯

icing signal switch　结冰开关

identification of friend-or-foe control box　敌我识别器控制盒

ignition switch　点火开关

indicated airspeed　指示空速

indicating error　指示误差

indicating light　指示灯

indication　读数

indicator　指示器

individual aircraft monitoring　单机监控

inertial force　惯性力

inertial navigation system（INS）　惯性导航系统

inertial sideslip angle　惯性侧滑角

inertial torque　惯性力矩

in-flight engine start switch　空中起动开关

initial position　起始位置

inlet guide heating　进气导向器加温

inlet guide icing signal　进气导向器结冰信号

INS auto navigation signal selection button　惯导自动导航信号选择按钮

INS battery switch　惯导电池开关

INS fault light　惯导故障提醒灯

INS state selector　惯导状态选择器

INS turn warning light　惯导转弯提醒灯

installation edge　安装缘

instrument brightness adjusting knob　仪表亮度调节旋钮

instrument brightness adjustment　仪表亮度调节

instrument indication　仪表指示器

instrument landing system（ILS） 仪表着陆系统

instrument panel 仪表板

integrated circuit 集成电路

intercom 机上通话

intercom control box 机通控制盒

interconnected bus 互连总线

interface board 接口板

label 标签

landing gear control 起落架操纵

landing gear extension 放起落架

landing gear signal 起落架信号

landing gear changeover switch 起落架转换开关

landing light control 着陆灯操纵

large storage module 大容量存贮模块

large-scale integrated circuit 大规模集成电路

large-storage data memory array 大容量参数存储器阵列

laser inertial/satellite integrated navigation system 激光惯性/卫星组合导航系统

lateral acceleration 横向加速度

latitude 纬度

layout 布局

left and right wing-tank cross-feed indicating light 左右连通指示灯

life of individual aircraft 单机寿命

light adjustment circuit 调光电路

light adjustment potentiometer 调光电位计

light shade 遮光罩

light signal 灯光信号

lighting power supply 照明电源

linear acceleration 线加速度

lithium battery 锂电池

load resistance 负载电阻

load spectra 载荷谱

local standard time 当地标准时间

locking wheel 锁紧轮

longitude 经度

longitudinal acceleration 纵向加速度

longitudinal level state 纵向水平状态

low power 低功率

low-fuel pressure indicator 低压燃油压力表指示器

low-temperature dedicated battery 低温专用电池

Mach indicator M 数表

magnetic heading 磁航向

main control panel 主控制板

main driving wheel system 主传动轮系

main icing warning light 结冰主告警灯

main recorder 主记录器

maneuverability 机动性

manual fuel consumption 人工耗油

manual fuel consumption pump switch 人工耗油油泵开关

mark line 标线

mass block 质量块

maximum indicating position 最大指示位置

maximum overload value 最大过负荷值

mean aerodynamic chord 平均气动力弦

measured time driving 测时传动

measured time handle 测时手柄

mechanical angular displacement 机械角位移

microcontroller 微控制器

microphone（MIC） 拾音器

microswitch 微动电门

middle position/neutral position 中间位置

minute hand 分针

misalignment signal 失调信号

monitor reset circuit 监控复位电路

motor 电机

mounting bracket 安装架

moveable pointer 活动指针

moveable travel range 可动行程范围

multimeter indicator 三用表指示器

multimeter switch　三用表开关

navigation cabin　领航舱

navigation instrument　航行仪表

navigation light　航行灯

navigator console　领航员操纵台

navigator fluorescent light　领航员荧光灯

navigator indicator　领航员指示器

navigator instrument panel　领航员仪表板

navigator oxygen annunciator　领航员用氧信号灯

navigator side cover plate　领航员侧面盖板

negative feedback brush　负回输电刷

negative feedback potentiometer　负回输电位计

negative thrust feathering check button　负拉力顺桨检查按钮

neutral gas explosion cap　中性气体爆炸帽

neutral gas fire extinguishing button　中性气体灭火按钮

neutral gas heating　中性气体加温

neutral gas indicating light　中性气体指示灯

neutral gas selection switch　中性气体选择开关

normal acceleration　法向加速度

normal cargo airdrop　正常投货

normal control circuit　正常操纵电路

northing acceleration　北向加速度

northing speed　北向速度

nose landing gear extension　前起落架放下

nose landing light　前着陆灯

nose wheelsteering　前轮转弯

off-course distance　偏航距

oil ejector radiation switch　滑油引射散热开关

oil quantity gauge indicator　滑油油量表指示器

oil quantity gauge switch　滑油油量表开关

oil radiating changeover switch　滑油散热转换开关

oil radiator shutter indicator switch　散热风门指示器开关

oil remaining warning light　滑油剩油警告指示灯

on-line programming　在线编程

on-site programming　现场编程

operating current　工作电流

operating depth　工作深度

operating frequency　工作频率

operating life　工作寿命

original position　原位

oscillation signal　震荡信号

output voltage　输出电压

over station beacon（IM）　过台信标（近）

over station beacon（MM）　过台信标（中）

over station beacon（OM）　过台信标（远）

over-current and overvoltage protection　过流、过压保护

overload　过载

overload meter power switch　过载表电源开关

oxygen instrument panel　氧气仪表板

oxygen relay box　用氧信号继电器盒

oxygen signal light　用氧信号指示灯

oxygen supply annunciator　用氧信号器

oxygen supply check button　用氧检查按钮

oxygen supply signal check button　用氧信号检查按钮

oxygen supply signal light　用氧信号灯

oxygen supply signal release button　用氧信号解除按钮

oxygen warning system　用氧警告系统

parachute cord recovery circuit　伞绳回收电路

parallel connection　并联

partial angle of attack　局部攻角

partial download　部分下载

partial feathering check button　部分顺桨检查按钮

pilot/copilot console　正/副驾驶员操纵台

pilot/copilot instrument panel　正/副驾驶员仪表板

pitch angle　俯仰角

pitot pressure adapter nozzle　全压转接嘴

pitot pressure hose　全压软管

pitot pressure union　全压接管嘴

pitot tube　全压管

pitot tube heating indicating light　空速管加温指示灯

pitot tube system tester　空速管系统试验器

pitot-static tube heating switch　动静压受感器加温开关

pointer pivot　指针轴

pointer setting driving　拨针传动

portable signal light socket　手提信号灯插座

position indicator　位置指示器

positive/negative acceleration　正/负加速度

potentiometer　电位计

potentiometer winding　电位器绕组

power plant　动力装置

power supply module　电源模块

power supply switch　供电开关

power-fail safeguard　掉电保护

power-on reset　上电复位

power-storage circuit　储能电路

preset dynamic pressure　给定动压

pressure indicator　压力表

pressure regulation safety valve　压力调节安全活门

pressure vacuum tester　压力真空试验器

processor　处理机

program mechanism indicating light　程序机构指示灯

propeller anti-icing　螺旋桨防冰

propeller heating indicating light　螺旋桨加温指示灯

propeller heating switch　螺旋桨加温开关

propeller ready indicating light　螺旋桨准备好指示灯

propeller stop release indicating light　螺旋桨解除制动指示灯

propeller stop release switch　螺旋桨解除限动开关

protection performance　防护性能

protective cover　保护罩

pulse circuit　脉冲电路

pulse relay　脉冲继电器

pulse signal　脉冲信号

pulse sound　脉冲声

pulse width　脉宽

push rod　顶杆

quick access recorder　快取记录器

quick-release screw　快卸螺钉

radio altimeter　无线电高度表

radio altimeter brightness adjusting knob　无线电高度表亮度调节旋钮

radio altimeter indicator　无线电高度表指示器

radio communication keying　无线电通话键控

radio magnetic indicator　综合航向指示器

radio talk button　无线电通话按钮

ramie thread　苎麻线

ramp door locked indicating light　货桥锁好指示灯

ramp dooropen indicating light　货桥门开指示灯

range-to-go　待飞距离

ratchet mechanism　棘轮机构

rated indicating light　额定指示灯

rated switch　额定开关

real time　实时

real-time clock　实时时钟

rear cargo door　后舱门

rear view　后视图

recorder ground check switch　记录器地面检查开关

recording capacity　记录容量

recording duration　记录时间

recording medium　记录介质

recording medium capacity　记录介质容量

recording mode　记录方式

reflective sticker　反光胶贴

relative displacement　相对位移

relative pressure altitude　相对气压高度

relay cover plate　继电器盖板

restraint　限制带

return-to-zero　回零

rheostat　变阻器

rigid connection　刚性连接

roll angle　横滚角

rubber vibration absorber　橡胶减震器

rudder　方向舵

sampling resistance　采样电阻

scale　刻度

screwjonit　螺纹接头

sealing tape　密封带

second hand　秒针

sensitive element　敏感元件

serial bus transmitting/receiving circuit　串行总线收发电路

serial data　串行数据

serial data bus　高速串行总线

service life　使用寿命

servo circuit　伺服电路

set altitude　设定高度

setter　给定器

shock absorbing device　减震装置

shock absorbing pad　减震垫

shutter position indicator　风门位置指示器

shutter switch　风门开关

signal box　信号盒

signal control box　信号控制盒

signal converter fault light　信号转换器故障灯

signal flare　信号弹

signal format　信号格式

signal gun control box　信号枪控制盒

signal gun control box cover plate　信号枪控制盒盖板

signal indicator　信号指示器

signal source　信号来源

signal window　信号窗

single pendulum　单摆

sink rate　下沉速率

sliding arm　滑动臂

sling　吊带

slotted screwdriver　一字螺刀

semi-conductor memory　半导体存储器

solid-state storage　固态存储

spring clip　弹簧夹子

spring vibration absorber　弹簧减震器

start fault light　起动故障灯

start indicating light　起动指示灯

starter anti-fire switch　起动机防火开关

starter changeover switch　起动机转换开关

starter fire warning button indicating light　起动机火警按钮指示灯

starter generator overload indicating light　起动发电机超负荷指示灯

starter oil pressure indicating light　起动机滑油压力指示灯

starter shutter switch　起动机风门开关

starter starting button　起动机起动按钮

starter starting indicating light　起动机起动指示灯

starter starting power switch　起动机起动电源开关

starter starting ready indicating light　起动机起动好指示灯

starting selector switch　起动选择开关

starting shutdown button　起动机停车按钮

starting time　起始时间

static air temperature　大气静温

static pressure sensor　静压受感器

static tube　静压管

status indicating light　状态指示灯

step motor　步进电机

stored program　存储程序

structural dimension　结构尺寸

synchronizer signal　同步器信号

tail anti-icing　尾翼防冰

tail heating indicating light　尾翼加温指示灯

tail heating switch　尾翼加温开关

temperature regulation selector switch　温度调节开关

terminal board　接线板

terminal switch　终点电门

terrain　地形

thermometer　温度表

three-axis accelerometer　三轴加速度计

throttle lever angle indicator switch　油门杆位置指示器开关

throttle lever assembly　油门杆组合件

throttle lever/power lever　油门杆

time delay switch　延时开关

time information　时间信息

time measurement　测时

time-to-go　待飞时间

timing circuit　走时电路

timing precision　走时精度

tolerance　公差

torquemeter　扭矩表

torque meter switch　扭力表开关

torquer　力矩器

total air temperature　大气总温

total sampling rate　总采样率

track angle　航迹角

traffic alert and collision avoidance system (TCAS)　空中交通告警和防撞系统

transceiver　收发机

transient (actual) angle of attack　瞬时(实际)攻角

transient acceleration　瞬时加速度

transient angle of attack voltage　瞬时攻角电压

transmission ratio　传动比

trim tab　调整片

true airspeed　真空速

true heading　真航向

TSI switch　转弯仪(通断)开关

turn and slip indicator　转弯(侧滑)仪

turn warning light　转弯提醒灯

typical value　典型值

underwater locator beacon　水下定位信标

unfeathering indicating light　反桨指示灯

upward acceleration　天向加速度

upward speed　天向速度

vacuum aneroid　真空膜盒

vane　风标

vertical acceleration　垂直加速度

vertical axis　垂直轴

vertical control　垂直控制

vertical speed　升降速度

vertical speed indicator　升降速度表

VHF/UHF radio set　超短波电台

voltage regulator　电压调节器

voltmeter　电压表

voltmeter selector switch　电压表选择开关

volume potentiometer　音量电位器

VOR　伏尔

WAIT/RUN indicating light　等待/运行指示灯

warning bell　警铃

warning bell ringing button　警铃断开按钮

warning light　告警灯

warning signal siren placard　警告信号鸣响标牌

washroom oxygen supply signal light　卫生间用氧信号灯

washroom oxygen supply warning bell　卫生间用氧警铃

water bubble　水泡

LANDING GEAR　起落架

water drain hole　排水孔

weather radar　气象雷达

weather radar display　气象雷达显示器

wheel brake　机轮刹车

wheel door control　轮舱门操纵

wind direction　风向

wind speed　风速

wind vane rotation angle　风标转角

windshield heating switch　风挡加温开关

windshield strong-weak heating selector switch　风档强、弱加温开关

windshield wiper switch　风挡刷开关

wing angle of attack　机翼攻角

wing fire-extinguish device　机翼灭火设备

wing heating indicating light　机翼加温指示灯

wing heating switch　机翼加温开关

wing plane　机翼平面

wing-tank cross-feed switch　油箱连通开关

wing-tank fuel quantity balancing switch　油箱油量平衡开关

white/fluorescent light switch　白灯/荧光灯开关

workshop check device　内场检测设备

zero point voltage　零点电压

zero position　零位

LANDING GEAR　起落架

abrade-proof sleeve　防磨套

accumulator　蓄压瓶

acoustic/optical signal electric device　灯光/声音电信号装置

actuator　作动筒

actuator casing　作动筒壳体

adapter　转接器

adjustable support arm　可调支臂

adjusting bushing　调节衬套

adjusting screw　调整螺钉

aerodynamic load　气动载荷

aero-washing gasoline　航空洗涤汽油

air pressure　气压

alloy steel　合金钢

aluminum bronze bush　铝青铜衬套

aluminum sleeve　铝轴套

ambient temperature　环境温度

annular chamber　环形腔

anti-skid braking device　防滑刹车装置

arc groove　圆弧形槽

area/access (channel)　区域/通道

armature　衔铁

arrow　箭头

auxiliary material　辅助材料

auxiliary travel　辅助行程

aviation hydraulic fluid　航空液压油

axial sleeve　轴套

ball joint　球形接头

bearing bush cover　轴瓦盖

blanking cap　堵帽

blanking plug　堵塞

bogie　车架

bogie fork ring　车架叉形耳环

bolt brace piece　螺栓撑片

box member/box-shaped member　盒形件

bracket　支架

brake accumulator　刹车蓄压瓶

brake balance mechanism　刹车平衡机构

brake pressure gauge sensor　刹车压力表传感器

braking flange/brake flange　刹车凸缘

braking wheel　刹车机轮

bulkhead　横隔板

bush cover　瓦盖

bushing dimension　衬套尺寸

bushing/bearing bush　轴瓦

butt landing gear mounting trolley　起落架安装车

butterfly spring　蝶形弹簧

— 101 —

button　按钮

cable　电缆

cable drum wheel　钢索鼓轮

cable plug　电缆插头

castle nut　槽形螺母

cause　故障原因

centering mechanism　纠偏机构

centrifugal acceleration　离心加速度

changeover valve　转换活门

charge valve　充气活门

charging pressure　充填压力

check valve　单向阀

chromium coating　镀铬层

circuit breaker　自动保护开关

clamp　卡箍/卡子

clearance/gap　间隙

cloth strip　布带

clutch　离合器

compression amount at parking　停机压缩量

compression quantity　压缩量

connection part　连接件

contacting switch　接触开关

control handle　操纵手柄

control stick　操纵杆

convex block　凸座

coordinating valve　协调活门

copper washer　紫铜垫圈

cotter pin　开口销

cotton cloth　棉纱

countersunk bolt　沉头螺栓

crank shaft　曲轴

cushion block/pad　垫块

cylinder seat　气缸座

damage　损伤

damper　缓冲器

damper strut　缓冲支柱

damping/absorbing function　阻滞缓冲作用

dead point　死点

dead weight　自重

deflection　挠度

deflection measuring device　挠度量具

deformation　变形

de-paint　脱漆

detective inspection　探伤检查

diameter　直径

differential unit　差动装置

dimmer tab　调光片

diode　二极管

disc part　盘形件

distribution cock　分配开关

docking car　对接车

door actuator　舱门作动筒

drifting amount　漂移量

driving lever　拨杆

driving piston/servo piston rod　主动/随动活塞杆

driving shaft/transmitting shaft　传动轴

duration　持续时间

dust cover　防尘盖

eccentric trunnion　偏心轴颈

elbow nipple　弯管嘴

elbowunion　弯管接头

electric load　电气负荷

electric winch　电动绞车

electrical pulse signal　电脉冲信号

electromagnet　电磁铁

emergency brake valve　紧急刹车活门

emergency unlocking rocker　应急开锁摇臂

end cover　端盖

equipment　设备

error　误差

excessive pressure/residual pressure　余压

exposure　外露量

extension/compression travel　拉伸/压缩行程

failure analysis　故障分析

feedback pull rod　反馈拉杆

feeler gauge　塞尺

felt ring　毡圈

felt ring groove　毡圈槽

filter　过滤器

fixed force　定力

fixing bracket　固定支座

fixing shaft　固定轴

fluid contamination　油液污染

fluid resistance　流体阻力

fluid-charging quantity　充油量

fluoroplastic ring　氟塑料圈

foldable brace rod　可折撑杆

follow-up system　随动系统

footpedal　脚踏板

forked lug　叉耳

forward/backward travel　正/反行程

friction　摩擦

friction block　摩擦块

front/rear axle　前/后轮轴

front/rear rocker　前/后摇臂

full travel　全行程

fuselage　机身

gauze　纱布

gear plate joint /toothed plate joint　带齿板接头

grease gun　油注枪

grease/lubricating grease　润滑脂

groove　油槽

ground power　地面电源

ground power cart　地面电源车

ground pump truck　地面油泵车

ground taxiing　地面滑行

guide clip plate　导向夹板

guide rail　导轨

hand-powered winch　手摇绞车

hook　挂钩

hose　软管

hydraulic braking unit　液压刹车装置

hydraulic filter　液压油滤

hydraulic fluid　液压油

hydraulic fluid quantity　油量

hydraulic handle valve　液压手动阀

hydraulic jack　液压千斤顶

hydraulic lock　液压锁

hydraulic pipe　液压导管

hydraulic pressure　油压

hydraulic solenoid valve　液压电磁阀

hydraulic test bed　液压试验台

impact　冲击

impact acceleration　冲击加速度

impurity　杂质

indicating light　指示灯

inductance load current　电感性负荷电流

industry nitrogen gas　工业用氮气

inertial force　惯性力

inertial load　惯性载荷

inertial sensor　惯性传感器

inertial wheel　惯性轮

inflation equipment　充气设备

initial pressure　始动压力

inlet/return line　来/回油路

inner/outer bushing　内/外衬套

inner/outer skin　内/外蒙皮

installation position　安装位置

interlock unit　联锁装置

isolation dielectric strength/insulation electrical shock strength　绝缘抗电强度

isolation resistance　绝缘电阻

jig/fixture　夹具

knuckle bearing　关节轴承

landing gear　起落架

landing gear observation window　起落架观察口

landing signal light box　着陆信号灯盒

lead sealing　铅封

leakage　渗漏

leather cup　牛皮碗

limit position　极限位置

lining disk　衬盘

linkage mechanism　连杆机构

load　载荷

load/unload　加载/卸载

load-bearing joint　承力接头

load-bearing part　承力构件

locating pin　定位销

lock hook　锁钩

lock key　锁键

lock ring　锁环

lock stopper/stop　限动件

locking screw　锁紧螺钉

locking wire　保险丝

louver　百叶窗

lug base/lug socket　耳座

magnesium alloy　镁合金

magnifier　放大镜

main joint　主接头

main landing gear　主起落架

main landing gear brace　主起落架撑杆

main landing gear inclined brace　主起落架斜撑杆

main wheel　主轮

main wheel brake　主轮刹车

maintenance access cover　维护口盖

maneuvering performance　机动性能

manual button　手控按钮

mark　记号

marking line　标志线

maximum deflection angle　最大偏转角

measuring instrument　测量工具

mechanical lock　机械构锁

mechanical-resistance action　抗机械作用

meter holder　表架

middle relay　中间继电器

moment　力矩

mould part　模压件

movable wheel rim　活动轮缘

movable/fixed pulley　动/定滑轮

movement clearance　运动间隙

moving valve　游动活门

moving/fixed member　运动/固定构件

neutral position　中立位置

nipple/union　接管嘴

nitrogen　氮气

nitrogen bottle　氮气瓶

nitrogen charging pressure　充氮压力

nomenclature　设备(元件)名称

nose landing gear　前起落架

nose landing gear well　前起落架舱

nose wheel steering　前轮转弯

observation access cover　观察口盖

oil chamber　油腔

oil container　贮油器

oil-air chamber　油气腔

ointment　油膏

oleo-nitrogen damper strut　油液氮气式缓冲支柱

oleo-nitrogen type　油液-氮气式

operating access cover/(working) access cover　工作口盖

operating chamber/working chamber 工作腔

operating fluid/working fluid　工作液体

operating gas/working gas　工作气体

operating light/working light　工作灯

operating line/working pipeline　工作油路

operating travel/working travel　工作行程

operation normalization/operation specification　工作规范

outer cylinder　外筒

outer cylinder joint　外筒接头

outer cylinder lug　外筒耳片

outer cylinder pipe　外筒管件

packing ring　涨圈

partition　隔板

pedal control　脚操纵

percentage meter 百分表

pin 销钉

pin hole 销钉孔

pin shaft 销轴

pipe head 导管头

piston rod 活塞杆

placard 标牌

planetary gear train system 行星轮系

plastic washer 塑料垫圈

plug connector 插销接头

plunger 柱塞

plunger pump 柱塞泵

positive/negative acceleration 正/负加速度

positive/negative angular acceleration 正/负角加速度

powder metallurgy braking disc 粉末合金刹车盘

preservation 油封

preservation paper 油封纸

press disc 压紧盘

press plate 压板

pressure check switch 气压检查开关

pressure difference 压力差

pressure of travel end 行程末端

pressure reducing valve 减压阀

pressure relief cock 泄压开关

pressure-bearing disc/support disc 承压盘

proportioner/quantilizer 定量器

protection panel 护板

push block 顶块

push rod 顶杆

pylon 吊挂

rail groove 导轨槽

rebounce 颠簸跳动

relay 继电器

reliability 可靠性

removable head 可卸头

removable sealing device 活动密封组

residual travel 剩余行程

resistance lad current 电阻性负荷电流

retraction circuit 收上电路

retraction/extension actuator 收放作动筒

retraction/extension position 收上/放下位置

return spring 回力弹簧

reverse pressure 反向压力

reversed procedure 相反程序

ring spring/spiral spring 环形弹簧

rivet 铆钉

roller 滚轮

roller gear plate/roller toothed plate 滚轮齿板

rotation shaft 转轴

rubber cup 胶碗

rubber ring groove 胶圈槽

safety cover 保险罩

safety nut 保险螺帽

safety pin 保险销

safety valve 安全阀

sandy soil 砂土

sealing device 密封装置

sealing guide sleeve 气密导套

sealing putty 密封腻子

sealing ring 密封圈

sealing rubber ring 密封胶圈

sealing tape 密封带

self-locking nut 自锁螺母

sensitivity 灵敏度

service ladder/working ladder 工作梯

service life 使用期限

shaft sleeve 轴套

shimmy device 减摆装置

shoulder 凸肩

shuttle valve 梭形活门

signal light 信号灯

signal system 信号系统

single metal braking disc 单金属刹车(动/静)盘

siren horn ring　警铃

skip time　跳越时间

slant surface　斜面

sleeve　套管

slide valve　滑阀

sling/hoisting ring　吊环

slotted screwdriver　一字螺刀

solenoid valve　电磁阀

solubility　可溶性

special-shaped sleeve　特型套筒

specified value　规定值

spline　花键

spring actuator　弹簧作动筒

stabilizing shock absorber　稳定缓冲器

stationary disc/moving disc　静盘/动盘

steel ball lock　钢珠锁

steel cable　钢索

steel drop-forging rod head　钢制模锻件杆头

stop nut　止动螺帽

stop ring　止动圈

stop watch　秒表

stopper　限制器/止动器

straight motion line　直线运动

straight nipple　直管嘴

stud　螺桩

support arm　支臂

support pipe　支管

support point　支撑点

suspension joint　悬挂接头

switching frequency times　转换次数

tail support　尾撑

takeoff/landing run　起飞/着陆滑跑

talcum powder　滑石粉

tapered needle bearing　圆锥滚针轴承

taxiing turning　滑行转弯

technological blanking cap　工艺堵帽

tensiometer　拉力计

tension spring　拉力弹簧

terminal switch　终点开关

throttle nipple　节流管嘴

throttle valve　节流活门

tool　工具

torque arm　防扭臂

torque spanner　定力扳手

transitional union　过渡管接头

triangle inclined strut　三角斜支柱

triangle truss　立体三角形架

tricycle retractable type　前三点可收放式

triode　三极管

troubleshooting/remedy　排故方法

twin-tandem landing gear　四轮小车式起落架

unlocking retraction/extension actuator　开锁收放作动筒

up/down vibration energy　上/下振动能量

uplock/downlock　上/下位锁

uplock/downlock mechanism　上/下位锁机构

uplock/downlock rocker arm　上/下位锁摇臂

upper/lower braking pull rod　上/下刹车拉杆

upper/lower chamber　上/下腔

upper/lower joint　上/下接头

upper/lower link　上/下连杆

upper/lower support ring　上/下撑圈

upper/lower torque arm　上/下防扭臂

upper/middle/lower universal joint　上/中/下万向接头

vacuum electron beam　真空电子束

variable section steel tube　变截面钢管

vibration acceleration　振动加速度

vibration frequency　振动频率

visual check　目视检查

warning board　警告牌

washing gasoline　洗涤汽油

wheel　机轮

wheel axle　轮轴

wheel cabin　机轮舱

wheel chock　轮挡

wheel deflection　轮偏转

wheel group　轮组

wheel hub　轮毂

wheel rim　轮缘

wheel tire　轮胎

wire　导线

working site　现场

NAVIGATION　导航

A/G distance measuring capacity　空/地测距能力

absolute barometric altitude　绝对气压高度

accelerometer failure　加速度计故障

additional resistance　附加电阻

address code　地址码

ADF compass　无线电罗盘

air damper　空气阻尼器

airdata　大气参数

air data computer　大气数据计算机

air data measurement control instrument　大气数据测控仪

air data system　大气数据系统

air pressure sensor　空气压力受感器

air traffic control transponder　空管应答机

air/air distance-measuring　空/空测距

air/air mode　空/空模式

air/ground mode　空/地模式

airborne equipment　机载设备

airborne marker panel　机载指示面板

airborne short-range navigation device　机载近程导航设备

aircraft symbol　飞机标记

airport database　机场数据库

airspeed indicator　空速表

airspeed probe　空速管

airtightness test　气密试验

alert area　告警区域

all-attitude combined gyro　全姿态组合陀螺

allowable error　允许误差

altitude code　高度编码

altitude indicator　高度指示器

altitude loss　高度损失

ambient temperature　周围温度

amplitude limiter　限幅器

analog signal　模拟信号

aneroid altimeter　膜盒式高度表

angle of attack(AOA)　攻角

angle of side slip　侧滑角

angle phase difference　角度相差

angular speed of turn　转弯角速度

angular speed rate　角速率

angular velocity annunciator　角速度信号器

annunciator　信号器

annunciator connected pressure　信号器接通压力

antenna gain　天线增益

antenna manual pitch angle　天线人工俯仰角

antenna stabilization mechanism　天线稳定机构

antenna stable range　天线稳定范围

antenna sweeping angle　天线扫掠角度

antenna transceiver changeover switch　天线收发转换开关

anti-transient power supply　抗瞬变电源

ANT-receiving　ANT 收讯

approachrate/proximity rate　接近速率

area/access No.　区域/口盖号

aspirating/charging/bleeding air　抽气/充气/放气

attenuator　衰减器

attitude　姿态

attitude azimuth system　姿态方位系统

attitude director indicator　地平指示器

aural message　语音信息

auto flight control system　自动飞行控制

系统

auto navigation　自动导航

auto transformer　自耦式变压器

auto rotation angular rate component　自转角速度向量

azimuth accuracy　方位准确度

azimuth mark　方位标志

azimuth motor　方位电机

azimuth signal　方位信号

balance modulator　平衡调制器

balance spring　平衡弹簧

balanced pulse　均衡脉冲

barometer altitude　气压高度

basic code　基本编码

beacon　信标

beacon antenna　信标天线

beacon subunit　信标分机

bell pulse　钟形脉冲

black epoxy nitroenamel　黑色环氧硝基磁漆

blanking cover/blanking cap　堵盖

broadcast station　广播电台

built-in test button/BIT button　自检按钮

buoy　浮标

cabin altitudepressure difference gauge　座舱高度压差表

caging error　锁定误差

calibration　标定

carrier　载波

carrier frequency　载波频率

carrier/subcarrier　载波/副载波

cathode-ray tube　阴极射线管

caution obstacle　注意障碍物

caution terrain　注意地形

centrifugal speed governor　离心式调速器

channel number　波道数

channel subunit　信道分机

chassis subunit　机架分机

check point　检查点

circuit board　电路板

circuit breaker　自保开关

circuit element　电路元件

circuit protection　电路保护

circular compass deviation　圆周罗差

clock circuit　时钟电路

clockwise　顺时针

clutter suppression ratio　杂波抑制比

code　编码

code pulse　信号脉冲

coincidence amplifier　重合放大器

color alphanumeric picture　彩色数字图象

compass deviation　罗差

compressed air　压缩空气

configuration module　配置模块

connector　连接器

connentional memory　常规存储器

continuous value　延续值

continuous wave　连续波

control box　控制盒

control display unit　控制显示器

control mechanism　控制机构

converter　转换器

coordinated resolver　同步分解器

correction mechanism　修正机构

correction potentiometer　修正电位计

corrective resolution advisory　修正性决断报告

counterclockwise　逆时针

coupler　耦合器

course setting　航线装订

course/heading mark　航向标记

critical angle of attack　临界攻角

critical angle of attack sensor　临界攻角传感器

cross track distance　偏航距

current frequency　当前频率

current stabilizing failure　稳流故障

cylinder　气缸

damping cylinder　阻尼筒

damping fluid 阻尼液

data card slot 数据卡插槽

data failure 数据错误

database searching area 数据库搜索区域

degree-of-freedom 自由度

delayer 延时器

diagnosis code 诊断码

dial 刻度盘

digital signal 数字信号

digital wheel window 数码轮窗口

direction scanning rate 方位扫描速率

directional sensitivity 定向灵敏度

direct-to flight plan leg 直达飞行计划航段

discrete magnitude signal 离散量信号

discrete signal 离散信号

display capability 显示能力

display radar 显示雷达

display range 显示范围

display window 显示窗口

distance advisory 距离通告

distance calculation 远距测距

distance mark 距离标志

distance measuring equipment 测距器

distance measuring response signal 测距应答信号

distance range 距离量程

distance/direction setting 距离/方位设定

distributor 分配器

drift 偏流

drift angle error 偏流角误差

duplexer 双工器

echo 回波

electric bridge theory 电桥原理

electric gyro 电动陀螺

electric heater 电加热器

electric thermometer 电动温度表

electrical heating device 电加温装置

electrical property 电性能

electromagnetic wave 电磁波

element panel 元件板

elevation motor 垂直电机

emergency horizon 应急地平仪

enhanced ground proximity warning computer 近地告警计算机

error 误差

evenness 均匀性

excitation interface 激励接口

expansion chamber 膨胀室

expectation value 期望值

failure in inertial navigation component 惯导部件故障

fast slaving speed 快协调速度

fault 故障

fault code 故障码

fault description 故障名称

fault flag 故障旗

fault isolation 故障隔离

fault level 故障级别

ferromagnetic object 铁磁性物质

field 外场

field atmospheric pressure setting knob 场压设定旋钮

field fault 外场故障

filling pulse 填充脉冲

fixed pointer 固定指针

flap control 襟翼操控

flat antenna 平板天线

flight code 飞行代码

flight distance 飞行区间

flight environment data system 飞行环境数据系统

flight path 航迹

flight route/flight course 飞行路线

frequency interval 频率间隔

frequency selector knob 频率选择旋钮

frequency stabilizing failure 稳频故障

frequency storing button 频率存储按钮

frequency synthesizer 频率合成器

frequency-synthesizing subunit　频合分机

function button　功能键

G/S antenna　下滑天线

general check　一般性检查

general-purpose tool　通用工具

geographic longitude/latitude　地理经/纬度

glide deviation　下滑偏差

glide slope cancel　取消下滑道

glide subunit　下滑分机

glideslope beacon　下滑信标

glideslope deviation　下滑道偏差

GPS navigator　GPS 导航仪

graduation value　分度值

Gray code data bus　格雷码数据总线

ground beacon marker　地面信标台

ground controller　地面管制员

ground object　地物

ground proximity warning system　近地告警系统

ground speed　地速

ground speed variance　地速超差

ground station　地面台

ground target　地物目标

grounding strip　接地条

guiding information　引导信息

gyro　陀螺

gyro azimuth　陀螺方位

gyro control task manager failure　陀螺控制任务机故障

gyro drift quantity　陀螺漂移量

gyro magnetic compass　陀螺磁罗盘

half compass drift error　半罗盘漂移误差

hard alert area　硬告警

heading　航向

heading angle setting　航向角设定

heading attitude system　航向姿态系统

heading attitude system indicator　航向姿态系统指示器

heating element　加温元件

heating power　加温功率

heterodyne radio　超外差式收音机

high-frequency socket　高频插座

horizon gyro platform　地平陀螺平台

horizontal distance threshold　水平距离门限

horizontalposition indicator　航向位置指示器

hose　软管

housing　表壳

hysteretic error/lagging error　迟滞误差

identification code　识别码

identification code pulse　识别信号脉冲

identification signal　识别信号

identification signal tone　识别信号声

indicated airspeed（IAS）　指示空速

indicating deviation（lag）　指示偏差（迟滞）

indication difference　示数差异

indicator pointer　指示器指针

information subunit　信息分机

initial elasticity　初始弹力

initial parameter setting　初始参数设定

initial position setting　初始位置设定

inner marker　近（内）距台

input/output　输入/输出

inquiry pulse　询问脉冲

installation bracket　安装托架

installation error angle　安装误差角

installation panel　安装板

instrument landing system　仪表着陆系统

integral amplifier　综合放大器

interface　接口

interface board　接口板

interface sub-unit　接口分机

interface unit failure　接口机故障

intermediate and image frequency rejection ratio　中/像频抗拒比

intermediate frequency/video circuit　中频/视频电路

intermediate-frequency　中频频率

interrogation/response　询问/应答

intruder　入侵飞机

kit　工具箱

knob　旋钮

landing direction　着陆方向

laser strapdown inertia/satellite integrated navigation system　激光惯性/卫星组合导航系统

latitude potentiometer　纬度电位计

latitude setting　纬度设定

Laval tube　拉瓦尔管

LCD screen　液晶显示屏

leaf spring/spring leaf　弹簧片

linear transition　线性过渡

link rod　连杆

LOC beacon　航向信标

local oscillator　本振器

location accuracy　定位精度

location beacon　定位信标

location correction　位置校正

location memory　位置记忆

location updating rate　位置更新率

longitude setting　经度设定

low-frequency socket　低频插座

low-noise amplifier　低噪声放大器

low-noise antenna　低噪声天线

low-noise receiver　低噪声接收机

low-pass filter　低通滤波器

low-speed data bus　低速数据总线

lubber line　航向标线

Mach meter　M 数表

magnetic compass　磁罗盘

magnetic compensator　罗差修正器

magnetic correction and slaving speed　磁修正协调速度

magnetic deviation/magnetic variation　磁差

magnetic heading　磁航向

magnetic heading sensor　磁航向传感器

magnetic heading swing　磁航向摆动

magnetic meridian　磁子午线

magnetic variation pointer　磁差针

magnetron　磁控管

mark　刻划

marker beacon　指点信标

master display　主指示器

measuring range　量程

medium marker　中距台

memory circuit　存储电路

menu button　菜单按钮

menu parameter　菜单参数

menu setting　菜单设定

meridian　子午线

meteorological target　气象目标

micro-ohm meter　微欧表

microprocessor board　微处理器板

microwave assembly　微波组件

minimum terrain clearance　最小地面间距

misalignment angle　失调角

mixer　混频器

mode button　方式键

mode selector failure　状态选择器故障

mode selector knob　模式选择开关

modulated frequency　调制频率

modulated wave　调制波

modulator　调制器

monitoring capability　监视容量

monitoring range　监视量程范围

Morse code　莫尔斯电码

mountain　山脉

mountain peak　山峰

navigation deviation　航向偏差

navigation indicator　领航指示器

navigation parameter　导航参数

navigation satellite　导航卫星

navigation source selection　导航源选择

navigation station　导航台

nickel-chrome resistance wire　镍铬电阻丝

non-volatile memory　非易失存贮

normal list　常规清单

normal slaving speed　正常协调速度

normal temperature　常温

notch　缺口

null position error　零位误差

observation value　观察值

obstacle ahead　前方障碍物

open circuit　开路

operating range　工作范围

operation button　操作键

outer marker　远/外距台

output interface format　输出接口格式

output parameter accuracy　输出参数精度

output range　输出范围

overload　过负荷

over-speed warning system　超速告警系统

over-temperature warning　超温告警

parallel altitude code　并行高度编码

percentile　百（分）位

phase relationship　相位关系

pin　引脚

pitot pressure　全压

pitot pressure input limit　总压输入限制

pitot tube　全压管

pitot-static system　全静压系统

planar slotted antenna（array）　平板峰阵天线

pointer　指针

position indicator　位置指示器

position longitude-latitude　位置经纬度

potentiometer　电位器

power filter　电源滤波器

power panel　电源板

power supply circuit　电源电路

power supply monitor　电源监控器

power supply sensitivity　电源灵敏度

power supply sub-unit　电源分机

power supply system　电源系统

power-on check　通电检查

power-supply fluctuation　电源波动

precipitator　沉淀器

preprocessor failure　预处理机故障

present flight course　当前飞行航线

present flight plan　当前飞行计划

preset flight route（course）　预定航线

preset frequency　预置频率

pressure setting　气压设定

preventive resolution advisory　预防性决断报告

prompt light　提示灯

prompt setting　提示设定

proximate traffic　邻近交通

pull up　拉起

pull-in bandwidth　捕捉带宽

pulse　脉冲

pulse peak　脉冲峰值

pulse peak power　脉冲峰值功率

pulse repetition frequency　脉冲重复频率

pulse width　脉冲宽度

pure inertial navigation　纯惯性导航

quadrant compass deviation　象限罗差

quantizer failure　量化器故障

quick alignment　快速对准

quick-release bolt　快卸螺栓

radio altimeter　无线电高度表

radio frequency assembly　射频组件

radio magnetic indicator　综合航向指示器/无线电磁指示器

radio navigation system　无线电导航系统

radio-frequency antenna　射频天线

rainfall rate　降雨率

range ring　距离环

receiving sensitivity　收讯灵敏度

reference code　基准编码

reference phase　基准相

register　寄存器

regular compass alignment　正常罗经对准

relative altitude　相对高度

relay box　继电器盒

remaining compass deviation　剩余罗差

rendezvous point　交会点

reply pulse　回答脉冲

resistance temperature　受阻温度

resolution advisory　决断报告

resolution alert　决断告警

reverser　换向器

right rotary knob　右旋钮

ring amplifier　环放大器

rotary knob　旋钮

rotor　转子

runway entry　跑道入口

runway extension line　跑道延长线

safety label　安全标签

sand paper　砂纸

scanner　扫描器

scanning driver　扫描驱动器

scanning driver base　扫描驱动器底座

screen intensity　显示屏辉度

sealing material　密封材料

select button　选择按钮

sensitive element　敏感元件

servo plate　伺服板

setting　装订，设定

shaking failure　抖动故障

shift lever drive mechanism　拨杆传动机构

shoreline　海岸线

short circuit　短路

sideslip indicator　侧滑仪

signal damper　信号阻尼器

signal operating point　信号工作点

signal point error　信号点误差

signal processing unit/signal processor　信号处理机

simulator　模拟器

sink rate　下降率

slant range　斜距

slave display　副指示器

slaving speed　协调速度

soft alert area　一级告警（软告警）

soft key　软按键

special anti-magnetic screwdriver　专用防磁螺刀

speed governor　调速器

spring ripple membrane　弹性波纹膜片

standard pressure altitude　标准气压高度

standby static changeover switch　备用静压转换开关

static air temperature　大气静温

static hole　静压孔

static hole for right system　右系统用静压孔

static pipeline system　静压管路系统

static pressure　静压

static pressure input limit　静压输入限制

static temperature sensor　静温传感器

static tube　静压管

stator　静子

status advisory　状态通告

stored flight leg　所存储的飞行计划航段

stored flight plan　所存储的飞行计划

strong magnetic field（source）　强磁场（源）

sub-unit　分机

supply voltage　电源电压

switch　开关

swivel joint　旋转接头

synchronization receiver　同步接收器

synchronizer　同步器

synchronous follow-up system　同步随动系统

synthetic envelope　合成包络

system weight　系统重量

tactical air navigation　战术空中导航系统/航空近程导航系统

tapered beam　锥形波束

tapered pin　锥形销

target echo　目标回波

temperature sensing resistance wire　感温电阻丝

ten percentile　十（分）位

terrain　地形

terrain display　地形显示器

terrain display inhibition　地形显示抑制

terrain failure　地形失效

test　试验

test socket　检测插座

test value　试验值

time threshold　时间门限

time-to-go　待飞时间

time-to-station　到台时间

total air temperature　大气总温

total temperature sensor　总温传感器

tracing rate　跟踪速率

tracking capability　跟踪能力

traffic advisory　交通报告

traffic alert　交通告警

traffic alert and collision avoidance system
空中交通告警和防撞系统

transceiver　收发机

transformer　变压器

trans-illuminated panel　导光板

transparent shroud　透明罩

transponder　应答机

trimmer/brightness regulator　亮度调节器

troubleshooting　处理方法

true/magnetic heading　真/磁航向

turn and slip indicator　转弯侧滑仪

turn indicator　转弯仪

turning meter　转弯表

two-coil logometer　双线圈电流比计

unevenness/heterogeneity　不均匀性

vacuum　真空

vacuum aneroid　真空膜盒

vacuum source　真空源

variable phase　可变相

velocity pressure　速压

velocity pressure annunciator　速压信号器

vertical speed　升降速度

vertical speed indicator　升降速度表

vertical speed pointer　垂直速度指针

vertical speed scale　垂直速度刻度

vibration mechanism　减振机构

volume control　音量控制

VOR/ILS system　伏尔/仪表着陆系统

VOR/LOC antenna　伏尔/航向天线

VOR/LOC sub-unit　伏尔/航向分机

warm-up time　预热期

warning altitude knob　警告高度旋钮

warning cursor　警告游标

warning flag　告警旗

warning light　警告灯

warning signal light　告警信号灯

warning signal light built-in test　告警信号
灯自检

wavebeam　波束

waveguide　波导

waveguide nipple　波导口

waypoint　航路点

waypoint number setting　航路点序号装订

weather radar　气象雷达

wind direction　风向

wind speed　风速

workshop fault　内场故障

OXYGEN　氧气

absolute temperature　绝对温度

adjusting nut　调整螺母

adjustment block　调节块

air inlet nipple　进气接嘴

air leakage　漏气

air leakage quantity　漏气量

air outlet nipple　出气接嘴

air tightness check/air tightness inspection
气密检查

aircrew oxygen system　空勤用氧系统

airtight cabin　气密座舱

aluminum liner/aluminum lining　铝内衬

aneroid diaphragm capsule assembly　真空膜盒组件

anti-oxidation grease　抗氧化润滑脂

ball valve　球形活门

blanking　堵漏

blanking cap　堵帽

blanking cover/blanking cap　堵盖

block　堵塞

bonding agent　粘接剂

bonding jumper　搭铁片

bruising　碰伤

bulge　鼓包

burst pressure　爆破压力

butterfly sheet　蝶形片

cabin　座舱

carbon fibre　碳纤维

cargo cabin　货舱

chromium-plating　镀铬

cockpit　驾驶舱

composite　复合材料

compressed washer　压缩垫圈

conical bracket　锥形架

crack　裂纹

demand mechanism flow　肺式机构流量

demand oxygen supply system　肺式自动型氧气系统

demand regulator　肺式调节器

demand valve　肺式活门

dent　压痕/压坑/压伤

depressuring　失密

diaphragm　膜片

direct proportion　正比例

dorsal fin　背鳍

emergency handle　应急手柄

emergency oxygen supply　应急供氧

emergency oxygen supply mechanism　应急供氧机构

felt　毛毡

filler block　垫块

fixed oxygen system　固定式用氧系统

fixed screw　固定螺钉

flat pipe connector　平管嘴

float　浮子

foam plastic filler　泡沫塑料填充物

foreign object　异物

gaseous oxygen　气态氧

gasket　垫片

glass fiber reinforced plastic oxygen bottle　玻璃钢氧气瓶

groove　凹槽

grounding wire/earthing wire　接地线

hand wheel　手轮

harmful gas　有害气体

hexagon flat nut　六角扁螺母

high-altitude flight　高空飞行

high-pressure inner cavity　高压内腔

high-pressure system　高压系统

high-pressure union　高压接嘴

high-stress glass fiber reinforced plastic　高强度玻璃钢

hypoxia　缺氧

inhalation resistance　吸气阻力

inner chamber　内室

landing gear well　起落架短舱

liner layer/lining layer　内衬层

locking wire　保险丝

low-pressure pipe　低压导管

low-pressure union　低压接嘴

main shaft　主轴

mask union　面罩接嘴

mechanical damage　机械损伤

mixed oxygen　混合氧

mounting plate　安装板

movable oxygen bottle　活动氧气瓶

neutral soap water　中性肥皂水

nickel-plating　镀镍

notch 切痕

operating altitude 使用高度

operating temperature 工作温度

original position 原位

outer chamber 外室

oxygen charging instrument panel 充氧仪表板

oxygen equipment accessory 氧气设备附件

oxygen filler 进氧接嘴/充氧咀

oxygen flow indicator 氧气示流器

oxygen high-pressure pipe 氧气高压导管

oxygen hose 氧气软管

oxygen instrument panel 氧气仪表板

oxygen mask 氧气面罩

oxygen percentage mechanism 含氧百分比机构

oxygen pressure gauge 氧气压力表

oxygen pressure reducer 氧气减压器

oxygen receiving valve 接氧活门

oxygen regulator 氧气调节器

oxygen signal light 用氧信号灯

oxygen source pressure gauge 氧源压力表

oxygen special wrench 氧气专用扳手

oxygen supply hose 供氧软管

oxygen supply valve 氧气源开关

oxygen system 氧气系统

oxygen-resistance 耐氧

parachute holder 伞托

pin 顶针

pipe joint 导管接头

plate assembly 板片组件

polytetrafluoroethylene sealing tape 聚四氟乙烯密封带

portable oxygen system 便携式用氧系统

pressure reducer 减压器

pressure reducer cover 减压器盖

pressure reducer regulator 减压调节器

pressure reducing mechanism 减压机构

pressure-regulating block 调压块

pure oxygen 纯氧

push rod 顶杆

raw rubber tape 生胶带

redundancy design 余度设计

regulator 调节器

rocker valve 摇杆活门

rotational angle 转动角

rough scrape/serious scratch 粗擦伤

screw-in elbow joint 拧入式弯管接头

screw-in union 拧入式管接头

sleeve coupling nut 外套螺母

special container 专用包装箱

steel ball 钢球

steel liner/steel lining 钢内衬

support component 配套产品

synthetic resin 合成树脂

taffeta 绢纺绸布

thread 螺纹

three-way union 三向接头

torque moment 扭转力矩

transparent mask 透明罩

union 接嘴

valve assembly 活门组件

valve base/valve seat 活门座

valve casing 开关壳体

valve lever 活门杆

vaporific water drop 雾状水滴

warning bell 警铃

washer 垫圈

welding seam 焊缝

BLEED AIR 引气

actuating mechanism 执行机构

adjusting bolt 调节螺柱

aerodynamic control structure 气动控制结构

aerodynamic force 气动力

air compressor 压气机

air filter 气滤

air horn system 气喇叭系统

air-delivery valve 输气管

airdrop 空投

bellow/corrugated tube 波纹管

bleed air over-pressure valve 引气超压开关

bleed air system 引气系统

butterfly plate 蝶板

bypass valve 分路活门

cabin oxygen supply 座舱用氧

casing assembly 壳体组件

channel 通道

connector 插头座

control air channel 控制气路

control cavity 控制腔

cooling accessory 制冷附件

ejection radiation airflow 引射散热气流

environment control system/air conditioning 环控系统

feedback cavity 反馈腔

filter sheet 滤片

general-purpose tool 通用工具

indication assembly 指示组件

inlet guide vane 进气道导向器

leading edge 前缘

main air channel 主气路

nipple 管嘴

oil ejector radiation valve 滑油引射散热活门

oil radiation ejector 滑油散热引射器

parachuting 空降

pilot valve 先导阀

pipe port 导管口

pressure regulating shutoff valve 压力调节关断活门

reducer union 变径接头

restricting venturi tube 限流文氏管

shaft base/axle base 轴座

shutoff valve 断流活门

small shaft/small axle 小轴

solenoid valve 电磁阀

specified value 规定值

spring force 弹簧力

structural part 结构件

supporting screw 支点螺钉

throttle valve 节流阀

universal compressed air source 公用压缩空气源

valve base/valve seat 活门座

valve leaf/flap 活门瓣

web 腹板

WATER /WASTE 供水/排污物设备

absolute teflon degreasing raw rubber tape 纯聚四氟烯脱脂生胶带

angle piece 角片

aviation chemical sanitary agent 航空化学卫生剂

balance weight 配重

blockage 堵塞物

boiled water 开水

braided rubber hose 编制胶管

ceiling light 顶灯

chain 链条

chute 滑槽

circulating toilet 循环马桶

clean water tank 净水箱

contactor 接触器

control steel cable assembly 控制钢索组件

decoration cover 装饰罩

double-cavity unidirectional flushing 双桶单向冲水

drain valve 排放阀

drain valve cover 泄放活门口盖

drinking water equipment/potable water equipment 饮用水设备

dustbin 垃圾箱

edible soda water 食用苏打水

electric heating pipe　电加热管

electric heating water tank　电加温水箱

electric oven　电烤箱

excessive water box　溢水盒

excrement pail　储便桶

felt　毛毡

filter　过滤器

flushing mechanism　抽水机构

flushing pipe　冲洗管

flushing toilet　抽水马桶

food cabinet　食品柜

funnel　漏斗

gas welding　气焊

glass wadding gasket　玻璃絮垫

handle　提手

hydrochloric acid　盐酸

lavatory　盥洗室

life raft cabinet　救生船柜

living facility cabinet　生活设备柜

mirror　镜子

mirror light　镜前灯

motor　马达

noninflammable varnished cloth　不燃漆布

oxygen flow indicator　氧气示流器

paper roll　纸卷桶

patch　补片

pipe clip　管卡

plug　插销

power distribution cabinet　配电柜

profile size/outline dimension　外形尺寸

pump　泵

punch　穿孔

quick-release stop pin　快卸止动销

rubber valve　橡胶阀门

rust-proof aluminum　防锈铝

sanitation system　卫生系统

sealing tape　密封带

sensor　感控器

signal pipe　信号管

sliding rail　滑轨

smoke detector　烟雾探测器

strap　卡带

stripping tape　捆扎带

sucking cylinder　吸筒

thermos bottle　保温水瓶

thermos bottle seat　暖水瓶座

timer assembly　计时器组件

toilet basin　马桶盆

toilet bottom hose　马桶底部短管

toilet device　马桶装置

toilet tub　马桶槽

washbasin　洗手盆

washing water　洗涤用水

waste disposal cart　排污车

waste water　污水

water charging cover　加水口盖

water charging nipple　加水管嘴

water drain pipe　放水管

water entering pipe　续水管

water faucet/water tape　水龙头

water jar　水壶

Z section　Z字型材

AIRBORNE AUXILIARY POWER
机载辅助动力装置

acceleration control　加速控制

acceptable fuel　可用燃油

access door　检修门

accessory cooling air duct　附件冷却空气管

accessory cooling air inlet duct　附件冷却空气进气孔

accessory cooling air inlet flange　附件冷却空气进口安装凸缘(法兰盘)

after fairing　后整流罩

air bleed valve　放气活门

air diffuser duct　空气扩压管

air inlet door　进气门

air inlet door actuator　进气门致动器

air inlet shutter electric mechanism　进气风门电动机构

air storage tank　空气存储器

air turbine starter　空气涡轮启动机

air/ground relay　空中/地面继电器

air-oil separator　空气/滑油分离器

aluminum forging　铝锻件

annular main fuel pipe　环形燃油总管

anticycle relay　反循环继电器

APU accessory drive case　APU 附件传动机匣

APU air inlet system　APU 进气系统

APU bleed valve　APU 放气活门

APU engine mount　APU 发动机安装座

APU engine shroud　辅助动力装置发动机罩

APU start switch　APU 启动开关

APU torque box cavity　APU 扭力盒内腔

armature　衔铁

atomized fuel　雾化燃油

atomizer　喷雾器

backup ring　密封支撑环

battery　蓄电瓶

battery charge relay　蓄电瓶充电继电器

battery timer　点火定时器

bleed air (dump) line　排气管路

bleed air duct　放气管

bleed air duct opening　排气管出口

bleed switch　放气开关

bleed-air ejector　抽气引射器

braid shield　编制套屏蔽

cable shield　电缆屏蔽

carbon pile　炭柱

centrifugal balancing weight　离心配重

centrifugal regulator　离心调节器

centrifugal switch　离心开关

chip detector　碎屑探测器

circuit breaker　断路器

clutch　离合器

coarse/fine filter　粗/细油滤

cold junction compensation　冷端补偿器

compensation wire　补偿导线

compressor air inlet duct　压气机进气管

conduit shield　导管屏蔽

conical bottom cover　锥体底盖

conical sleeve　锥形套筒

constant pressure control air line　恒压控制空气管路

constant speed control　恒速控制

contact　接点

cooling air collector　冷却空气收集器

cooling air crossover　冷却空气交输管

cooling air fan　冷却空气风扇

cooling air shroud　冷却罩

cooling air shutoff valve　冷却空气关断活门

cooling air valve　冷却空气活门

cooling fan shut off valve limit switch　冷却风扇关断活门限制开关

crank/cold run　冷转

current divider contactor　分流接触器

damping time　阻尼时间

DC motor　直流马达

de-oil solenoid valve　除油电磁活门

differential air pressure regulator　差动空气压力调节器

dip stick　量油尺

direct-current (DC) electronic system　直流电气系统

door open switch　进气门开启开关

door position switch　进气门位置开关

drain fitting　排泄管接口

drain line　排泄管路

drain plug　排放塞

drain stand pipe　排泄竖管

drip pan　承屑盘

dry sump　干油槽

eject-radiating air inlet window　引射散热进气窗

elapsed time indicator access　经（过的）时

（间）指示器进口

electric erosion ignition plug　电蚀电嘴

electrode　电极

electromagnet　电磁铁

electromagnetic interference　电磁干扰

electromotive force　电动势

electronic control unit　电子控制装置

electronic over temperature switch　电子过温开关

exhaust duct　排水管

exhaust duct shroud　排气管套罩

exhaust gas temperature probe　排气温度探头

exhaust muffler　排气消音器

exhaust pipe　尾喷管

exhaust pipe adapting ring　排气管转接环

external DC supply　外部直流直源

fault relay　故障断电器

felt metal liner　毡金属衬套

filter bowl　滤杯

filter bypass valve　油滤旁通活门

filter element　油滤芯

filter sheet　滤片

fire repeat relay　防火重复继电器

fire shutdown relay　防火断路继电器

fire truck　消防车

firewall　防火墙

flame igniter　火舌点火器

flow limiter　限流器

flush opening　齐平式进气口

flush scoop　埋入式进气口

fuel additives　燃油添加剂

fuel atomizer　油雾喷射器（喷油器）

fuel boost pump　燃油增压泵

fuel contamination　燃油污染

fuel control unit　燃油控制装置

fuel drainage　燃油排放设备

fuel filter cap　油滤盖

fuel heater　燃油加热器

fuel heater check valve　燃油加温器单向活门

fuel icing　燃油结冰

fuel pump　燃油泵

fuel shutoff valve　燃油关断活门

fuel solenoid valve　燃油电磁阀

fuel starvation　燃油不足

fuel supply joint　供油接头

fuel supply pressure　供油压力

fuel tank　燃油油箱

fuel valve actuator　燃油活门致动器

full authority electronic controller　全权式电子控制装置

gas turbine　燃气涡轮

gas turbine engine　燃气涡轮发动机

gear box　齿轮箱

gear fuel pump　齿轮式油泵

generator　发电机

generator control unit　发电机控制装置

generator excitation winding　发电机激磁绕组

hairspring　游丝

heat shield　隔热屏

heat shroud　隔热套

heatproof asbestos ring　石棉隔热圈

heat-resistant alloy steel plate　耐热合金钢板

helical electromagnet　螺管式电磁铁

high oil temperature driver　滑油高温驱动器

high oil temperature switch　高滑油温度开关

high permeability shield　高导磁屏蔽

high pressure fuel filter　高压燃油滤油

high pressure fuel line　高压燃油管路

high pressure relief valve　高压释放活门

high tension distributing system　高压点火分配系统

hydraulic accumulator　液压蓄压器

igniter　点火器

igniter boss　点火器衬块

igniter cam　点火凸轮

igniter lead　点火导管

igniter lead connector　点火导管接头

igniter plug　点火塞

ignition exciter　点火励磁机

ignition switch　点火开关

immersion thermocouple　浸入式热电偶

induction sheet　感应片

inlet guide valve　进气导向活门

integral amplifier　积分放大器

latch　扣锁

limit switch　限制开关

link axial reverse load　联杆轴向反向负荷

loop resistance　电路阻值

low oil pressure driver　滑油低压驱动器

low oil pressure shut down relay　滑油低压断路继电器

low oil pressure switch　低滑油压力开关

low oil pressure timer　滑油低压定时器

low oil quantity switch　低滑油量开关

low pressure fuel filter　低压燃油油滤

low pressure fuel line　低压燃油管路

lower shroud　下罩

lubricating　润滑油黏性

magnetic particle inspection　磁力探伤检查

master control switch　主控制开关

mat　衬边

moisture collector　水分收集器

mounting bracket　安装支架

muffler　消音器

multiplier　乘法器

nickel-cadmium battery　镍镉电瓶

oil cooler　滑油冷却器

oil cooler inlet line　滑油冷却器进口管路

oil cooler outlet line　滑油冷却器出口管路

oil drain cock　放油开关

oil filter　滑油油滤

oil level sensor　滑油油面传感器

oil lever switch　滑油油面开关

oil pressure pump　滑油增压泵

oil pump　滑油泵

oil sampling　滑油抽样

oil scavenge pump　滑油回油泵

oil sump　滑油槽

oil tank　滑油箱

oil tank filer port　滑油箱加油口

oil-return valve　回油活门

oil-to-fuel heat exchange　滑油-燃油热交换器

on-board battery　机载电瓶

over current driver　过电流驱动器

over speed driver　过速驱动器

over speed anticycle relay　过速反循环继电器

over speed shutdown relay　过速断路继电器

over temperature driver　过温驱动器

overspeed reset switch　过速复位开关

over temperature switch　过热开关

packing　密封圈

paraffin paper　石蜡纸

permanent magnet generator　永久磁产生器

pneumatic solenoid valve　气动电磁活门

positive displacement hydraulic motor　变容液压马达

positive displacement pneumatic motor　变容气动马达

power on relay　电源接通继电器

pre-amplifier　前置放大器

pressure reducing valve　减压活门

pressure regulating valve　调压活门

primary capacitor　初级电容器

proportional amplifier　比例放大器

proportional control valve　比例调节活门

ram scoop　冲压式进气口

rate control valve　速率控制活门

rejector pipe upper holder　引射排气管上支座

reverse-flow check valve　逆流单向活门

routing of the fuel supply line 燃油管路敷设

screen 滤网

seal drain port 封严排气口

selector solenoid valve 选择器电磁活门

sensor rotor 传感器转子

sensor stator 传感器静子

sequencing oil temperature switch 程序滑油压力开关

shroud ring 箍环

shut down 停车

shut down relay 停车继电器

sight oil separator 目测油表

signal lose detector 信号损失探测器

signal monitor 信号控制器

silica gel 硅胶

silicon rectifier 硅整流器

sleeve strut 套管式支柱

speed governor 调速器

speed reducer casing 减速器机匣

start contactor 启动接触器

start interlock relay 启动联锁继电器

start latch relay 启动锁继电器

start motor 启动马达

start relay 启动继电器

starter 启动机

starter clutch 启动离合器

starting main fuel pipe 起动燃油总管

starting torque 启动扭矩

start-stop relay 启动停车继电器

start-up period timer 试车周期计时器

surg 喘振

surg bleed air exhaust 防喘放气口

surg bleed valve 防喘振放气活门

switcher valve 开关活门

tachometer 转速表

tangential hole 切向孔

thermal pressure relief 热压力释放

thermal shielding/heat shield 隔热屏

thermoelectric instrument 热电式仪表

thermostat 温度自动调节器/恒温器

time delay 延时

timed acceleration fuel control unit 加控加速燃油系统装置

transformer 变压器

transformer rectifier 变压整流器/变流器

turbine exhaust drain cup 涡轮排气排泄座

turbine exhaust port 涡轮排气口

upper shroud 上罩

upper shroud mount bracket 上罩安装支架

valve actuator 活门致动器

vibration isolator 隔振器

vibrator contact 振荡器触点

voltage relay 电压继电器

wet sump 湿油槽

DOORS 舱门

access cover/hatch 口盖

access frame 口框

accessory board 附件板

acid solution 酸类溶液

actuator body 筒体

actuator union 筒管嘴

adaptability 贴合度

adaptor 转接嘴

adjustable screw 可调螺钉

adjusting thread rod 调节螺杆

aero-washing gasoline 航空洗涤汽油

aging 老化

air pressure reducer 空气减压器

aircraft aerodynamic contour/configuration 飞机气动外形

aircraft axis 飞机轴线

aircraft jacking state 飞机顶起状态

aircraft parking state 飞机落地状态

airdrop operator 空投员

airdrop parachuting console 空投空降操

纵台

airdrop pull rod 空投拉杆

airframe 机体

airtight band 气密带

airtight load 气密载荷

airtight riveting 气密铆接

airtightness 气密

alignment 对齐

aliquation/peel-off 起层

alkali solution 碱类溶液

allowable difference 允许偏差

aluminum alloy forging 铝合金锻件

anchor nut 托板螺母

angle 角材

anti-shedding shaft 止脱轴

arrow 箭头

assembly 装配

assembly technique 装配工艺

aviation kerosene 航空煤油

axle pin 轴销

bail-out 应急跳伞

base 基座

basin doorframe 盆形门框

bearing 承受

bent edge 弯边

blade spring 板簧

blanking cap 堵帽

blind nut 螺盖

bonding surface 搭接面

box element 盒形件

box-shape framework structure 盒形框架结构

bracket 托架

brake 刹车

breaking load 破坏载荷

broken 折断

broken wire 断丝

cargo 货物

cargo cabin 货舱

cargo floor axis 货舱地板轴线

cargo ramp fluid line 货桥油路

casting 铸件

center distance 中心距

central rocker 中央摇臂

circular arc 圆弧

clamp 卡带

clearance/gap 间隙

closed state 关闭状态

cloth cover 布盖

cloud-shape lock spring/turn-lock spring 云形锁弹簧

coat 涂覆

cockpit 驾驶舱

coincidence 重合度

cold-work hardening 冷作硬化

compression state 压缩状态

cone screw 锥螺纹

connecting plate 连接片

console panel 操纵台面板

contour/configuration 外形

control pull rod 操纵拉杆

conventional frame 普通框

copper part 铜制件

cotter pin 开口销

counter-clockwise 逆时针

crack 裂纹

crane 吊车

crane sliding rail 吊车滑轨

crescent groove 月牙形槽

cross beam 横梁

cross section 截面

damage 破损

damper 阻尼器

decorative inner skin 装饰内蒙皮

deformation 变形

delay valve 延迟活门

dent 凹陷

deposit valve 沉淀开关

depth 深度

dipping 蘸

distortion 扭曲

distribution changeover valve 分配转换活门

door bracket 门托架

door edge 门缘

door sealing system 舱门密封系统

doorframe 门框

doorframe edge strip 门框缘条

double rocker arm 双摇臂

double-side hydraulic lock 双面液压锁

downlock cable rocker arm 下位锁钢索摇臂

drop-press forming part 落压成形件

dust-proof sleeve 防尘套

elasticity 弹性

electrical console 电气操纵台

electrical control 电气控制

electrical control switch 电气操纵开关

electrical element 电气元件

emergency brake handle 应急刹车手柄

emergency control 应急操纵

emergency control hydraulic fluid line 应急操纵油路

emergency control system 应急操纵系统

emergency exit 应急出口

emergency hydraulic fluid line 应急油路

emergency opening 应急开启

empty cavity 空腔

enclosure 封闭

end cover eyelet 端盖耳环

entrance ladder 登机梯

entrance ladder hanging ring 门梯挂环

entry door 登机门

extension 伸出

extension/retraction mechanism 收放机构

external handle 外部手柄

extrusion/section 型材

eyelet bolt 耳环螺栓

failure 失效

fastening screw 紧钉螺钉

faying surface 贴合面

feeler 塞尺

felt ring 毡圈

fixing strap 固定条带

flange disc union 法兰盘接管嘴

flat-head shaft 平头轴

floor door 地板门

floor door lock 地板门锁

flow division valve 分流活门

flow regulating valve 流量调节阀

flow restricting window 节流窗口

flow restriction hole 节流小孔

fluctuation 波动

foreign force 外力

fork 叉子

fork lug 叉耳

fracture surface 断面

frame 框

frame door 框门

frame edge/flange 框缘

frame plate 框板

framework 骨架

friction 摩擦力

front support point 前端支撑点

fuselage theoretical contour/configuration 机身理论外形

galvanization 镀锌

gas path 气路

glass assembly 玻璃组件

graphite lime grease 石墨钙基润滑脂

gravity 重力

groove/notch 沟槽

ground air pressurizing unit 地面空气增压设备

ground air-charging equipment 地面充气设备

ground console 地面操纵台

ground DC power supply　地面直流电源

ground hydraulic pump cart　地面油泵车

guide plate　导向板

guide sleeve　导套

guide sliding rail　导向滑轨

height　高度

hexagonal socket nut　内六角螺母

high pressure pipeline　高压管路

high-pressure cavity　高压腔

hinge　铰接

hinge hole　铰接孔

hook　挂钩

hydraulic actuator　液压作动筒

hydraulic control　液压操纵

hydraulic control valve　液压控制阀

hydraulic fluid return chamber　回油腔

hydraulic fluid return line　回油路

hydraulic fluid return union　回油管嘴

hydraulic lock valve　液压锁活门

hydraulic manual valve　液压手动阀

hydraulic pipe　液压导管

hydraulicpressurization system　液压增压系统

hydraulic system　液压系统

hydraulic tank/hydraulic reservoir　液压油箱

included angle　夹角

inner skin　内蒙皮

inner surface　内表面

insert hole　钥匙插孔

internal handle　内部手柄

jam　卡涩

landing light　着陆灯

left/right reservoir　左/右油箱

light signal device　灯光信号装置

lightening hole　减轻孔

link rod　连杆

load intensity test　载荷强度试验

loading ramp door　货桥大门

loading ramp hydraulic system　货桥液压系统

loading ramp lock mechanism　货桥锁机构

locating pin　定位销

lock actuator　锁作动筒

lock base　锁座

lock core　锁芯

lock groove　锁槽

lock hook　锁钩

lock mechanism　锁机构

lock pin　锁销

lock pin hole　锁座插销孔

lock position indicating line　锁闭位置指示线

locking mechanism　锁死机构

locking nut　锁紧螺母

longeron　纵梁

longitudinal partition　纵隔板

lower joint　下接头

low-pressure cavity　低压腔

lug　耳片

lug bolt　耳片螺栓

magnesium alloy　镁合金

magnifier　放大镜

main landing gear　主起落架

main landing gear door　主起落架舱门

main load-bearing member　主要承力构件

manual valve　手动阀

marking line　标记线/刻线

maximum opening　最大开度

mechanic indicating rod　机械指示杆

mechanical damage　机械损伤

molding　模锻件

neutral position　中立位置

normal control line　正常操纵油路

normal opening　正常开启

nose landing gear door　前起落架舱门

nose pliers　尖嘴钳

observation access cover　观察口盖

observation window　观察窗

oil tank　油池

oil tray　贮油盘

one-way buffer　单向缓冲器

one-way delay valve　单向延迟活门

one-way restricting function　单向节流作用

operating chamber　工作腔

operating oil line　工作油路

operation position　工作部位

oriented organic glass　定向有机玻璃

original position　原始位置

outer cylinder　外筒

outer skin　外蒙皮

outer surface　外表面

overall load　总体受力

overpressure protection　超压保护

paint layer　漆层

parking　停机

partition　隔板

passage/channel　通道

piston full travel　活塞全行程

piston rod　活塞杆

piston rod sleeve　活塞杆套

piston travel　活塞行程

pit　压伤

placard　标牌

plate/sheet　板材

plate-bent part　板弯件

plunger　栓塞

pneumatic control valve　冷气操纵开关

polyurethane adhesive fluid　聚酸脂胶液

pop rivet/hollow rivet　空心铆钉

preservation paper　油封纸

pressure holding return line　保压回路

pressure loss　压力损失

pressure reducing union　减压管嘴

pressure relief valve　减压阀

pressurized hydraulic fluid way　压力油路

pre-stretched　预拉伸

pre-tightening force　预紧力

protective layer　保护层

protective sleeve　包覆物

pull rod　拉杆

pulley　滑轮

pulley bracket　滑轮支架

pulley groove　滑轮槽

pulley rim　滑轮缘

pump union　泵管嘴

push rod mechanism　顶杆机构

quick-release lock　快卸锁

rear cargo door　货舱后大门

red enamel　红色磁漆

reinforced edge strip　加强条

reinforced groove　加强槽

reinforcement　加强

release pressure/depressuring　卸压

removable floor　可卸地板

reset paw　复位爪

residual pressure　剩余压力

residual travel　剩余行程

restraint　约束

restricting orifice　节流孔

retainer　挡盘

retraction　缩回

retraction/extension space　收放空间

reversal installation　反向安装

reverse sequence　反顺序

rib　肋

rigid support　硬支撑

rivet　铆接

riveted structure　铆接结构

rocker arm　摇臂

rotating shaft　转动轴

rubber section　橡胶型材

rust spot　锈蚀斑点

safety device　保险装置

safety unit　保险机构

sand paper　砂纸

scratch　划伤

sealing 密封件

sealing ring 封严圈

sealing rubber tube 密封橡胶管

sealing strip 密封条

seat 底座

secondary load-bearing member 非主要承力构件

self-locking nut 自锁螺母

separation point 分离点

sequential action 顺序动作

service ladder 工作梯

sheet-metal part 钣金成形件

shock 冲击

side panel 侧壁板

single actuator 单向作动筒

single-side hydraulic lock 单面液压锁

skin 蒙皮

sleeve 套筒

slide valve 滑阀

sloping floor 斜地板

small cover 小盖

spherical surface cover 球面盖

spherical surface skin 球面皮

spherical thin-wall structure 球面形薄壁结构

sponge rubber plate 海绵橡胶板

spot welding 点焊

spring 弹簧

spring handle 弹簧手柄

spring washer 弹簧垫圈

stainless steel ball handle lock 不锈钢弹子执手门锁

steel bushing 钢制衬套

steel cable tension 钢索张力

step 台阶

stop block 挡块/限动块/止动块

stop pin 挡销

stop ring 挡圈

stop screw 止动钉

stopper 止动器

stopper block/stopping block 止动器挡块

strand 股

stringer 长桁

structure fixing part 结构固定件

structure height 结构高度

structure type 结构形式

support abutment 支座

support point 支点

suspension joint 悬挂接头

symmetry plane 对称平面

system pressure 系统压力

tail opening 尾段大开口

tail rod 尾杆

technological hole 工艺孔

tensiometer 张力计

thermal safety valve 热安全活门

thickness 厚度

thread broken 滑丝断扣

three-position four-way control valve 三位四通方向控制阀

top emergency exit 上应急口盖

torsional spring 扭簧

transverse partition 横隔板

transverse positioning 侧向定位

triangular spar 三角大梁

turnbuckle 松紧螺套

unreasonable bending 不合理弯曲

upper joint 上接头

varnish 清漆

ventilation pressurizing joint 通气增压接头

vibration absorber 减振器

visual check 目视检查

washer 垫圈

web 腹板

weld 焊接

wheel chock 轮挡

white gauze 白纱布

width 宽度

window frame　窗框

FULSELAGE　机身

absorbent cotton　脱脂棉

accelerant　促进剂

access cover　口盖

accommodation　容纳

aerodynamic load　气动载荷

air conditioning bay　空调舱

air conditioning equipment bay　空调设备舱

air delivery pipeline　输气管道

air leakage　漏气量

aircraft base level line/aircraft level reference datum　飞机水平基准线

airflow　气流

airframe structure　机体结构

airtight cover　气密罩

airtight line direction　机身气密线走向

airtight load　气密载荷

airtight riveting　气密铆接

airtight sealing　气密密封

airtight section　密封型材

airtight test bed　气密试验台

airtightness test　气密性试验/密封性检查

all-metal semi-monocoque structure　全金属半硬壳式结构

angle　角材

annular section　环向型材

annular-groove screw　环槽钉

anodizing coating　阳极化层

anti-corrosion performance　防腐性能

anti-fatigue performance　抗疲劳性能

anti-sliding　防滑

armrest box　扶手盒

automatic drain valve　自动排水活门

auxiliary frame/secondary frame　辅框

average reading/arithmatic average value　算术平均值

aviation organic glass　航空有机玻璃

balance tab/compensating plate　补偿片

beam　梁

bearing　承受

bevel　斜削

bonding layer　胶合层

bottom　底部

bottom beam　底梁

bow connecting section　弓形连接型材

box beam　盒形梁

box body　盒体

box corrugation plate　盒形波纹板

boxcover　盒盖

bridge plate　桥形板

bulge　鼓包

butt joint　对接接头

butt-joint frame　对接框

butt-joint plate　对接板

butt-joint seam　对缝

button-head rivet/mushroom-head rivet　扁圆头铆钉

carbon spring steel wire　碳素弹簧钢丝

cargo　货物

cargo compartment floor　货舱地板

cargo load　货载

casting aluminum alloy　铸铝合金

casting magnesium alloy　铸镁合金

caution zone　警戒区

ceiling　顶棚

center wing　中央翼

center wing fairing　中央翼整流罩

centerline　中线

certificate　合格证

chemical-milling skin　化铣蒙皮

circular　圆形

cockpit　驾驶舱

cockpit canopy　驾驶舱盖

cockpit floor　驾驶舱地板

colorized flare　彩标弹

colorized signal flare 彩色信号弹

common brush 普通毛刷

compound bristle brush 合成硬毛刷

compressed air 压缩空气

concentrated load 集中载荷

contacting surface 接触面

conventional frame 普通框

corner fitting joint 角盒接头

correction 校正

countersunk rivet 埋头铆钉

craze 银纹

cross section 横截面

cross section area 剖面面积

crossbeam 横梁

cutout 缺口

decoration panel 装饰板

diameter 直径

distance between stringers 桁距

distribute 扩散

divergent 发散

dorsal fin 背鳍

dorsal fin platform 背鳍形状

double-curved 双曲度

down stream butt-joint seam 顺气流对缝

drain valve 排水活门

draining mechanism 排水机构

duralumin 硬铝合金

electrical property 电性能

emergency door 应急舱门

emergency exit hatch 应急口盖

emergency pressure release switch 紧急卸压开关

emergency steel cable 应急钢索

emergency up/downaccess cover 应急收放口盖

epoxy clearcoat 环氧清漆

equal distance 等距

equipment bay cover 设备舱口盖

extending section 延伸段

extending section of tailplane 平尾延伸段

extrusion 挤压型材

faying surface sealing 缝内密封

fillet sealing 缝外密封

fine sand paper 细砂纸

fine white cloth 细白布

flange 框缘

flat-head rivet 平头铆钉

floor crossbeam 地板横梁

floor of navigation cabin 领航舱地板

flowmeter 流量计

fluorescent inspection 萤光检查

foot pedal 脚踏板

force transfer member 传力件

forging aluminum alloy 锻铝合金

forming 成形

frame plane 框平面

front fairing 前整流罩

front fuselage 机身前段

front longeron 前纵梁

front side panel 前侧壁

funnel 漏斗

fuselage 机身

fuselage planform 机身几何形状

fuselage skeleton 机身骨架

fuselage skin distribution 机身蒙皮分布

glass cloth 玻璃布

glass fiber reinforced plastics honeycomb 玻璃钢蜂窝

glass of navigator's compartment 领航舱玻璃

grease base 基膏

grid plate 方格板

grind 抛光

guide rail 导向滑轨

guider 导向器

hard anodizing treatment 硬阳极化处理

hard-anodizing 硬阳极化

heavy vehicle axle load 重车轴载

hinge 合页

hoist beam 吊车梁

ice-throwing 抛冰

illustration placard 说明标牌

inclined floor 斜地板

industrial organic glass 工业有机玻璃

inertial load 惯性载荷

inner edge/flange 内框缘

intersection point 交点

jogging 下陷

key part/critical part 关键件

lap joint 搭接处

lapping paste 研磨膏

lateral member 横向构件

lateral section 横向型材

leakage quantity 泄漏量

lightening hole 减轻孔

lighting hole 透光孔

load 载荷

load of inner and outer pressure difference
内外压差载荷

load per area unit 单位面积载重量

load per length unit 单位长度载重量

loading ramp 货桥

log book 履历本

longitudinal member 纵向构件

longitudinal section 纵向型材

low-carbon alloy structure steel 低碳合金结
构钢

lower edge strip 下缘条

lower front panel 前下壁

lower longeron 下零纵

lower panel 下壁板

lower rear panel 后下壁

lowersemi-frame 下半框

luggage 行李

magnesium alloy framework 镁合金框架

magnesium alloy front beam 镁合金前梁

magnesium alloy side beam 镁合金侧梁

main component 主要构件

main landing gear well 主起落架舱

maximum cross section 最大截面

mechanical system 机械系统

middle fuselage 机身中段

middle side panel 中侧壁

milling angle piece 铣切角片

moulding part 模压件

moulding rubber ring 模压橡胶圈

navigation cabin 领航舱

navigation radar bay 导航雷达舱

non-circular cross section 非圆剖面

non-pressurized cabin 非气密舱

non-pressurized panel 非气密壁板

nose dome/nose fairing 机头罩

nose landing gear well 前起落架舱

observation window 观察窗

opening 开孔

outer edge 外框缘

overall length 全长

overall load 总体载荷

overlap 搭接

over-pressure 超压

paint 漆

panel 壁板

parallel arrangement 平行排列

partition/bulkhead 隔框（板）

pillips screwdriver 十字螺刀

pilot's stand 驾驶台

plate-bent frame 板弯框

point dent 点状凹陷

polysulfide 聚硫胶

press plate 压板

pressure hose 充压软管

pressure increasing-decreasing ratio 压力升
降速率

pressure releasing 释压

pressurized cabin 气密舱

pressurized panel 气密壁板

protection panel　防护板

protective layer/coating　保护层

protective skin　防护蒙皮

putty cloth　腻子布

quench　淬火

quick-release lock　快卸锁

radar antenna bay　雷达天线舱

radar bay　雷达舱

radar fairing　雷达罩

rear cargo door　后大门

rear end frame plate　后端框板

rear fairing　后整流罩

rear fuselage　机身尾段

rear longeron　后纵梁

rear reserved cabin　后备用舱

rectangle strip　矩形带板

reinforced frame　加强框

reinforced plate　加强板

reinforced stringer　加强长桁

reinforced strip　加强条带

repair　修理

residual pressure　余压

resin　树脂

ripple　波纹

roller　滚棒

roller device　滚棒装置

rubber gasket　橡胶垫片

rubber pad　橡胶垫

rubber sealing ball　橡胶密封球

rubber section　橡胶型材

rubber tube　橡胶管

rudder gasket　胶垫

safety device　安全装置

seal element　气密元件

sealant　密封胶

sealing element　密封元件

sealing plate　封缘板

sealing putty　密封腻子

sealing tape　密封胶带

semi-frame　半框

serrated pad　齿形垫板

serrated plate　齿形板

service ladder　工作梯

shot peening strengthening treatment　喷丸强化处理

side beam　侧梁

side fairing　侧整流罩

side guide rail　侧导轨

side panel　侧壁板

side rear panel　后侧壁

sign　迹象

silicon rubber　硅橡胶

simple toilet　简易厕所

skin　蒙皮

skylight framework　天窗骨架

sloping frame　斜框

sloping rib　斜肋

sloping support abutment　架斜支座

solvent　溶剂

sound/noise　响声

spinning　旋压

sponge rubber　海棉橡胶

sprayer　淋雨器

spraying water　洒水

staggered arrangement　交错排列

steel bolt　钢制螺栓

steel section　钢制型材

stiffener hole/strengthened pit　加强窝

stiffness　刚度

stop joint　制动接头

stress concentration　应力集中

stretch-wrap section/bent　拉弯型材

stringer　长桁

stringer axis　长桁轴线

strip plate　带板

structure height　结构高度

support abutment　支座

supporting frame　支承框

supporting partition　支撑隔板

surface treatment　表面处理

tail cabin　尾舱

tail fairing　尾罩

tail section　机身尾部

talcum powder　滑石粉

tapered basin shape　锥形盒体

tensile strength limit　抗拉强度极限

tension bolt　受拉螺栓

tension-shear bolt　拉剪螺栓

test equipment　试验设备

test room　试验间

tiedown ring/mooring ring　系留环

torque spring　扭簧

total cargo load　货运总载重量

transverse butt-joint seam　横向对缝

trapezoid　梯形

triangle box beam　三角形盒形梁

triangle partition　三角形隔板

uniform airtight load/mean airtight load　均布气密载荷

uplock/downlock　上位锁/下位锁

upper edge strip　上缘条

upper front panel　前上壁

upper longeron　上零纵

upper panel　上壁板

upper rear panel　后上壁

uppersemi-frame　上半框

upwarp　上翘

ventilation hole　通气孔

ventilation window　通风窗

vertical partition　垂直隔板

vibration　振动

vulcanizer　硫化剂

wadding-free cotton cloth　无絮棉布

washing agent　清洗剂

water leakage pipe/water-leaking pipe　漏水管

watertight sealing　水密密封

weather radar antenna　气象雷达天线

web　腹板

wetsealing　湿密封

white acrylic enamel　白色丙烯酸磁漆

windshield　风挡玻璃

windshield glass framework　风挡玻璃骨架

wood powder　木粉

wooden hammer　木锤

ENNGINE NACELLES　发动机短舱

access cover lock　口盖锁

access frame　口框

accessory casing　附件机匣

aging　老化

air inlet　进气道

air inlet/air intake　进气口

anchor nut　托板螺帽

angle　角材

annular inlet　环状进气道

attachment/connection　连接

axis　轴线

baffle　挡板

ball-end bracing　球头支撑杆

bolt　螺栓

bonding jumper　搭铁线

bracing　撑杆

bracing abutment　撑杆支座

bracket　支架

bulkhead　隔框

casting　铸件

caution　注意

clamp　卡箍

clearance/gap　间隙

connecting pin　连接插销

cool　冷却

corrosive object　腐蚀物

corundum/carborundum　金刚砂

cover plate　盖板

cowling/fillet　包皮

crack　裂纹

damage　损坏

decelerator fairing/speed reducer fairing　减速器整流罩

deformation　变形

defuel/drain　放（油）

dent　凹坑

diagonal bracing　斜撑杆

duralumin alloy plate　硬铝板材

ear-end cable　耳环绳带

ejection pipe　引射管

elbow bracket　弯管支架

electric deicer　电防冰装置

electro-corundum sandblast　电刚玉砂喷砂

engine exhaust pipe　发动机尾喷口

engine mounting　发动机架

engine turbine casing　发动机涡轮机匣

equipment　设备

erode　腐蚀

exhaust　喷出

extend　延伸

fairing part　整流部件

fault　故障

fine sandpaper　细砂纸

firewall　防火墙

fitting　接头

fixing ring　固定环

fluorescent-penetrant inspection　荧光渗透检验

frame　框

frame web　框板

front section　前段

frontal drag/frontal resistance　迎面阻力

groove ring　槽形环

heating window access cover　加温窗口盖

heat-resistance shield　耐热护板

heat-resistant stainless steel　耐热不锈钢

hinge　铰链

hook　吊钩

hot air stream　炽热气流

hub spinner/hub fairing　桨毂整流罩

inboard wing　中外翼

inertial force　惯性力

inner/outer frame edge strip　内/外框缘条

inspection/check　检验/检查

installation　安装

key part/critical part　关键件

lath　边条

leading edge skin　前缘蒙皮

link rod　连杆

load　载荷

load-bearing section　承力段

locating pin　定位销

lock　锁紧

lock seat　锁座

lower panel　下壁板

lower surface　下表面

magnesium alloy　镁合金

magnetic particle inspection　磁力探伤

maintenance practice　维修实施

mid section　中段

moulding　模压件

moulding rubber　模压橡胶

mounting edge　安装边

oil radiator　滑油散热器

outer sidewall　外侧壁

outer skin　外蒙皮

outlet　喷气口

pad　垫板

paint　油漆

paint coat　油漆层

partition　隔板

periphery　周缘

pin groove　锁销槽

pipeline　管路

polishing　打磨

prevent　预防

propeller 螺旋桨

propeller root fairing 桨根整流包皮

protection 保护

quick-release lock 快卸锁

reinforced edge 加强边

reinforced groove 加强槽

reinforced ring 加强环

removal 拆卸

removal/eliminating 排除

repair 修理

replacement 更换

rigidity/stiffness 刚度

rivet 铆钉

rockerarm bracket 摇臂支架

rotary configuration 旋转体形

rotating handle shaft 旋转柄轴

rubber air-proof section 橡胶密封型材

rubber sealing pad/rubber gasket 橡胶密封垫

rubber tape/rubber sealant 胶带

rust/corrosion 锈蚀

scheduled maintenance 定期维修

screw 螺钉

separation 分开

shed 脱落

shock-absorbing bushing 减震衬套

shutter 风门

silicone rubber section 硅橡胶型材

spacer 垫片

spar 大梁

special washer 专用垫圈

spot-welding riveting assembly 点焊铆接件

steel tube 钢管

steel wire 钢丝绳

stringer 长桁

strip 条带

structure 结构

stud 螺桩

support 支撑

support arm 支臂

suspension 悬挂

tail cowling 尾罩

tail nacelle 尾舱短舱

thrust 推力

turbine ventilation shroud 涡轮吹风罩

upper fairing 上包皮

upper/lower access cover 上/下口盖

ventilation outlet 通风出气口

visual check 目视检查

web 腹板

weld 电焊

welding seam 焊缝

X-ray X光

STABILIZERS 尾翼

adjusting thread sleeve 调整螺套

aerodynamic contour 气动外形

anchor nut 托板螺帽

anti-icing leading edge 防冰前缘

balance tab 补偿片

balance weight 平衡配重

bending 弯制

bulkhead 隔板

butt joint 对接接头

cable pipe 电缆导管

channel section 槽型截面

chemical etching/chemical milling 化学铣切

common rib 普通肋

composite 复合材料

connecting bolt 连接螺栓

countersunk head rivet 埋头铆钉

double-spar structure 双梁式结构

electrical heating anti-icing device 电加温防冰装置

elevator 升降舵

end rib 断面连接肋

extrusion/section 型材

fairing 整流罩

fairing envelope 整流罩包皮

fitting 接头

flange 缘条

front spar 前梁

fuselage axis 机身轴线

hinging bolt 铰链螺栓

horizontal stabilizer 水平安定面

knuckle bearing 关节轴承

metal structure 金属结构

mold 模压

overlap 搭接

panel 壁板

rear spar 后梁

rectify 整流

reinforced pad 加强垫板

reinforced partition 加强隔板

reinforced rib 加强肋

removable leading edge 可卸前缘

root rib 根肋

rudder 方向舵

spacer block 垫块

spring tab 弹簧补偿片

stamping 冲压

stiffening 加强

stiffening strut 加强支柱

stringer 长桁

strut 支柱

supporting rib 支撑肋

suspension abutment 悬挂支座

suspension arm 悬挂支臂

suspension fitting 悬挂接头

suspension spar 悬挂小梁

suspension support 悬挂支臂

tensile strength 抗拉强度

trailing edge 后缘

transverse load-carrying member 横向承力构件

trim tab 调整片

trimmer 修正片

tubular beam 管梁

universal joint 万向接头

universal partition 普通隔板

vertical stabilizer 垂直安定面

water drain hole 漏水孔

web 腹板

wing rib 翼肋

wing tip 翼尖

WINDOWS 窗

adhere 粘合

air bag 气囊

air-charging nipple 气嘴

airtightness 气密性

aluminum plate 铝板

angular sheet 角片

assembly area 装配区域

axial force 轴向力

basin type part 盆形件

bolt 螺栓

bubble 气泡

cabin 座舱

canopy 座舱盖

cargo cabin/cargo compartment 货舱

cavity 空腔

chain crack 链状裂纹

clearance 间隙

cockpit 驾驶舱

compression bolt 压紧螺钉

conductive film 导电膜

connecting bolt 连接螺栓

conventional frame 普通框

cork gasket 软木垫片

cotton thread 棉丝线

craze 银纹

curved glass 曲面玻璃

decoration panel 装饰板

defect　缺陷

deformation　变形

delaminate　脱胶

depth　深度

diagonal　对角线

diameter　直径

electric heating glass　电加温玻璃

eliminate　消除

ellipse　椭圆形

emergency door　应急门

entry door　登机门

extruding　挤出

fastening bolt　紧固螺栓

fastening screw　紧固螺钉

frame edge　框缘

framework　框架

fuselage　机身

fuselage front section　机身前段

gasoline　汽油

glass assembly　玻璃组件

handle　手柄

hinge　铰接

humidity　湿度

lead sealing　铅封

length　长度

link rod/link　连杆

locking wire　保险丝

longitudinal partition　纵向隔板

moment　力矩

mounting framework　安装框架

natural emulsion　天然乳胶

navigation cabin　领航舱

navigator　领航员

nitrogen gas　氮气

nose fairing glass　机头罩玻璃

nozzle　管嘴

observation window　观察窗

organic glass　有机玻璃

oriented organic glass　定向有机玻璃

panel　壁板

pilot　驾驶员

plane glass　平面玻璃

plate-bent part　板弯件

polyvinyl butyral rubber sheet　聚乙烯醇缩丁醛胶片

press block　压块

press plate　压板

pressing block assembly　压块组合件

pre-tightening force　预紧力

rivet　铆钉

rocker arm　摇臂

rubber gasket　橡胶垫片

rubber section　橡胶型材

rubber sheet　橡胶板

scratch　擦伤

sealant　密封胶

sealing cap　密封帽

sealing putty　密封腻子

sealing tape　密封带

seam　对缝

shave　削刮

silicate glass　硅酸盐玻璃

skin　蒙皮

skylight framework　天窗骨架

sliding rail　滑轨

spanner/wrench　扳手

spline　软木条

sponge rubber sealing tape　海绵橡胶密封带

stamping part/pressing part　冲压件

stringer　长桁

symmetry axis　对称轴线

terminal post　接线柱

thickness　厚度

tipping material　包边材料

torque wrench　定力矩扳手

ventilation window　通风窗

visibility　能见度

vision area　视野区

wedge 压座

window frame 窗框

wire 导线

wooden putty knife 木制腻子刀

WINGS 机翼

access cover 口盖

access cover chain 口盖链条

access frame 口框

adapter 转接器

adhesive 胶

adhesive trace 胶痕

aerodynamic vane 导流片

aging 老化

aileron 副翼

aileron bay 副翼舱

airflow 气流

airtight line 气密线

airtight load 气密载荷

airtight riveting 气密铆接

airtight riveting structure 气密铆接结构

airtightness requirement 气密性要求

all-metal structure 全金属结构

aluminum plate 铝板

anchor nut 托板螺帽

angle 角材

angle box 角盒

angle fitting 角接头

angle piece 角片

angle protractor 测角仪

anti-icing system 防冰系统

antimagnetic bolt 防磁螺栓

anti-rust aluminum 防锈铝

arc-shape angle 弧形角材

aviation balloon cloth 航空气球布

aviation rubber plate 航空橡胶板

aviation sponge rubber plate 航空用海绵橡胶板

axial compensation 轴向补偿

axial load (pulling) 轴向（拉）载荷

backward difference 逆差

bag tank 软油箱

bakelized paper 胶纸

bolt 螺栓

bond 粘结

bonding jumper 搭铁

bonding strip 搭接片

box member 盒形件

breakdown plate 分离面平板

bulkhead 隔板

butt bolt 对接螺栓

butt cover plate 对接盖板

butt profile 梳状件

butt rib 对接肋

butt seam 对接缝

butt strut 对接支柱

butting mode 对接形式

cable bundle 电缆束

cable shroud 电线罩

cadmium-plating 镀镉

cantilever high-wing 张臂式上单翼

cathedral angle 下反角

cement ground 水泥地

center wing 中央翼

chamfer 倒角

check/inspection 检查

chemical etching/chemical milling 化学铣切

chord length 弦长

clamp 卡箍

clamping plate 夹板

clearance/gap 间隙

cloth cover 布盖

compass sensor 罗盘传感器

concentrated load 集中载荷

connecting angle 连接角材

contact resistance 接触电阻

control cable 操纵钢索

conventional rib　普通肋

corrosion　腐蚀

corrugated plate　波形板

countersunk rivet　埋头铆钉

crack　裂纹

curvature　曲率

defect　损伤

deflection　偏转

deformation　变形

defueling access cover　放油口盖

defueling port　放油口

degumming/unbonding　脱胶

dehydrated kerosene　脱水煤油

initia breakdown　设计分离面

deteriorating　变质

diameter　直径

dihedral angle　上反角

dirigible fabric/balloon fabric　气球布

disengagement　脱落

double-slotted extension flap　双缝后退式襟翼

double-spar wing box　双梁翼箱

dust cover　防尘罩

dye inspection　着色检查

end bulkhead　端隔板

end thread　尾扣

engine mounting　发动机架

epoxy zinc yellow primer　环氧锌黄底漆

equal section　等截面

equidistant template　等距样板

exhaust gas/waste gas　废气

exhaust port　排气口

exposing　暴露

extruded section　挤压型材

extrusion　型材

fatigue-resistant performance　抗疲劳性能

faying surface sealing　缝内密封

fence-frame　围框式

field　外场

file　锉刀

filler port plug　注油口塞

fillet sealing　缝外密封

fine plain　细平布

fixed screw　固定螺钉

flap　襟翼

flap bay　襟翼舱

flap retraction/extension actuator　襟翼收放作动筒

flashlight　手电

floating anchor nut　游动托板螺帽

flutter　颤振

foamed plastic block　泡沫塑料块

foldable cover plate　可折动盖板

formation light　编队灯

forward difference　顺差

fracture　断裂

frame　框

front chamber　前室

front spar　前大梁

fuel filling port　燃油注油口

fuel quantity gauge　油量表

fuel quantity gauge sensor　油量表传感器

fuse (electric)　保险丝(电)

fuselage　机身

gas-injecting zone　喷流区

geometric twist　几何扭转

glass cloth laminate　玻璃布层压板

gravity center　重心

gravity filler opening　重力注油口

greasy dirt　油污

guide plate　导板

gun oil　炮油

hammer　榔头

handling fitting　搬运接头

hard aluminum plate/duralumin　硬铝板

hard chromium　硬铬

hard inorganic glass　硬料无机玻璃

heat source　热源

high lift device　增升装置

hi-lock bolt　高锁螺栓

hi-lock nut　高锁螺帽

hinge　铰链

hinge pin　通条

hollow shaft　空心轴

horizontal flow direction　横气流方向

hot air　热空气

huck rivet　环槽铆钉

inboard engine　内发动机

inboard wing　中外翼

incidence angle　安装角

indication　指示

inner skin　内蒙皮

installation　安装

integral rib　整体肋

integral tank　整体结构油箱

jack　千斤顶

jack fitting　千斤顶接头

jack socket　千斤顶窝

lateral balance　横向配平

lateral displacement　横向位移

lead brass　铅黄铜

leading edge　前缘

left/right wing　左右机翼

lift　升力

lightening hole　减轻孔

limiter　限动器

link rod/link　连杆

load-bearing structure　承力部件

locking wire（mechanic）　保险丝（机械）

longitudinal partition　纵隔板

looseness　松动

low temperature lubricating grease　低温润滑脂

lower cover plate　下盖板

lower edge strip　下缘条

lower panel　下壁板

machining　机械加工

magnesium alloy extruded section　镁合金挤压型材

magnifier　放大镜

main flap　主襟翼

maintainace　维护

mark line　标记线

mass balance test　质量平衡检验

material　材料

measuring tool/measuring equipment　测量工具

member　构件

metal dust　金属微尘

mooring fitting　系留接头

moulding　模压件

moulding fitting　模压件接头

moulding flange　模压件法兰盘

movable component　可动部件

navigation light　航行灯

needle bearing　滚针轴承

non-metallic scraper　非金属刮刀

nut　螺帽

nylon cushion block　尼龙垫块

opening　开口

operating light　工作灯

original position check　原位检查

outboard aileron　外副翼

outboard engine　外发动机

outboard wing　外翼

overlap　搭接

phenolic adhesive　缩醛胶液

plane form/flat shape　平面形状

plate/sheet　板材

pliers　钳子

polyurethane adhesive　聚氨脂胶

polyurethane adhesive liquid　聚氨脂胶液

polyvinyl chloride foam plastics　聚氯乙烯泡沫塑料

preparations　准备工作

pressure oil filler　压力注油嘴

process panel　工艺壁板

propeller hub spinner　螺旋桨桨帽

pulley support　滑轮架

pump　油泵

quenching　淬火

quick-release lock　快卸锁

reamer　铰刀

rear chamber　后室

rear engine mount bracing　后段撑杆

rear section　后段

rear spar　后大梁

rectangle　长方形

reinforced access frame　加强口框

reinforced groove　加强槽

reinforced plate　加强板

reinforced rib　加强肋

reinforced strut　加强支柱

removable access cover　工艺口盖

removable leading edge　可卸前缘

removable panel　可卸壁板

removal　拆卸

replacement　更换

rib plate　肋板

rigidity　刚度

rocker arm　摇臂

rolling　滚制

rotary axis　旋转轴线

rubber　摩擦

rubber pad　橡胶垫板

rubberized fabric　胶布

rubber-sealing pad　橡胶密封垫

rust　锈蚀

scissor　剪刀

scissors difference　剪刀差

scratch　划伤

screw　螺钉

screw joining　螺接

screw rod　螺杆

screwdriver　螺丝刀

sealant　密封胶

sealant stem　胶梗

sealing grease　密封润滑脂

sealing plate　封严板

sealing plug/airtight-plug　气密油塞

sealing rib　密封肋

sealing rubber ring　密封胶圈

sealing tape　密封带

section　截面

sediment　沉淀物

self-tapping screw　自攻螺钉

serrated gasket　齿形垫

serrated pad　齿形垫板

service ladder　工作梯

servicing　保养

sharp edge　锐边

shotpeening treatment　喷丸强化处理

side plate　侧板

skin　蒙皮

sled　滑车

sliding rail　滑轨

sliding rail channel　滑轨槽

sloping rib　斜肋

small beam　小梁

socket　插座

solidification　固化

span　翼展

span length　展长

spar　大梁

spoiler access cover　扰流板口盖

stainless steel plate　不锈钢板

stainless steel rivet　不锈钢铆钉

stamp/press/punch　冲压

standard part　标准件

station　站位

step connection　台阶式连接

step difference　阶差

stringer　长桁

stringer jaw　长桁爪

stripping 剥落

structure 结构

super duralumin 超硬铝

support arm 支臂

support plate 撑板

suspending cable 吊索

suspension arm 悬挂支架

suspension joint 吊挂接头

symmetry axis 对称轴线

tank bay 油箱舱

taper surface 圆锥面

tearing 破裂

technical requirement 技术要求

technological bolt 工艺螺栓

theoretical configuration 理论外形

thickness 厚度

thread 螺扣

tightening torque 拧紧力矩

tightness 拧紧度

tongue piece 舌状片

tool 工具

tooth strip 齿条

torque wrench 定力板手

total load bearing 总体受力

trailing edge 后缘

transverse partition 横隔板

trapezium shape 梯形

trim tab 调整片

truss head bolt/mushroom-head bolt 扁圆头螺栓

tube passing 筒道

unpressurized area 非密封区

upper airfoil 上翼面

upper butt fitting 上对接接头

upper edge strip 上缘条

upper joint 上接头

upper panel 上壁板

upper skin 上蒙皮

upsetting 镦锻

vent hole 通气孔

vent pipe fairing/vent pipe shroud 通气管罩

vertical beam 竖梁

washer 垫圈

water drain hole 漏水孔

wavecrest 波峰

wave trough 波谷

web 腹板

weight balance 配重

weld 焊接

wet riveting 湿铆

wheel chock 轮挡

wind speed 风速

wing anti-icing pipe 防冰导管

wing bracket/wing stand 机翼托架

wing rib 翼肋

wing tip 翼尖

wing tip fairing 翼尖整流罩

working surface 工作表面

PROPELLERS 螺旋桨

accessory gear box 附件传动机匣

actual speed 实际转速

adapter unit/adapting cylinder 转接器组件

adjusting gasket 调整垫片

aerodynamic configuration 气动外形

aerodynamic performance 气动性能

aerodynamic variable pitch propeller 气动式变距螺旋桨

airborne tool 随机工具

allowable maximum rotating speed 允许最大转速

allowable speed deviation 允许转速偏差

altitude 高度

aluminum-cased holder 铸铝支架

ambient temperature 外界温度

ammeter 电流表

annular groove 环形槽

annular lock　环型锁

anti-icing system　防冰系统

armature circuit　电枢电路

armature core　铁芯

asbestos pad　石棉垫

assembly workshop　装配车间

auto-feathering circuit breaker　顺桨自动保护开关

auto-feathering　自动顺桨

auto-feathering check switch　自动顺桨检查开关

auto-feathering relay　自动顺桨继电器

auto-feathering system　自动顺桨系统

automatic timing mechanism　自动定时机构

automatic torque feathering　扭矩自动顺桨

auto-pitch variation　自动变距

auxiliary motor　辅助电机

auxiliary oil pump　辅助滑油泵

axial propeller　共轴反转螺旋桨

axis of propeller　螺旋桨轴线

balancing speed　平衡转速

blade　桨叶

blade angle　桨叶角

blade heater　桨叶加热器

blade heating component　桨叶加温元件

blade hinge　桨叶铰

blade incidence angle　桨叶安装角

blade profile　桨叶型面

blade tipping　桨叶包边

blade unit　桨叶组件

blanking plug　堵塞

boost-up cavity　增压腔

butt nut　对接螺母

cam/bulging　凸轮

canvas jacket　帆布套

carbon brush box　碳刷盒

centering muff　定心轴套

centrifugal balancing weight　离心配重

centrifugal force　离心力

centrifugal mechanism　离心机构

centrifugal pitch　离心定距

centrifugal speed regulating mechanism　离心调速机构

check feathering relay　检查顺桨继电器

clamp bolt　夹紧箍

clamp lock　卡锁

coil　线圈

combustion chamber　燃烧室

communication oil line　沟通油路

communicator circuit breaker board　通讯员自动开关板

component　元件

compressor blade　压气机叶片

concentrated warning light box　集中告警灯盒

conducting ring　导电环

connecting area　结合面

connecting pipe　连接导管

constant pitch line/fixed pitch line　定距油路

constant pressure valve　恒压活门

constant rotating speed　恒定转速

constant-speed propeller　恒速螺旋桨

contact　触点

contact current　接触电流

contactor　接触器

continuous flying　连续飞行

control circuit　控制电路

control device　控制装置

control lever　操纵杆

control switch　控制开关

control system　操纵系统

counter propeller　共轴反转螺旋桨(同一台发动机驱动)

counter clockwise rotation　处右向左旋转

counter-force　反向作用力

cowling plate　整流盘

crane　吊车

crane hook　吊钩

crank/cold run　冷转

current cross　串电

curve of engine running　试车曲线

deicer boot　除冰套(带)

deicing switch　除冰开关

deicing timer　除冰定时器

delay time　滞后时间

dent　凹痕

depreservation　启封

detachable blade propeller　可拆桨叶螺旋桨

detachable connector　可拆接头

detachable cowling　可卸整流罩

diagonal　对角线

distribution board　配电盘

distributor valve　分配活门

diverter valve　分流活门

drain pipe　漏油管

driven gear　从动齿轮

driving element　传动机件

driving gear　主动齿轮

ducted propeller　函道螺旋桨

dummy propeller　试车桨

duration　持续时间

duration of starting pulse　起动脉冲持续时间

dynamometer　测力计

EICAS indicator/display　EICAS 显示器

elbow/elbow bend pipe　弯管

electric bridge　电桥

electric motor　电机

electric plug　电插头

electric propeller　电动变距螺旋桨

electrical cavity　电气腔

electrical control system　电气操纵系统

electrical heating　电热防冰

electrical heating anti-icing device　电防冰装置

electrical heating anti-icing mechanism　电加温防冰机构

electrical heating de-icing device　电加温除冰元件

electrical heating system　电加温系统

electrical part　电气部件

electrical pitch-control mechanism　电气式变距机构

electro-hydromatic propeller　电动液压式自动变距螺旋桨

emergency feathering　应急顺桨

emergency feathering shutdown　应急顺桨停车

emergency feathering switch　应急顺桨开关

emergency hydraulic feathering　紧急液压顺桨

end brush box bracket　端面刷盒支架

end brush holder　端面刷架

end cover　端盖

end face teeth　端面齿

end spline　端面花键

engage　接通/吸合

engaged voltage　吸合电压

engine fault signal relay　发动机故障信号继电器

engine in-operation　工作着的发动机

engine nacelle　发动机短舱

engine oil line　发动机滑油路

engine out-of-operation　不工作的发动机

engine parameter　发动机参数

engine power output shaft　发动机功率输出轴

engine rotating speed　发动机转速

engine shutdown　发动机停车

engine shutdown relay　发动机停车继电器

engine speed reducer　发动机减速器

engine warm-up　暖机

enter feathering relay　进入顺桨继电器

environment temperature　环境温度

exit feathering relay　退出顺桨继电器

failure indicating light　故障指示灯

faired over hub　整流式桨毂

false feathering　假顺桨

faring cap　整流罩

fault indicating light of engine　发动机故障指示灯

feathering angle　顺桨角

feathering circuit　顺桨电路

feathering control panel　顺桨操纵板

feathering cycle　顺桨循环

feathering hub　顺桨机构桨毂

feathering pipeline　顺桨管路

feathering procedure　顺桨程序

feathering propeller　可顺桨螺旋桨

feathering pump　顺桨泵

feathering ready　顺桨准备

feathering ready circuit　顺桨准备电路

feathering ready relay　顺桨准备继电器

feathering stop　顺桨止动

feathering system　顺桨系统

feathering system circuit　顺桨系统电路

feathering time　顺桨时间

feathering timing mechanism　顺桨定时机构

fiber glass propeller　玻璃钢螺桨

fine-pitch oil chamber　小距油腔

fine-pitch oil line　小距油路

fine-pitch oil line pressure annunciator　小距油路压力信号器

fine-pitch stop　小距止动

fixed pitch oil line pressure annunciator　定距油路压力信号器

fixed-pitch propeller　定距螺旋桨

fixing bracket　固定支架

flange　法兰盘

flight condition　飞行条件

flight day　飞行日

four-blade propeller　四叶螺旋桨

front casing　前机匣

frontal drag　迎面阻力

fuel control unit　燃油调节器

full feathering　全顺桨

full-feathering propeller　全顺桨螺旋桨

fuse　熔断器

gear pump　齿轮油泵

gear vale　齿谷

general current　总电流

given speed　给定转速

glue line　胶缝

go-around　复飞

governor synchronize　调速器同步器

ground power supply　地面电源

head airflow　迎面气流

head assembly　头部组件

heating valve　加温活门

hexagonal head screw　六角头螺钉

high pitch line　大距油路

high-pitch oil chamber/cavity　大距油腔

high-pressure oil　高压滑油

hoisting weight　起吊重量

horizontal position　水平位置

hose　软管

hot oil　热滑油

hub　桨毂

hub assembly　桨毂组件

hub case　桨毂体

hub spinner/hub fairing　桨毂整流罩

hydraulic centrifugal type　液压离心式

hydraulic fixed pitch control valve　液压定距活门

hydraulic pitch control　液压定距

hydraulic pitch-control mechanism　液压变距机构

hydraulic signal　液压信号

hydropitch propeller　液压自动变距螺旋桨

idling speed　慢车转速

ignition coil　点火线圈

indicating light　指示灯

indicating light of engine unfeathering　发动

机回桨指示灯

inductive load　感性负荷

inlet oil temperature　进油温度

inlet pressure　进口压力

insensitiveness　不灵敏度

installation　组装

installation disc　安装盘

insulation resistance　绝缘电阻

interlock device　联锁装置

lacquer surface　涂漆面

landing taxiing distance　着陆滑跑距离

lateral centrifugal force　横向离心力

lateral component force　横向分力

lateral component moment　横向分力矩

left-hand rotation propeller　左旋螺桨（反转桨）

left-hand screw　左螺纹

limit speed　极限转速

link rod unit　连杆组件

load condition　负载状态

locking gasket　锁紧垫片

log book　履历本

lower left bracing rod　左侧下撑杆

lubricating oil　润滑剂

lug　凸耳

magnetic sensor　磁传感器

manual feathering　人工顺桨

manual feathering button　人工顺桨按钮

manual unfeathering　人工回桨

match　匹配

measuring　测定

mechanical damage　机械损伤

mechanical pitch stop　机械定距

mechanical pitch-control mechanism　机械式变距机构

mechanical propeller　机械式变距螺旋桨

megger　兆欧表/高阻表

metal propeller　金属螺桨

microswitch　微动开关

mid-pitch stop angle　中距限动角

mid-pitch stop mechanism　中距限动机构

mid-pitch stop switch　中距限动电门

motor　电动机

multi-blade propeller　多叶螺旋桨

mutual effects　双向作用

navigator circuit breaker box　领航员断路器箱

negative pole　负极

negative thrust　负拉力

negative thrust auto-feathering　负拉力自动顺桨

negative thrust auto-feathering circuit　负拉力自动顺桨电路

negative thrust check device　负拉力检查装置

negative thrust feathering check switch　负拉力顺桨检查开关

negative thrust feathering sensor　负拉力顺桨传感器

negative travel　反向行程

net weight　净重

normal temperature　常温

nose cap/dome　螺旋桨桨帽

notch　槽口

offset propeller hub　偏置螺旋桨轴

oil cavity　油腔

oil cylinder　油缸

oil distribution bush　分油衬筒

oil distribution valve　分油活门

oil drain pipe　放油管

oil drain sleeve　放油套

oil filter　滑油滤

oil groove　油槽

oil hole　通油孔

oil inlet　进油口

oil inlet cavity　进油腔

oil line　滑油管路

oil pipe unit　滑油管组件

oil pressure　油压

oil pump　滑油泵

oil pump assembly　油泵组件

oil tank　滑油箱

operating current　工作电流

operating flow　工作流量

operating liquid　工作液

operating performance　工作性能

operating state　工作状态

operating voltage range　工作电压范围

output　输出

overhead console　顶部操纵台

overload　过载

packing box　包装箱

partial feathering　部分顺桨

partial feathering button　部分顺桨按钮

partial shedding　局部脱落

periodical inspection　定期工作

pin　销子

pin valve　针形活门

piston cavity　活塞腔

piston unit　活塞组件

pitch change actuator　变距作动筒

pitch lock mechanism　距桨锁定机构

pitch variation mechanism　变距机构

pitch variation mode　变距方式

pitch variation range　变距范围

pitch-reversing mechanism　逆桨机构

pitch-setting mechanism　距桨调节机构

plug connector　插销接头

plug screw　螺塞

plumbing　管路

positive pole　正极

positive travel　正向行程

post-flight inspection　飞行后检查

power distribution box　电源配电盒

power-on check　通电检查

pre-flight inspection　飞行前检查

preservation　油封

preservation oil　油封用油

pressure annunciator　压力信号器

pressure fluctuation　压力波动

pressure reducing valve　减压活门

pressure regulating valve　调压活门

propeller　螺旋桨

propeller blade　桨叶

propeller blade icing　桨叶结冰

propeller blade root　桨根

propeller blade root cowling　桨根整流罩

propeller brake mechanism　螺旋桨制动机构

propeller casing　桨壳

propeller casing unit　桨壳组件

propeller digital synchronize　数字式螺旋桨同相器

propeller governor　螺旋桨调速器

propeller over-speed　飞转

propeller pitch-control mechanism　螺旋桨变距机构

propeller root fillet　桨根整流包皮

propeller rotating speed　螺旋桨转速

propeller shaft　螺旋桨桨轴

propeller stop release switch　螺旋桨解除限动开关

propeller synchronizer　螺旋桨(转速)同步器

pulling propeller　拉进式螺旋桨

pushing propeller　推进式螺旋桨

quick-release screw　快卸螺钉

radial brush box bracket　径向刷盒支架

radial brush holder　径向刷架

rated DC voltage　额定直流电压

rated operation rotating speed　额定工作转速

rated power　额定功率

readout　读数

recover　复原

reducer casing　减速机匣

regular distribution　均匀分布

regulating screw　调节螺钉

reinforcing access frame　加强口框

relative thickness　相对厚度

relay　继电器

reliability　可靠性

resistor/resistance　电阻

restrictor　节流器

reverse-pitch propeller　可逆桨螺旋桨（反距桨）

right-hand rotation propeller　右旋螺桨

rotating spced　转速

rotation direction　旋转方向

rotation inertia　转动惯性

rubber cup　胶碗

rubber fabric hose　夹布胶管

rubber seal washer　橡胶密封垫

rubber-asbestos washer　橡胶石棉垫圈

rudder buffer bushing　橡胶减震衬套

safety washer　保险片

scratch　擦伤/划伤

screw lock　螺钉锁

sealing compound　封口胶

sealing ring　封严圈

selector valve　选择活门

self-lock　自锁

semi-feathering propeller　半顺桨螺旋桨

semi-variable pitch propeller　半变距螺旋桨

sensor　传感器

sequential engaging　循序接通

service ladder　发动机工作梯

shaft coupling　联轴节

shaft power　轴功率

shear　剪力

short circuited　短路

shoulder　凸肩

shutdown in-air　空中停车

shutdown solenoid valve　停车电磁活门

shutdown system　停车系统

shutdown valve　停车活门

signal light　信号灯

single oil line　单油路

single-blade propeller　单叶螺旋桨

slide valve　滑阀

slip ring　滑环

slotted teeth to cover the plate　槽齿封闭盖板

socket　插座

solenoid valve　电磁活门

spare part　备件

speed bias torque motor　速度偏转扭矩电机

speed regulating assembly　调速器组件

speed regulator　调速器

spinner　桨罩

spinner centering muff unit　整流罩定心轴套组件

spring lock　弹簧锁

starting　起动

starting in-air　空中起动

starting in-air switch　空中起动开关

steel cable　钢丝绳

stop releasing indicating light　解除限动指示灯

stop releasing switch　解除限动开关

storage　贮存

stud　螺桩

subsonic propeller　亚音速螺旋价格

supersonic propeller　超音速螺旋桨

support component list　配套单

supporting bracket　托架

swivelling propeller　可偏转螺旋桨

tandem propeller　纵串螺旋桨

tension/compression load　拉/压载荷

terminal　接线柱

terminal voltage　端电压

three-blade propeller　三叶螺旋桨

throttle lever　油门杆

thrust　拉力

tightening torque　拧紧力矩

torque feathering check switch　扭矩顺桨检

查开关

torque feathering interlock throttle angle 扭矩顺桨联锁油门角度

torque measuring mechanism 测扭机构

torque measuring sensor 扭矩测量传感器

torque moment measuring gauge 扭转力矩测量器

torque oil pressure 扭矩油压

torque pressure 扭矩压力

torque spanner 定力扳手

total time 总时间

tractor 拉进式

transmission shaft 传动轴

transonic propeller 跨音速螺旋桨

turbine blade 涡轮叶片

turbine exhaust temperature 涡轮后温度

two-blade propeller 双叶螺旋桨

two-mode propeller timer 双模式螺旋桨定时器

unfeathering circuit 回桨电路

unfeathering indicating light 回桨信号灯

unfeathering pressure annunciator 回桨压力信号

unfeathering solenoid valve 回桨电磁阀

unfeathering time 回桨时间

union 接管嘴

unshrouded propeller 无函道螺旋桨

upper cavity 上腔

variable-pitch propeller 变距螺旋桨

vent plug 放气塞

voltage 电压

voltmeter 电压表

warm-up 预热

wide-blade propeller 宽叶螺旋桨

wire 导线

wire fixing clamp 导线固定夹

wooden propeller 木质螺桨

POWER PLANT 动力装置

A/C base level line 飞机水平基准线

A/C symmetry axis 飞机对称轴线

ability of "three defenses" "三防"能力

AC generator 交流发电机

accelerating lamination limiter 加速层板限制器

accelerating nozzle 加速喷油嘴

acceleration performance 加速性

accessory casing 附件机匣

accessory section cowling 附件整流罩

accidental ignition 意外点火

accumulative fuel 积油

acoustic panel（liner） 消声板（承受声载荷板）

activated agent 活性剂

active combustion 有效燃烧

actuator lock assembly 整流罩制动锁组件

adiabatical combustion 绝热燃烧

adjustable inlet 可调进气道

advanced ignition 提前点火

after fan 后风扇

afterburner diffuser 加力燃烧室扩压器

afterburner fuel manifold 加力燃烧总管组件

afterburner igniter 加力燃烧室点火器

afterburner ignition 加力燃烧室点火

afterburner jet pipe 加力燃烧室喷管

afterburner nozzle 加力燃烧室喷嘴

afterburner test 加力燃烧室试验

air bleed unit 放气机构

air compressor 压气机

air compressor blade 压气机叶片

air impingement starting system 双击式启动系统

air passage 气流通道

air scoop 进气嘴

air separator/air deaerator　空气分离器

air starter/air-injection starter　空气启动机

air stream　空气流

air turbine starter　空气涡轮启动机

airborne starter/on-board starter　机上启动机

air-driven turbine　空气驱动涡轮

air-mixing chamber　空气混合室

air-rectifier　整流器

alcohol　酒精

allowable bleed airflow　允许抽气量

altitude limit　限制高度

ambient temperature　外界温度

angle piece　角片

annular air inlet　环形进气道

annular cascade test　环形叶栅试验

annular combustion chamber　环形燃烧室

annular combustion chamber with centrifugal fuel combustor　甩油盘式环形燃烧室

annular diffuser　环形扩压器

annular nozzle　环形喷嘴

annular pipeline　环形管路

annular reverse-flow combustor　环形回流式燃烧室

anti-icing pressure annunciator　防冰压力信号器

appearance check　外观检查

arc chamber　电弧(加热的预燃)室

atomizer chamber　雾化室

atomizing nozzle　雾化喷嘴

attach fitting　安装接头

attachment ring　连接环

augmented turbine　加力式涡轮

automatic feathering　自动顺桨

automatic ignition　自动点火

automatic ignition-advance device　自动提前点火装置

automatic pitching　自动变矩

autotransformer starter　自耦变压器式启动机

auxiliary igniter　辅助点火器

auxiliary inlet　辅助进气口

auxiliary power unit　辅助动力装置

aviation kerosene　航空煤油

axial diffuser　轴向式扩压器

axial load　轴向载荷

axial-flow compressor　轴流式压气机

axial-flow compressor passage　轴流式压力机通道

axial-flow turbine　轴流式涡轮

axial symmetry inlet　轴对称进气道

baffle-type chamber　阻流式燃烧室

balancing plate　平衡板

base profile　基础叶型

basket-tube combustion chamber　联筒燃烧室

battery ignition　电瓶点火

battery timer　点火定时机构

battery-powered starter　蓄电池电动启动机

blade angle　桨叶角

blade flutter　叶片颤振

blade loading limit　叶片载荷极限

blade shaping　叶片造型

blade with vortex generator　带旋涡发生器叶片

blade/vane　叶片

blanking cover　堵盖

bleed airflow　引气量

blocker door　折流门

blowout characteristic　熄火特性

bonding jumper　搭接线

booster blades (vane)　增压叶片

booster compressor　增压压力机

booster inlet guide vane　增压进口导流叶片

booster rotor assemble　增压转子组件

booster stator assemble　增压静子组件

boostor coil　启动点火线圈

boundary-layer bleed inlet　有附面层吸除装

置的进气道

buffer　缓冲器

buried ducting　暗道

buried engine duct　埋入式进气导管

by-pass door　放气门

cable control　钢索操纵

cambered blade profile　有弯度叶型

cannular combustion chamber　筒环形燃烧室

cantilever moment　悬臂力矩

can-type combustion chamber　管型燃烧室

capture area　捕获面积

carbon brush　碳刷

carbon brush box　碳刷盒

cartridge turbo-starter　火药涡轮启动机

cascade nozzle　串联式喷嘴

casing treatment　机匣处理

casing/case　机匣

catalytic igniter　催化点火器

center console　中央操纵台

center hole　中央小孔

central-body inlet　有中心体的进气道

central-body supersonic inlet　有中心（锥）体的超音速进气道

centrifugal compressor　离心式压气机

centrifugal compressor diffuser　离心式压气机扩压器

centrifugal mechanism　离心机构

centrifugal nozzle　离心式喷嘴

ceramic-lined chamber　陶瓷（衬垫）燃烧室

chloridion　氯离子

circulation flow zone　环流区

clamp ring　卡圈

cleaning agent　清洗剂

cleaning fluid　清洗液

cleaning fluid formulation　清洗液配方

cleaning nozzle　清洗喷嘴

cleaning system　清洗系统

cleaning type　清洗性质

cloth cover　蒙布（罩布）

coarse filter　粗油滤

coefficient of combustion effectiveness　燃烧安全系数

coefficient of static pressure recovery　静压恢复系数

cold air　冷空气

cold start/dry motoring　冷启动

cold stream　冷气流

cold-weather start　冷天启动

combination cooling blade　复合式冷却叶片

combines compressor　混合式压气机

combustion chamber/combustor　燃烧室

combustion efficiency　燃烧效率

combustion in parallel layes　平行层燃烧

combustion intensity　容热强度

combustion product　燃烧产物

combustion theory　燃烧理论

combustor casing　燃烧室机匣

combustor cooling　燃烧室冷却

combustor outer casing　燃烧室外机匣

combustor wall temperature　燃烧室壁温

communication valve　连通开关

compositor　复激器

compressed-air starter　压缩空气启动机

compression nut　压紧螺母

compression pad　抗压垫

compressor adjustment　压气机调节

compressor air flow　压气机空气流量

compressor blade root damper　压气机叶片根部减震器

compressor case/compressor casing　压力机机匣

compressor characteristic　压气机特性

compressor efficiency　压气机效率

compressor element stage　压气机基元级

compressor fan cowling　压气机风扇整流罩

compressor front roller bearing　压气机前滚棒轴承

compressor rotor 压气机转子

compressor rotor blade 压气机转子叶片

compressor stage 压气机的级

compressor stall 压气机失速

compressor stator 压气机静子

compressor stator vane 压气机静子叶片（整流叶片）

compressor surging 压气机喘振

compressor test 压气机试验

conical shaping 锥面造型

conical-spike inlet 锥形中心体进气道

connecting disc 结合盘

constant bleed airflow 经常性抽气量

constant equivalent power regulation 等当量功率调节

constant rotating speed 恒定转速

constant rotating speed regulating 等转速调节

constant speed device 恒速装置

constant speed drive pneumatic starter 恒速驱动气动启动机

constant turbine inlet gas temperature regulation 等涡轮前燃气温度调节

constant-pressure combustion 定压燃烧

constant-pressure combustor 定压燃烧室

constant-temperature combustor 等温燃烧室

constant-volume combustion 定容燃烧

control system 操纵系统

convection cooling blade 对流冷却叶片

convergent-divergent inlet 收敛扩散型进气道

cooled combustion chamber 冷却式燃烧室

cooling air shroud 冷气罩

cooling effectiveness test of turbine blade 涡轮叶片冷却效果试验

core compressor 核心压气机

cowl access doors 整流罩检修门

cowl accessory 整流罩附件

cowl flap 整流罩鱼鳞片

cowl hinge 整流罩铰链

cowl panel 整流罩板

cowl safety latch 整流罩安全锁

cowl sling point 整流罩掉点

cowl support 整流罩支架（撑）

cowl support strut 整流罩撑杆

crane 吊车

crank/cold run 冷转

cranking 发动机冷转

cross-generator starting 发动机交叉启动

current collector 集流器

curve test 试车曲线

dangerous area 危险区

DC generator 直流发电机

DC starter/generator 直流起动发电机

dead start 重载启动

dead-center ignition 死点点火

deflector 导流门

delay lamination limiter 延迟层板限制器

delayed ignition 延迟点火

delayed start 延迟打火启动

design altitude 设计高度

design point and off design point 设计点与非设计点

diagonal order 对角线顺序

diffuser 扩压器

diffuser case 扩压器机匣

diffuser efficiency 扩压器效率

diluent stator combustor 掺混静子式燃烧室

dilution zone 掺混区

direct ignition 直接点火

direct-flow combustion chamber 直流式燃烧室

disc-drum rotors 盘鼓式（混合式）转子

discharge coefficient 流量系数

disc-type annular combustor 盘式环形燃烧室

disc-type rotor 盘式转子

distilled water 蒸馏水

distributed combustion 分散燃烧

diverter 隔道

double annular combustion chamber 双层环形燃烧室

double-annular ram-induction combustor 双环脉冲压进气燃烧室

double-entry compressor 双面进气压气机

double-flow combustion chamber 双路式燃烧室

double-oblique shock inlet 双斜激波进气道

double-stage compressor 双级压气机

drag 阻力

drag coefficient 阻力系数

drain fuel pipe 余油管

driving shaft 传动轴

droplet combustion 液滴燃烧

drum-type rotor 鼓筒式转子

dual action starter 双作用启动机

dual energy supply starter 双能源启动机

dual fuel nozzle 双路燃油喷嘴/双孔燃油喷嘴

dual-circuit ignition exciter 双偶电路式点火激发器

duct-cover 防尘罩

dust-net 防尘网

dynamo starter 启动发电机(启动和发电两用)

eight-unit-can-annular combustion chamber 八个火焰筒的环形燃烧室

ejector heat-radiation bleed airflow 引射散热抽气量

ejector type nozzle 引射喷嘴

electric ignition 电点火

electric mechanism 电动机构

electrical harness 电器配线

electrical inertia starter 电动惯性启动机

electrical starter 电动启动机

electromagnetic ignition 电磁点火

emergency air inlet 应急空气进口

emergency hydraulic feathering 紧急液压顺桨

emergency ignition 应急点火

end face slotted tooth 端面槽齿

engine altitude chamber 发动机高空试验室

engine appearance 发动机外形

engine axis 发动机轴线

engine continuous operation duration 发动机连续工作时间

engine efficiency 发动机效率

engine inner cavity 发动机内腔

engine instantaneous vibration value 发动机瞬时振动值

engine life 发动机寿命

engine link 发动机支撑连杆

engine main/auxiliary mounting joint 发动机主/辅安装节

engine model 发动机型号

engine mount 发动机安装支架

engine nacelle 发动机短舱

engine nacelle-wing fairing 发动机短舱与机翼接合整流罩

engine oil tank/pressure relief door 发动机滑油箱/压力释放门

engine operation parameter 发动机工作参数

engine pylon/strut 发动机吊架(挂)

engine shutdown 发动机停车

engine start 发动机启动

engine start lever 发动机启动手柄

engine starting 发动机起动

engine tail cone 发动机尾锥

engine type 发动机类别

engine vibration damper 发动机减震器

engine-mounting trunnion 发动机安装轴颈

equivalent divergent angle 当量扩张角

equivalent shaft horse power（ESHP） 当量功率

erosion igniter 电馈电嘴

excess air coefficient　余气系数

excessive fuel　富油

excessive fuel supply　过度富油

excessive fuel system　余油系统

exhaust and ventilation cooling system　排气和通风冷却系统

exhaust emission　排气发散

exhaust gas temperature　排气温度

exhaust gas temperature sensor　排气温度传感器

exhaust pipe　排气管

exhaust pipe blanking cover/blanking cap　尾喷管堵盖

exhaust-gas turbine　排气涡轮

exit temperature distribution　出口温度分布

exit temperature distribution coefficient　出口温度分布系数

extension pipe　延伸管

extension pipe shroud　延伸管外罩

external compression inlet　外压式进气道

external start　外部启动

external-internal compression supersonic air inlet　外内混合(压缩)式超音速进气道

extinguishing agent　灭火液

fabric hose connector　夹布胶管接头

fairing support plate　整流支板

false start　假开车

false start/wet motoring　假启动

fan　风扇

fan blade spacer　风扇叶片调整垫片

fan blade　风扇叶片

fan disc　风扇安装盘

fan outlet guide vane　风扇出口导向叶片

fan shaft　风扇轴

fan trim balance screw　风扇平衡微调螺钉

fan turbine　风扇涡轮

fastening belt/fixing strap　固定带

feathering position　顺桨位置

feathering system　顺桨系统

film cooling blade　气膜冷却叶片

film-cooled flame tube　气膜冷却式火焰筒

filtering area　过滤面积

filtering pore　滤孔

fine filter　细油滤

fine filter blocking annunciator　细油滤堵塞信号器

fire extinguishing system　灭火系统

fire-proof wall　防火墙

fireseals　防火隔板

firtree inline with scallop　枞树式排列的扇形孔

fitting clearance　配合间隙

fixed cowling　固定整流罩

fixed inlet　固定唇部的进气口

fixing clamp　固定卡

flame chamber　火焰室/燃烧室

flame holder test　火焰消除器

flame igniter　火舌点火器

flame tube/burner (liner)　火焰筒

flame velocity　火焰传播速度

flameholder　火焰稳定器

flap inlet　鱼鳞板式的进气道

flight direction　飞行方向

flight envelope　飞行包线

fluid type　溶液类型

flushing fluid　冲洗液

force-bearing frame　承力框

forward inboard cowling　前内侧整流罩

forward outboard cowling　前外侧整流罩

four-blade propeller　四叶螺旋桨

free turbine　自由涡轮

free-stream tube area　自由流管面积

free-vortex compressor　自由涡轮压力机

friction surface　摩擦面

front fan　前风扇

front inlet　前进气道

fuel consumption　燃油消耗量

fuel consumption meter sensor　燃油耗量表

传感器

fuel control unit 燃油调节器

fuel control unit inspection cover 燃油调节器检查口

fuel filter 燃油滤

fuel nozzle inlet pressure 工作喷嘴前压力

fuel pressure 燃油压力

fuel supply quantity 供油量

fuel system 燃油系统

fuel valve shutoff switch 燃油活门关断开关

fuel-air combustion starter 油气燃烧启动机

fuel-air ratio 油气比

fuel-rich combustion 富油燃烧

full size turbine test 全尺寸涡轮试验

gas generator 燃气发生器

gas stream 燃气流

gas turbine 燃气涡轮

gas turbine starter 燃气涡轮启动机

gas-generator turbine 燃气发生器涡轮

gas-turbine air starter 燃气涡轮供气的启动机

gas-turbine compressor 燃气涡轮压力机

generator cooling air overboard port 启动发电机冷却空气进口

grease/lubricating grease 润滑脂

greasy stain 油垢

ground cleaning truck 地面清洗车

ground start 地面启动

ground starter/ground-powered starter 地面启动机

ground starting envelope 地面起动包线

guide vane 导向叶片

hanging start 迟滞启动

headwind 迎风

heat loss 热阻

heater sleeve pipe 加温机套管

heating element 加温元件

heating wire plug 加温导线插头

hemispherical combustion chamber 半球形燃烧室

high energy (spark) igniter 高能(火花)点火器

high energy ignition exciter 高能点火装置

high loading combustor 高负荷燃烧室

high loading turbine 高负载涡轮

high pressure annular combustor 高压环形燃烧室

high temperature turbine 高温涡轮

high temperature turbine test 高温涡轮试验

high tension distributing system 高压点火分配系统

high tension leads 高压引线

high voltage air igniter 火花电嘴

high voltage surface gap igniter 沿面电嘴

high-altitude start 高空启动

high-center ignition 高压磁电机

high-pressure (HP) compressor 高压压气机

high-pressure combustion 高压燃烧

high-pressure ratio axial-flow compressor 高增压比轴流式压气机

high-pressure turbine (HPT) 高压涡轮

high-speed and high-temperature gas 高速高温燃气

high-stage compressor 高压级压力机

high-temperature part 高温零件

high-tension ignition 高(电)压点火

high-turbulence combustion 高紊(流度)燃烧

high-turbulence combustion chamber 高紊流燃烧室

high-velocity combustion 高速燃烧

hinged cowling 铰接式整流罩

hoist fitting 发动机提升附件(接头)

homogeneous combustion 均匀燃烧

hot stream 热气流

hot-air pipe 热空气导管

hub spinner/hub fairing   桨毂整流罩

hydraulic starter   液压启动机

hypersonic combustor   高超音速燃烧室

hypersonic inlet   高超音速进气道

ideal combustion   理想燃烧

idling   慢车

igniter   点火器/电嘴

igniter/igniting/ignition chamber   点火燃烧室

ignition adjusting lever   点火调节杆

ignition boss   点火器衬块

ignition cable   点火电缆

ignition cam   点火凸轮

ignition characteristic   点火特性

ignition coil   点火线圈

ignition delay   点火延迟

ignition distributor   点火分配器

ignition energy   点火能量

ignition exciter   点火激发器

ignition harness   点火导线束

ignition lead   点火引线

ignition lead connector   点火引线接头

ignition magnetos   点火磁电机

ignition plug   电嘴

ignition relay   点火继电器

ignition switch   点火开关

ignition time   点火时间

ignition time lever   点火定时杆

ignition transformer   点火变压器

immediate relight   空中瞬时点火

impact load   撞击载荷

imperfect combustion/poor combustion   不完全燃烧

impingement impulse starter   空气冲击启动机

impurity   杂质

indirect ignition   间接点火

induction vibrator   感应振动器

inertia starter   惯性启动机

in-fight start   空中启动

in-flight ignition   空中点火

in-flight starting envelope   空中起动包线

initial ignition   起始点火

initiating combustion   起始燃烧

inlet additive drag   进气道附加压力

inlet buzz   进气道嗡鸣

inlet casing/inlet case   进气机匣

inlet cowling   进气道整流罩

inlet dynamic characteristic   进气道动态特征

inlet guide vane   导向器叶片

inlet guide vanes   进气道流叶片

inlet lip   进气道唇口

inlet mass-flow coefficient   进口流量系数

inlet operating condition   进气道工作状态

inlet starting   进气道启动

inlet surg   进气道喘振

inlet test   进气道试验

inlet throat   进气道喉道

inlet total pressure recovery coefficient   进气道总压恢复系数

inlet/outlet pressure   进口/出口压力

inner/outer point   内/外点

input/output power   输入/输出功率

instantaneous combustion   瞬时燃烧

instantaneous ignition   瞬时点火

intermediate zone   过渡区

intermediate-pressure (IP) compressor   中压压气机

intermediate-pressure turbine (IPT)   中压涡轮

internal compression inlet   内压式进气道

irregular combustion   反常燃烧

isobaric combustion   等压燃烧

isolation valve connector   隔断活门接头

jet pump/injector pump   隐射泵

jet starter   喷气启动机

jet stream   喷气流

jet stream/jet efflux   射流

jet turbine 喷气涡轮

jet-engine compressor 喷气发动机压气机

jet-flag blade 喷气叶片

knocking combustion 爆震燃烧

late ignition 过迟点火

lead filter 引线滤波器

lead sealing 铅封

leaking liquid 泄漏液体

leaking port 泄漏口

lean fuel 贫油

leaning-flow turbine 斜流式涡轮

level measurement 水平测量

linkage mechanism 联动机构

lock plate 保险片

locking wire 保险丝

log book 履历本

low calorific value 低热值

low pressure rotor 低压转子

low pressure start 低压空气启动

low pressure stator 低压静子

low-drag inlet 低阻力进气道

lower pressure cases 低压机匣

low-pressure (LP) compressor 低压压气机

low-pressure air starter 低压空气启动机

low-pressure diffuser 低压扩压器

low-pressure diffuser pipe 低压扩压管

low-pressure pneumatic turbine starter 低压气动涡轮启动机

low-pressure turbine (LPT) 低压涡轮

low-stage compressor 低压级压气机

low-tension ignition 低(电)压点火

lubricating system 润滑系统

lug strut 叉耳支柱

magnetic system of high-tension magnetic 高压磁电机磁系统

magneto (impulse) starter 启动磁电机

magneto ignition 磁电机点火

main performance 主要性能

main technical data 主要技术数据

main vibration shock absorber 主减震器

main/auxiliary oil pump 主/辅助滑油泵

main/auxiliary fuel pump 主/辅助燃油泵

main/upper/middle/lower link 主/上/中/下撑杆

manual feathering 人工顺桨

manual unfeathering 人工回桨

mass 质量

match of inlet and engine 进气道与发动机匹配

maximum torque 最大扭矩

mechanical damage 机械损伤

metal chip 金属屑

metal sheet 金属片

mid-rear bearing 中后轴承

mixed combustion chamber 混合燃烧室

mixed compression inlet 混合压缩式进气道

mixed-flow compressor 混流式压气机

mixed-flow turbine 混流式涡轮

mixer 混合器

mixture (fuel) 混合(燃)油

mixture (oil) 混合(滑)油

mould fungus 霉菌

mounting bolt 安装螺栓

mounting bracket 安装支架

mounting joint 安装节

movable cowling 可移动整流罩

muffler 消音器

multiple combustion chamber 分管燃烧室

multiple starts 多次启动

multiple-shock inlet 多激波进气道

multiple-stage compressor 多级压气机

multi-stage pressure compounded turbine 多压力级涡轮

multi-stage turbine 多级涡轮

multi-stage velocity compounded turbine 多速度级涡轮

multistep ignition step start 多级点火气动

NACA-type inlet/NACA scoop NACA 型进

气道

nacelle access cover　短舱口盖

negative thrust auto-feathering　负拉力自动顺桨

no-load start　空载启动

nonsteady combustion　不稳定燃烧

nonsustaining combustion　不持续燃烧

normal atmospheric pressure　标准大气压力

normal combustion　正常燃烧/充分燃烧

normal start　正常启动

normal-shock inlet　正激波进气道

nose ring cowling　前环整流罩

nozzle cowling　喷管整流罩

nozzle pressure　喷嘴压力

nut　螺母

off-axis inlet　侧面进气道:偏离(发动机)轴线的进气道

oil barrel　接油桶

oil consumption　滑油消耗量

oil drain valve/oil drain cock　放油开关

oil filler access　滑油注入口

oil filtering accuracy　滑油过滤精度

oil flow　滑油流量

oil heat dissipation capacity　滑油散热量

oil pressure sensor　滑油压力传感器

oil quantity indicator sensor　滑油油量表传感器

oil scavenge pump　抽油泵

oil system　滑油系统

oil tank　滑油箱

oil temperature sensor　滑油温度传感器

oil-mist separator　油雾分离器

on start　连通启动

one-side-entry combustor　单侧进气式燃烧室

operating condition/operating state　工作状态

operating fluid　工作液

operating rotating speed　工作转速

operation characteristic　工作特性

operational fuel　使用燃油

operational oil　使用滑油

optimum oil temperature　最佳滑油温度

original profile　原始叶型

oscillatory combustion　振荡燃烧

overall dimension　外廓尺寸

oxidizer-rich combustion　富氧化剂燃烧

packing box　包装箱

packing box bracket　包装箱托架

parking apron　停机坪

partial combustion　局部燃烧

penetration zone　穿透深度

perfect combustion　完全燃烧

performance data　性能数据

periodical bleed airflow　周期性抽气量

piloted air-blast fuel nozzle　气动雾化式燃油喷嘴

pin　销子

pipe　导管

piston aero-engine ignition system　活塞式航空发动机点火系统

piston aero-engine starting ignition device　活塞式航空发动机启动点火装置

piston compressor　活塞式压气机

plane cascade test　平面叶栅试验

pneumatic starter　气动启动机

porous gas turbine blade　多孔燃气涡轮叶片

potable water　可饮用自来水

power lever angle　油门角度

power recovery turbine　功率恢复涡轮

precombustion-chamber ignition　预燃室点火

pre-compression　预压缩

preliminary start　预先启动

premature ignition　过早点火

pressure balance cavity　卸荷腔

pressure ratio of compressor　压力机增压比

pressure regulating valve　调压阀

pre-start/before starting　启动前

primary air　一股流/主空气

primary igniter　主点火器

primary nozzle　主喷嘴

primary zone　主燃区

profile in cascade　叶栅叶型

propeller hub　桨毂

propeller shaft　螺旋桨桨轴

purified water　纯净水

pylon aprons　吊架护罩

pyrogen igniter　高温点火器

quick-release pin　快卸销

quick-release self-sealing coupling　快卸自封接头

radial diffuser　径向扩压器

radial load　径向载荷

radial-flow compressor　径流式压气机

radiator shutter　散热器风门

ram compressor　冲压式压气机

ram-air turbine　冲压空气涡轮

ramjet inlet　冲压(式)喷气发动机进气道

rapid repeat start　快速重复启动

ratchet　棘轮装置

ratchet clutch　棘轮离合器

rated state　额定状态

rear inboard cowling　后内侧整流罩

rear outboard cowling　后外侧整流罩

recirculation flow zone　回流区

recommended cleaning cycle　推荐清洗周期

redial-flow turbine　径流式涡轮

regulating system　调节系统

regulator　调节器

reheat combustion chamber/afterburner　加力燃烧室

reheat ignition/after burning ignition　加力点火

reigniter chamber　再次点火式燃烧室

reliable ignition　正常点火

relight button　空中启动按钮

remaining deformation　剩余变形

retarded combustion　延迟燃烧/减速燃烧

reverse-flow combustion chamber　回流式燃烧室

rocker switch　摇臂开关

rotating cascade test　旋转叶栅试验

rotating speed oscillation　转速摆动

rotating stall　旋转失速

rotation direction　旋转方向

rotational resistance angle　旋转阻力角

rotatory sealing device　旋转封严装置

rotor blade　转子叶片

rubber cup　皮碗

rubber hose　胶管

rubber ring　胶圈

rubber sealing ring　橡胶封严圈

running start　运转启动

rust　锈蚀

safety coupling　保险联轴节

safety requirement　安全要求

salt　盐分

scallop　扇形空

secondary air　二股流

secondary igniter　副点火器

secondary nozzle　副喷嘴

second-class bolt　2级螺栓

sediment　沉积物

self-piloting start　自动控制循环启动

semiconductor igniter　半导体电嘴

sequence start　顺序启动

service ladder　工作梯

sharp-lip inlet　尖唇进气道

shielded igniter lead　带屏蔽的点火器导线

shock-wave ignition　激波点火

short annular combustion chamber　短环形燃烧室

shroud　外罩

shrouded turbine blade　带冠的涡轮叶片

shutdown　停车

shutter-controlled air inlet　节气门控制的进气道

simultaneous ignition　同时点火

single-entry compressor　单面进气压力机

single-flow combustion chamber　单路式燃烧室

single-rotor engine　单转子发动机

single-spool axial compressor　单转子轴向式压气机

single-spool compressor　单转子压气机

single-stage compressor　单级压力机

single-stage turbine　单级涡轮

slab-shaped combustion chamber　扁平式燃烧室

slider　滑板

sling/suspension　吊挂

slow start　慢启动

smoke number　发烟数

smooth start　平稳启动

softening coefficient　软化系数

sonic inlet　音速进气道

spacer bushing　间隔衬套

spacer/washer　垫片

spark duration　火花持续时间

spark energy　火花能量

spark igniter　点火电嘴

spark plug　火花塞

spark-plug ignition　电嘴点火（火花塞点火）

speed regulator　调速器

spinner　整流罩

spinner nose cap　整流罩盖

split three-stage turbine　双转子三级涡轮

spray ring　喷射环

stage ignition　分级点火

standard atmosphere condition　标准大气（条件）

start bleed valve　启动引气活门

start time　启动时间

starter button　启动按钮

starter generator power cutoff speed　起动发电机断开转速

starter oil discharge duct　启动机空气排放管

starter switch　启动机开关

starter/relight selector switch　启动/再点火选择开关

starter-generator　启动发动机

starting box　起动箱

starting fuel nozzle　启动喷嘴

starting ignition　启动点火

starting power supply　起动电源

starting program/starting procedure　起动程序

starting system　起动系统

state operation area　状态工作区

static pressure tube　静压管

stationary stator vane　固定静子叶片

steady-flow combustion　定常流燃烧

storage capacitor　存储变容器

storage period　贮存期

straightened blade　整流叶片

subsonic cascade　亚音叶栅

subsonic combustion　亚音速燃烧

subsonic compressor　亚音速压力机

subsonic inlet　亚音速进气道

subsonic profile　亚音叶型

subsonic stage　亚音级

subsonic turbine　亚音速涡轮

supercharger impeller　增压器叶轮

supercritical combustion　超临界燃烧

supersonic cascade　超音叶栅

supersonic combustion　超音速燃烧

supersonic compressor　超音速压力机

supersonic cruise inlet　超音速巡航进气道

supersonic inlet　超音速进气道

supersonic profile　超音叶型

supersonic stage　超音级

supersonic turbine　超音速涡轮

support　支撑件

surface combustion　表面燃烧

surg　喘振

surg boundary　喘振边界

surg limit　喘振边界

surg margin　喘振裕度

sustained combustion　持续燃烧

sweat-cooled flame tube　发汗冷却式火焰筒

sweat-cooled thrust chamber　发汗冷却燃烧室

swirl chamber/whirl chamber　旋流室

swirl combustion chamber　旋流式燃烧室

swirl-can type combustor　涡流罐式燃烧室

swirler　涡流器

synthetic hydrocarbon oil　合成烃滑油

tachometer　转速表

tachometer sensor　转速表传感器

tail cone　尾锥

tail cowling　尾部整流罩

takeoff envelope　起飞包线

takeoff state　起飞状态

tandem blade　串列式叶片

tandem journal　前后轴颈

tangential blowing blade　切向喷气叶片

temperature ratio　加温比

test run　试车

testing similar condition　试验相似条件

thermal load　热负荷

thermal part　热部件

three-dimensional flow　三元流

three-dimensional inlet　三元进气道

three-spool compressor　三转子压气机

three-stage compressor　三级压力机

three-stage turbine　三级涡轮

three-way joint/three-port connector　三通接头

throttle control handle　油门操纵手柄

throttle lever position sensor　油门杆位置表传感器

thrust producing combustion　产生推力的燃烧

thrust reverser cowling　反推力装置整流罩

thrust ring　安装发动机承力隔框

time interval　间隔时间

timing mechanism　定时机构

torch igniter　火炬电嘴

torque　扭矩

torque pressure gauge sensor　扭矩压力表传感器

torque wrench　定力扳手

total bleed airflow　总抽气量

total temperature tube　总温管

track/slider fairing　导轨/滑板整流罩

train-type ignition　循序点火

transitional combustion　过渡燃烧

transonic cascade　跨音叶栅

transonic compressor　跨音速压力机

transonic profile　跨音叶型

transonic stage　跨音级

transonic turbine　跨音速涡轮

transpiration cooling blade　发散冷却叶片

transportation trolley　运输车

tube-annular combustion chamber　联管燃烧室

turbine blade　涡轮叶片

turbine blowing cover　涡轮吹风罩

turbine cover　涡轮罩

turbine efficiency　涡轮效率

turbine engine ignition system　涡轮发动机点火系统

turbine guide vanes　涡轮导向器叶片

turbine inlet gas temperature　涡轮前燃气温度/涡轮进口温度

turbine outer case　涡轮外机匣

turbine rotor　涡轮转子

turbine shaft　涡轮轴

turbine starter generator　涡轮发电装置

turbine starter/vane starter　涡轮启动机

turbine stator（nozzle guide）　涡轮静子（导

向器）

turbine stator cane（nozzle guide vane）　涡轮静子叶片（导向器叶片）

turbine-driven compressor　涡轮压气机

turbo expander　涡轮冷却器

turbo-accelerator　涡轮加速器

turbocharger，turbo-supercharger　涡轮增压器

turbocompressor　涡轮压力机

turbojet combustion chamber　涡轮喷气发动机燃烧室

turboprop engine　涡轮螺旋桨发动机

turbulent boundary layer combustion　紊流附面层燃烧

twin inlet　双进气道

twin shaft compressor　双轴压力机

twin-spool axial compressor　双转子轴向式压气机

twin-spool compressor　双转子压气机

two-dimensional flow　二元流

two-dimensional inlet　二元进气道

two-stage turbine　两级涡轮

underslung inlet　吊挂式进气道

upper/lower air bleed pipe　上/下放气管

vane compressor　叶片式压力机

vane diffuser　叶片式扩压器

vaneless diffuser　无叶片式扩压器

vaporizing combustor　蒸发式燃烧室

vaporizing fuel nozzle　蒸发管燃油喷嘴

variable geometry turbine　变几何涡轮

variable guide vane　可调安装角叶片

variable inlet guide vane　可调进气导流叶片

variable pitch blade　可调静子叶片

variable rotating speed regulating　变转速调节

variable stator vane　可整流叶子（可转动静子叶片）

variable-area turbine　变面积涡轮

variable-discharge turbine　变流量涡轮

variable-geometry combustion chamber　变几何燃烧室

variable-pitch turbine　变叶距涡轮

variable-stator compressor　可调静子（叶片）压力机

variation law　变化规律

vent tank　通气油箱

vibration disc /shock absorber disc　减震盘

vibration overload coefficiency　振动过载系数

vibration pad/shock absorber pad　减震垫

vibration ring/shock absorber ring　减震圈

vibration speed indicator sensor　振动速度表传感器

vibration strut/shock absorber strut　减震支柱

viscosity　黏度

vortex control device　涡流控制装置

vortex generator　涡流发生器

vortex generator with flap　带鱼鳞片的涡流发生器

vortex shed　涡束

vortex spoilers　涡流扰流器

warm-up　暖车

warning signal light　警告信号灯

water injection　注水

wedge-shaped blade　楔形叶片

wet start　湿启动（燃油加水启动）

wheel chock　轮挡

windmilling start　风转启动

wing leading edge　机翼前缘

wing trailing edge　机翼后缘

ENGINE FUEL AND CONTROL
发动机燃油及控制

accelerating valve　加速活门

adapter　转接头

adjust control tube　可调整的控制管

adjuster　调节器

afterburner fuel regulator　加力燃烧室燃油调节器

afterburner fuel-flow regulator　加力燃烧室燃油流量调节器

air bleed valve　放气活门

air compressor　压气机

air inlet temperature　进气温度

air intake　进气道

air-bleed valve　放气活门

airflow control regulator　空气流量调节器（防喘调节器）

amplifier　放大器

amplifying element　放大元件

atmospheric environment　大气条件

automatic accelerator　自动加速器

automatic starter　自动启动器

auxiliary fuel pump　辅助燃油泵

back pressure　背压

balk lever　限动杆

barostat　气压调节器

bellcrank　曲柄

bimetallic screen filter plate　双金属网状滤片

block　堵塞

bracing　撑杆

buffer　缓冲器

bullet shaped throttle　锥尖形油门

butterfly throttle　蝶形油门

bypass valve　旁通活门

cable quadrant　钢索扇形件

calculating cam　计算凸轮

calculating element　计算元件

calculating lever　计算杠杆

calibration　标定

capacitance-type fuel lever transducer　电容式油面传感器

casing　壳体

cavitation　气穴

centrer instrument panel　中央仪表板

centrifugal hydraulic tachogenerator　液压离心式转速表传感器

centrifugal mechanical tachogenetator　机械离心式转速表传感器

circular main fuel pipe　环形燃油总管

clamping hoop　夹紧箍

coarse filter　粗油滤

combustion mixture-ratio regulator　燃烧混合比调节器

compact picture　紧凑画面

compensate rod　补偿杆

compensation　补偿

compensator　补偿器

constant fuel-flow regulator　恒量供油（等流量）调节器

constant pressure drop valve　等压差活门

constant pressure valve　定压活门

container　接油盒

control element　控制元件

correcting element　校正元件

crossfeed manifold pressure switch　交叉输油管压力开关

cutoff valve　断油活门

dial/scale　刻度盘

dowel　定位销（暗销）

driven gear　从动齿轮

driving gear　主动齿轮

dynamic balance　动平衡

dynamic pressure regulator　速压调节器

eccentric bushing　偏心衬套

elastic element　弹性元件

electric bridge　电桥

electrical pulse　电脉冲

emergency feathering shutdown　应急顺桨停车

emergency fuel regulator　应急燃油调节器

emergency reducing valve　急降活门

engine combustion chamber/engine combustion combustor　发动机燃烧室

engine indication and crew alert system 发动机指示和空勤告警系统

engine shufoff valve 发动机燃油关断活门

engine shutdown 发动机停车

engine shutdown solenoid valve 停车电磁活门

engine start 发动机开车

engine start fuel pipe 起动油管

engine start fuel reducing solenoid valve 起动切油电磁阀

engine start fuel supply solenoid valve 起动供油电磁活门

engine start nozzle 起动喷咀

engine throttle 发动机油门

engine throttle control lever 发动机油门杆

equivalent shaft horse power（ESHP） 当量功率

exhaust gas temperature 排气温度

exhaust hole 排气孔

expanding temperature transmitter 膨胀式温度传感器

eyelet bolt 活节螺栓

filler temperature transmitter 充填式温度传感器

filter block pressure difference annunciator 油滤堵塞压差信号器

filter element 过滤元件

filter paper 滤纸

filter plate 滤片

filtering mesh 滤孔

filtering precision 过滤精度

fine filter 细油滤

fine filter blocking annunciator 细油滤堵塞信号器

flexible connector 柔性接头

floating shaft 浮动轴

float-type fuel lever transducer 浮子式油面传感器

flow divider 流量分配器

flow equalizer 流量均衡器

fluid resistance 流体阻力

fluidic amplifier 滑阀式液压放大器

forward thrust lever 前推力杆

fuel consumption gauge sensor 耗量表传感器

fuel consumption sensor 耗量传感器

fuel control amplifier 燃油控制放大器

fuel control unit 燃油调节器

fuel density compensator 燃油密度补偿器

fuel distribution system 燃油分配系统

fuel distributor 燃油分配器

fuel drain valve/fuel drain cock 放油开关

fuel filter bypass warning switch 燃油滤旁通告警开关

fuel filter differential pressure switch 燃油滤压差开关

fuel flow proportioner 燃油流量比例器

fuel flow regulator 燃油流量调节器

fuel flow transmitter 燃油流量传感器

fuel governor reset arm 燃油调节器复位臂

fuel heater 燃油加温器

fuel inlet nipple 进油接管咀

fuel inlet union 进油接管嘴

fuel lever transducer 油面传感器

fuel low pressure warning switch 燃油低压告警开关

fuel outlet union 出油接管嘴

fuel overfill control 过量加油控制

fuel pressure indicating system 燃油压力指示系统

fuel pressure indicating system 燃油压力指示系统

fuel pressure transmitter 燃油压力传感器

fuel pump 燃油泵

fuel quantity compensator 燃油油量补偿器

fuel quantity indication 燃油油量指示

fuel quantity probe 燃油油量传感器

fuel return check valve 回油单向活门

fuel shutoff valve　防火开关

fuel supply and delivering pump　燃油供输油泵

fuel supply quantity　供油量

fuel tank　燃油箱

fuel-delivery regulator　输油调节器

fuel-flow equalizer　（多油箱的)均耗调节器

fueling/defueling hose receptacle　加油/放油管接口

fuel-line stabilizer　燃油管路稳定器

full throttle　油门全开

gear selector　变速选择杆

gear-type high-pressure fuel pump　齿轮式高压燃油泵

gross weight indictor　飞机总重指示器

handle　手柄

high fuel pressure system　高压系统

holding stand　保持架

hydraulic delay unit　液压延迟器

hydraulic delayer　液压延迟器

hydraulic signal　液压信号

idle speed governor　慢车转速调节器

idle valve　慢车活门

idling　慢车

impeller　叶轮

indicating device　指示装置

indicating system　指示系统

inlet pressure/intake pressure　进口压力

intake　进口

interference source　干扰源

jet-baffle type amplifier　喷嘴-挡板式液压放大器

jettison pump pressure switch　放油泵压力开关

kinetic energy　动能

laminated restrictor　层板式限流器

limiter　限制器

linear voltage signal　线性电压信号

lock bar　锁杆

locking collar　锁圈

locking washer　保险垫圈

low fuel pressure system　低压燃油压力传感器

lower cover　下盖

low-lever warning sensor　低油面告警传感器

low-pressure system　低压系统

magnetic amplifier　磁放大器

magnetic dumping generator　磁阻发生器

main fuel pump　主燃油泵

main tank boost pump pressure switch　主油箱增压泵压力开关

max/min fuel pressure limiter　最大/最小燃油压力限制器

maximum speed governor　最大转速调节器

measuring accuracy　测量精度

mechanical energy　机械能

mechanical impurity　机械杂质

mechanical rotation signal　机械转动信号

modular structure　模块式结构

negative thrust auto feathering　负拉力自动顺桨

nozzle　喷嘴

operating nozzle　工作喷咀

outer casing nut　外套螺母

outlet pressure　出口压力

overfill float switch　过量加油浮子开关

overfill reset switch　过量加油复位开关

over-temperature　超温

pickle (package) an engine　油封发动机

place the throttle off　收油门到底

port throttle　左发动机油门

power lever angle　油门杆角度

power limiter　功率限制器

power regulator　功率调节器

power-on　通电

precooler　预冷器

pressing strip　压条

pressure drop regulator　压差调节器

pressure fueling panel　压力加油控制板

pressure ratio regulator　压比调节器

pressure ratio sensor　压力比传感器

pressure reducing chamber　减压室

pressure reducing valve　减压活门

pressure regulator　压力调节器

pressure sensor　压力传感器

pressurization　增压

printed circuit board　印制电路板

reheat regulator　加力调节器

replace the engine　更换发动机

resistance temperature transmitter　电阻式温度传感器

retard the throttle to idle　收油门到慢车

reverse thrust control link　反推力操纵连杆

reverse thrust lever　反推力杆

reverse thrust lever interlock actuator　反推力杆联锁调节器

reverser control rod　反推力操纵拉杆

rigging pin hole　定位销孔

roller bearing　滚子轴承

rotor　转子

safety valve　安全活门

scavenge pump　回油泵

sealing cup ring　封严皮碗

sealing ring　密封圈

seal-off　脱焊

seat cover　座盖

sediment　沉淀物

shaft horse power (SHP)　轴功率

shield　屏蔽

side picture　副画面

signal detection amplifier　信号检测放大器

signal processor　信号调理器

signal switch assembly　信号变换组件

signal volume　信号量

silicon pressure sensor　硅压力敏感元件

slant　扭斜

speed governor　转速调节器

spherical collar　球形环

sprayer　喷雾器

starboard throttle　右发动机油门

start lever　启动杆

start lever assembly　启动杆组件

start system　起动系统

starter power-off switch　起动机断电开关

stress　应力

supercharger pressure regulator　增压器调节器

tacho sensor　转速传感器

tacho-generator　测速发电机

tacho-limiter/speed limiter　转速限制器

tandem rectifiers　前后整流器

tank pressure regulator　油箱压力调节器

teleflex　软套管/转套

temperature compensator　温度补偿器

temperature limiter　温度限制器

temperature regulator　温度调节器

temperature sensor　温度传感器

thermal bimetal　热双金属

thermocouple　热电偶

thinner lead mat　铅薄垫

throttle angle/power lever angle　油门角度

throttle at idle　慢车油门

throttle element　节流元件

throttle orifice　节流嘴

thrust drum　推力鼓轮

thrust lever　推力杆

thrust lever control switch　反推力控制开关

thrust lever position resolver　推力杆位置分解器

timer adjuster　定时装置调节器

torque auto feathering circuit　扭矩自动顺桨电路

total fuel indicator　燃油总量指示器

total temperature　总温

transient fuel consumption　瞬时油耗

transmitting valve type amplifier　射流式放

大器

vane-type boosting pump 旋板式增压泵

volumetric shutoff compensator 油满关断补偿器

volumetric shutoff unit 油满关断装置

weak DC voltage signal 弱直流电压信号

Wheatstone bridge 惠斯登电桥

IGNITION 点火

afterburner 加力燃烧室

airborne tool 随机工具

aluminum alloy 铝合金

amplitude 幅值

anti-corrosion 防腐

argon welding 氩弧焊接

boundary position 边界位置

bushing clamp 衬套箍

carbon deposit 积碳

case type assembly 盒式组件

casting magnesium alloy 铸镁合金

center electrode 中心电极

circuit breaker board 自保开关板

complex shape 形状复杂

connection nut 连接螺母

contactor 触头

continuous ignition 持续点火

copper bushing 铜衬套

copper sealing ring 铜封严圈

cross section 截面

decoration panel 装饰板

duralumin alloy 硬铝合金

electric energy 电能

electric erosion layer 电蚀层

electrical treatment 电处理

engine airborne tool 发动机随机工具

engine compressor casing 发动机压气机机匣

engine ignition contactor 发动机点火接触器

engine nacelle 发动机短舱

engine propeller 发动机螺旋桨

engine service ladder 发动机工作梯

engine start fuel solenoid valve relay 起动燃油电磁开关继电器

engine starting stage 发动机启动阶段

extinguishing bottle 灭火瓶

fixed screw 固定螺钉

fluorine wire 氟导线

follow-up 随动

fuel supply starting solenoid valve 启动供油电磁活门

fuel-air mixture 油气混合气

galvanic corrosion spark igniter 电蚀电嘴

graphite grease 石墨润滑脂

ground condition 地面条件

ground power unit 地面电源车

ground starting ignition 地面启动点火

heat-resistant enamel paint 耐热磁漆/瓷漆

heat-resistant non-alkali strap 耐热无碱带

hexagonal screw 六角螺钉

high voltage ignition cable 高压点火电缆

high-strength low-carbon alloy structure steel 高强度底碳合金结构钢

high-voltage radio shield harness 高压防波导线

high-voltage wire plug 高压导线插头

horizontal stabilizer 水平安定面

hydraulic power-off switch 液压断电开关

igniter 点火器

ignition cable 点火电缆

ignition circuit breaker 点火自保开关

ignition coil 点火线圈

ignition plug 点火电嘴

ignition power supply system 点火电源

in-flight ignition 空中点火

in-flight starting ignition 空中起动点火

inseam 内缝

insulating tube 绝缘管

insulator surface 绝缘体表面

iron-shimmed asbestos pad 夹铁丝石棉垫

loop ceramic insulator 环形陶瓷绝缘体

low-energy ignition plug 低能电嘴

magnetic oscillator 磁振荡器

maintenance access cover 维护口盖

manufacturer 承制厂

maximum cross section 最大截面

mechanical system 机械系统

megger 兆欧表

metalized silver ionic layer 金属化银离子层

micro-capacitor 微电容器

navigation radar bay 航行雷达舱

neutral gas bottle 中性气体瓶

non-pressurized cabin 非气密舱

nose landing gear well 前起落架舱

operation end surface 工作端面

outer electrode 外电极

outseam 外缝

peak voltage 峰值电压

plate bond face 平板结合平面

pre-energized exercise 预先赋能锻炼

primary current intensity 初级电流强度

primary winding 初级绕组

radar bay 雷达舱

radial burning damage 径向烧损

rear combustor casing 燃烧室后外套

rear end frame 后段框

rectifier 整流器

roller device 滚棒装置

sealed type induction coil 密封式感应线圈

secondary winding 次级绕组

serial connection 串联

side electrode 侧电极

simplify start procedure 简化启动程序

single line system 单线制

start fuel spray 起动喷油

start relay box 起动继电器盒

starting spray nozzle 起动喷嘴

surg hose 防波软管

thiokol sealant 多硫密封胶/聚硫橡胶

tidy and beautiful 整齐美观

tiedown ring/mooring ring 系留环

time procedure annunciation 时间程序信号

timing mechanism 定时机构

vertical stabilizer 垂直安定面

vibrator connector 振动器接点

voltage peak value 电压峰值

warm-up 预热

watertight sealing 水密密封

wind force 风力

wire core section 线芯截面

wrought aluminum alloy 锻铝合金

AIR 进气

AC/DC generator 交/直流发电机

acceleration 加速

access cover latch 口盖锁

accessory 附件

accumulation 积聚

adjust/test 调整/试验

adjustment gasket 调整垫

affect 影响

aging 老化

airadapter 空气管接头

air bleed valve 放气开关

air bleed valve housing 放气活门壳体

air flow 气流

air inlet 进气道

air pipe 空气导管

air separator 空气分离器

air volume flow 空气容积流量

air-bleed valve 放气活门

airspeed indicator 空速表

airtightness 密封性

alignment 对准

aluminum sheet 铝片

ambient temperature　外界气温

angular box　角盒

annular chamber　环形腔

annunciator　信号器

anti-icing pressure annunciator　防冰压力信号器

aperture/hole diameter　孔径

asbestos gasket　石棉垫

axis　轴线

bearing　轴承

bi-switch/two-way switch　双向开关

blade journal　叶片轴颈

blanking cap　堵头

blink/flash　闪亮

blowing hose　吹风软管

blowing pipe　吹风管

bolt　螺栓

bonding jumper　搭接线

butt bolt　对接螺栓

cabin　座舱

cable plug　电缆插头

cable-insulating layer　电缆绝缘层

cam　凸轮

cause　故障原因

caution　注意

centrer instrument panel　中央仪表板

change　变化/转换

circuit breaker　自动保护开关

circuit breaker board　自保开关板

clamp　卡箍

clearance/gap　间隙

clip　夹子

cockpit　驾驶舱

coil　线圈

coil circuit　线圈回路

combustion chamber casing　燃烧室机匣

combustion chamber/combustor　燃烧室

compensation　补偿

compress　压缩

compressed air　压缩空气

compression spring　受压弹簧

compressor control　压气机控制

connecting nut　连接螺母

connecting pipe　连接管

contact pointoff value　接点断开值

control switch　操纵开关

cool　冷却

cooling air shroud　冷却罩

cooling assembly　冷却组件

copilot console　右操纵台

copilot instrument panel　右驾驶员仪表板

correctness　正确性

cotter pin　开口销

counter-clockwise　反时针

crack　裂纹/龟裂

damage　损坏

DC power supply　直流电源

deaerating ability　抽气能力

decreasing　减少

defect　缺陷

deformation　变形

diaphragm　膜盒

disbond　脱胶

discharge　放（气）

disconnecting　拆下/断开

displacement　位移

dynamic hole　动压孔

dynamic pressure　动压

dynamic-static pressure calibrator　动静压校验器

elasticity　弹性

elbow　弯管

electric mechanism　电动机构

electric mechanism valve　电动机构活门

elimination　排除

energizing contact　通电节点

engine air intake system　发动机进气系统

engine airborne tool　发动机随机工具

engine anti-icing system　发动机防冰系统

engine high-temperature area　发动机高温部位

engine inlet hot air　发动机进气道热空气的

engine inlet icing signal system　进气道结冰信号系统开关

engine nacelle　发动机短舱

engine nacelle theoretical configuration　发房理论外形

equipment　设备

evaporate　蒸发

exhaust gas temperature　排气温度

exhaust pipe　排气管/尾喷口

extension pipe　延伸管

extinguishing　灭火

failure analysis　故障分析

fairing support plate　整流支板

fastener　紧固件

fire-warning light　火警信号灯

first calibration　初次校验

fixed nut　固定螺母

fixed screw　固定螺钉

flexible connection　软式连接

foreign object　外来物

front bearing of propeller axis　桨轴前轴承

fuel control unit　燃油调节器

gasket　垫片

ground power unit（GPU）　地面电源车

half-round head rivet　半圆头铆钉

head wind　迎面风/逆风

heading　航向

heating and air supply electric mechanism　加温供气电动机构

heating component　加温元件

heating resistance wire　加温电阻丝

hollow support plate cover　空心支板盖

horizontal axis　水平轴线

hot air duct　热空气导管

hot oil　热滑油

icing annunciator relay　结冰信号器继电器

icing detection　结冰探测

icing indicating light　结冰指示灯

icing region　结冰区域

icing signal control box　结冰信号控制盒

idling　慢车

illuminating　点亮

increasing　增大

indicating light　指示灯

injection power　引射力

inlet guide vane　进气导向器叶片

inlet pipe　进气导管

inner labyrinth　内篦齿

inspection/check　检查

installation　安装

intake port　进气孔

jam/block　卡死

jointing surface/faying surface　结合表面

lateral surface　侧面

lead screw　丝杠

lead seal　铅封

leading edge　前缘

leak　泄漏

local atmospheric condition　当地大气条件

locking block　锁块

locking wire　保险丝

loosen　拧松

main warning light　主告警灯

maintenance access cover　维护口盖

maintenance practice　维修实施

material　材料

measure　测量

melt　融化

metal bellow　金属波纹膜盒

microswitch　微动开关

middle-rear bearing　中后轴承

motor　电机

motor ventilating hood　电机通风罩

mounting base　安装座

mounting flange　安装边

movement　移动

multimeter　万用表/三用表

negative pole　负极

nomenclature　名称

non-metal part　非金属件

normal operation　正常工作

nozzle guide vane　涡轮导向叶片

nut gear　螺母齿轮

Ohmmeter　欧姆表

oil eject-radiating electric valve　滑油引射散热电动活门

oil inlet temperature　滑油进油温度

oil line　滑油道

oil lubrication system　滑油润滑系统

oil mist separator　油雾分离器

oil radiator　滑油散热器

oil sealing device　滑油封严装置

operating revolutions per minute　工作转速

operation status　工作状态

original position　原始位置

outlet tube　出气管

output shaft　输出轴

part　零件

pin　插销/插针

pipe　导管

pipe joint/union　管接头

piston　活塞

piston ring　涨圈

plug　插头

plug point　插钉

positive pole　正极

power　功率

power on contact point　通电接触点

power supply changeover switch　电源转换开关

precaution　预防

preparations　准备工作

press　按压

pressure bleed valve　放压开关

pressure signal sensor　信号压力传感器

pressure-test hose　测压软管

procedure　步骤

propeller brake　螺旋桨制动

quantity　数量

radial groove　径向槽

radiation　散热

radiator shutter　散热器风门

rated status　额定状态

recover/restore　恢复

regulating screw　调节螺钉

regulating shim　调整垫片

relay　继电器

release　放开

reliability　可靠性

remark　备注

removal　拆卸/去除

replacement　更换

resistor　电阻器

restricting orifice　限流孔

retraction　收起

return　返回

return pipe　回流管

revolutions per minute　转速

rib　肋

rib-sealing cavity　肋形封严腔

rotating　转动

round nut　圆螺母

rubber bellow/corrugated tube　橡胶波纹管

rubber pad　胶垫

rubber ring　胶圈

rust　锈蚀

safety nut　保险螺母

scavenge stage of main oil pump　主滑油泵回油级

schematic diagram　原理图

screw　螺钉

screw rod　螺杆

sealing　封严

sealing cavity　封严腔

sealing ring　封严圈

seat disk　座盘

sensitive area　受感区域

sensitive element　敏感元件

sensor　传感器

serial number（S/N）　序号

service ladder　工作梯

shaft end long groove　轴端长槽

shaft end section/shaft end surface　轴端面

shake　摇动

signal circuit　信号电路

signal light　信号灯

sleeve　套管

smoke　烟

socket/receptacle　插座

spare part　备件

speed pressure/dynamic pressure　速压

speed reducer　减速器

spring seat disc　弹簧座盘

spring stop rod assembly　弹簧限动杆组件

spring tab　弹簧片

spring washer　弹簧垫圈

start　（发动机)开车/起动

static chamber　静压室

static port　静压孔

static pressure　静压

stop pad　止推盘

strut　支板

stud　螺桩

surg　喘振

suspension　悬挂

symmetry axis　对称轴线

symptom　故障现象

tail wind　顺风

takeoff and maximum operating status　起飞和最大工作状态

taxi　滑行

temperature sensing probe　感温棒

tenth-stage compressor　第十级压气机

terminal switch　终点开关

test equipment　试验设备

throttle　油门

throttle nozzle/restricting nozzle　节流嘴

throttle washer　节流垫

tightening moment　拧紧力矩

tool　工具

torque automatic feathering sensor　扭矩自动顺桨传感器

total-pressure probe　总压测头

transmission air-sealing cavity　传动封严腔

travel　行程

troubleshooting　排除方法

tube root heater　管根加热器

turbine blade　涡轮工作叶片

turbine blowing cover cooling air inlet　涡轮吹风罩冷却进气口

turbine casing　涡轮机匣

turbine disk　涡轮盘

turbine roller bearing　涡轮滚子轴承

type No.　型号

valve closing rotating speed　关闭转速阀门

valve link　活门连杆

valve shaft　活门轴

ventilation　通风

ventilation pipe　通气管

ventilation system　通气系统

vibration　振动

warm-up　暖机

warning　警告

welding part/weld assembly　焊接件

welding seam　焊缝

wheel chock　轮挡

wing anti-icing valve　机翼防冰活门

wrench　扳手

yellow sealing pound/yellow adhesive sealant　黄色封口胶

ENGINE CONTROLS
发动机操纵系统

acceleration control　加速控制

air idling　空中慢车

airflow control　空气流量控制

annular potentiometer　环形电位器

augmentation control/boost control　加力控制

automatic control system　自动控制系统

automatic regulating system　自动调节系统

automatic shutdown　自动停车

automatic temperature control system　自动温度控制系统

axial clearance　轴向间隙

blockage　挡块

braking mechanism　制动机构

comb plate　梳状板

control stand　操纵台

control surface lock　舵面锁

controlled object　被控对象

crank arm　曲臂

crosswind　侧风

digital-analog hybrid computer　数-模混合计算机

dirty fuel test　脏油试验

distributing plunger　分流柱塞

drain cup　漏油杯

dust-proof cover　防尘盖

dynamic test　动态试验

elevator　升降舵

emergency feathering switch　应急顺桨开关

emergency shutdown　紧急停车

engine close-loop control　发动机闭环控制

engine control rob　发动机操纵拉杆

engine electronic control　发动机电子控制

engine emergency shutdown system　发动机应急停车系统

engine open-loop control　发动机开环控制

engine power control system　发动机功率操纵系统

engine vibration monitoring system　发动机振动监视系统

fan pressure ratio control　风扇压比控制

fixed screw　固定钉

flap control switch　襟翼收放开关

flow divider　分流活门

fluorescent light　荧光灯

formation light　编队灯

friction brake piece　摩擦制动片

full electronic control system　全电子控制系统

hand wheel disk　手轮盘

hydraulic control　液压操纵

hydraulic pump cart　液压油泵车

hydromechanical control system　液压机械式控制系统

integrated auto throttle servo mechanism　综合自动油门伺服机构

landing gear up/down switch　起落架收放开关

locking piece　锁紧片

magnet rotor　磁铁转子

main engine control　主发操纵

manual shutdown　人工停车

mechanical control　机械操纵

mixed electronic-hydromachenical control system　电子-液压机械混合式控制系统

piston rod　活塞杆

placard　标牌

pneumatic control　气动操纵

power control　动力操纵

power lever　油门杆

product quality test　产品质量试验

push rod　推杆

radial clearance　径向间隙

rectangle joint　直角接头

regulated parameter　被调参数

reinforced pad　加强垫

remote control　遥控

restrictor　节流活门

return line　回油管路

rheostat　变阻器

rolling sleeve　滚套

slide valve　滑阀

speed control　转速控制

stall　失速

start control drum　启动操纵钢索鼓轮

start control push-pull cable　启动操纵推拉钢索

three-way joint/three-port joint　三通接头

throttle control handle　油门操纵手柄

trim tab　调整片

turbine temperature control　涡轮温度控制

variable geometric control　可变几何形状控制

warning signal siren　警告信号鸣响

ENGINE INDICATING　发动机指示

access cover　口盖

accuracy　精度

actual value　实际值

add　增加

additional error　附加误差

additional resistance　附加电阻

adjusting resistance　调整电阻

adjustment　调整

aircrew　空勤

alignment　对准

allowable error　允许误差

alternating current（AC）　交流电

altitude　高度

ambient temperature　环境温度

amplifier　放大器

amplifier transformer　放大变压器

analysis system　分析系统

area/channel　区域/通道

assembly　零组件

attenuate　衰减

automatic fuel-reducing device　自动切油装置

axial clearance　轴向间隙

balance　平衡

bearing　轴承

bolt　螺栓

bonding jumper　搭铁线

bronze alloy　青铜合金

buffer　缓冲器

bulb　灯泡

bushing　衬套

cable plug　电缆插头

calibration　校验

capacitance　电容

carbon　碳

cause　故障原因

center instrument panel　中央仪表板

check point　检查点

circuit　电路

circuit breaker board　开保关板

clamp　卡箍

coil　线圈

coil resistance　线框电阻

coil winding　线圈绕组

collection　采集

compensating wire　补偿导线

compensating wire bundle　补偿导线束

conformance　符合

conformity　符合度

constant　常数

contact resistance　接触电阻

control circuit　控制电路

converter　变换器

cover plate　盖板

crack　裂纹

current consumption　消耗电流

damp time　阻尼时间

deceleration　减速

deceleration ratio　减速比

deflection　偏转

deformation　变形

delay error　迟滞误差

deterioration　（绝缘电阻）变坏

deviation　偏差

diaphragm　膜片

direct current（DC）　直流电

direct proportion　正比例

direction　方向

displacement　位移

distinguish　区别

distribution　分布

double T frequency selective network　双T型选频网络

electrical part　电子器件

electromagnetic sensitive element　电磁式敏感元件

electromotive force　电动势

electronic tube　电子管

elimination　排除

engine exhaust gas thermometer　发动机排气温度表

engine first running　发动机初次试车

engine nacelle　发动机短舱

engine tachometer　发动机转速表

engine vibration G-load indicator　发动机振动过载表

equipment name　设备名称

equipment rack　设备架

error　误差

extending connecting wire　延接线

external resistance　外线电阻

failure analysis　故障分析

fastening nut　紧固螺母

filament circuit　灯丝电路

fixing iron core　固定铁芯

fixing sensor　固定传感器

flange　框缘

floor　地板

floor access cover　地板口盖

fluid　流体

fragile element　易损元件

frame　框

free end　自由端

frequency　频率

friction resistance　摩擦阻力

fungus　霉菌

fuse　保险丝

gap　缺口

gasket/spacer　垫片

gear　齿轮

germanium rectifier　锗整流器

glass fiber　玻璃纤维

hairspring　游丝

hairspring resistance　游丝电阻

humidity　湿度

illumination　燃亮

indicating scale　指示刻度

indicator　指示器

induction coil　感应线圈

induction current　感应电流

induction electromotive force　感应电动势

induction sheet　感应片

inductive resistance　感抗

inlet　入口

input　输入

installation　安装

installation angle　安装角

installation position　安装位置

instantaneous electric break　瞬时断电

insulation material　绝缘材料

insulation resistance　绝缘电阻

inverse proportion　反比例

iron-nickel alloy　铁镍合金

junction surface　对接面

lamp holder 灯座

lead-out end 引出端

locating pin 定位销

locking wire 锁紧丝

low air pressure tolerance 耐低气压要求

lug 接线片

magnet assembly 磁铁组合件

magnetic circuit system 磁路系统

magnetic delay disc 磁带圆盘

magnetic pole 磁极

magnetism 磁性

main shaft 主轴

measure 测量

measuring range 测量范围

mechanic adjusting value 机械指标调定值

mechanical property 机械性能

metal chip 金属屑

metal disc 金属盘

meter connecting wire 接表线

meter test 试表

millivoltmeter 毫伏计

moisture 潮气

monitor 监视

monostable circuit 单稳态电路

mounting bracket 安装架

movable armature 活动衔铁

movable coil 活动线圈

movable system 活动系统

nozzle 喷嘴

oil pressure 滑油压力

operating principle 工作原理

operating range 工作范围

operating temperature 工作温度

operation altitude 使用高度

operation state and parameter 工作状态和参数

output 输出

oven 烘箱

over temperature protection 超温保护

percentage 百分比

permanent magnet 永久磁铁

phase voltage 相间电压

pin 插销

platform 平台

plug 插头

pointer 指针

positive pulse signal 正脉冲信号

positive/negative error 正/负误差

positive/negative pole 正/负极

positive/negative travel 正反行程

power 功率

power source 电源

power spike 尖峰电源

power supply 供电

power supply transformer 电源变压器

preparations 准备工作

pressure measuring range 压力测量范围

primary-standby power supply changeover switch 主备电源转换开关

procedure 程序

pulse current 脉动电流

quantity 数量

rain protection 耐淋雨要求

raster circuit 屏板电路

raster voltage 屏压

rated voltage 额定电压

relay 继电器

remark 备注

removal 拆卸

remove 去除

repair 修复

replacement 调换

resonance operating frequency 共振工作频率

revolutions per minute 转速

rotating magnet group 转动磁铁组

rotation angle 旋转角度

rotation magnetic field 旋转磁场

rotation moment　转动力矩

rotation speed signal converter　转速信号转换器

rotation speed transfinite fuel-reducing　转速超限切油

rotor　转子

sandwich　夹层

scale　刻度

schematic diagram　原理图

screwdriver　螺刀

secondary winding　次级绕组

sensitivity　灵敏度

sensor　传感器

serial number（S/N）　序号

series connection　串联

signal light　信号灯

silicon rubber sleeve　硅橡胶套管

sleeve coupling nut　外套螺母

specified value　规定数值

spring　弹簧

stator　静子

stator coil　定子线圈

storage temperature　贮存温度

symptom　故障现象

synchronous motor　同步电机

technical data　技术数据

temperature compensating line　温度补偿线路

temperature difference　温差

temperature system　温度系统

terminal　接线端

terminal board　接线板

terminal box　接线盒

thermistor　热敏电阻

thermocouple　热电偶

thermo-joint/hot junction　热接点

three-phase AC generator　三相交流发电机

tolerance of acceleration　耐加速度要求

tolerance of vibration　耐振动要求

torque meter　扭转力矩表

transfer　传递

transmission gear　传动齿轮

transmission shaft　传动轴

transmission system　传动系统

triangular mark　三角指标

trouble shooting　排故方法

true value　真实值

two-pointer tachometer　双针指示器

type　牌号

type No.　型号

uniformity　均匀性

variable resistor　可变电阻

vibration amplifier　振动放大器

vibration sensor　振动传感器

voltage regulator tube　稳压管

warning　警告

warning light　警告灯

washer　垫圈

weld　焊接

weldspot　焊点

whole set　全机

zero position　零位

EXHAUST　排气系统

adjustable nozzle　可调节式尾喷管

aero-washing gasoline　航空洗涤汽油

aluminum-color heat-resistant paint　铝色耐热漆

annular expanding channel　环形扩散通道

argon arc welding　氩弧焊

baffle　挡板

bifurcation assy　反推装置双叉管组件

blocker door　反推装置折流门

box-shaped piece　盒形件

bullet nozzle　有整流锥喷管

butt flange edge　对接法兰边

butt step difference　对接阶差

cascade support ring　叶栅支承环

cascade vane　反推装置叶栅

cascade vane segment　叶栅扇形体

clamp　箍圈/卡圈

clamshell　蚌壳式折流门

clamshell-type thrust reverser　蚌壳型反推
装置(双折流板式)

collector ring　集流环

conical-wall nozzle　锥形喷管

convergent nozzle　收敛喷管

convergent-divergent nozzle　收敛-扩散喷管

crack-arrest hole　止裂孔

curling　卷边

deflecting nozzle　偏转喷管

diagonal　对角线

divergent nozzle　扩散形喷管

drag link　反推装置牵引连杆

ejector nozzle　引射喷管

elbow　弯管

exhaust case　排气机匣

exhaust gas system　排气系统

exhaust gas temperature sensor　排气温度传
感器

exhaust jet test of simulation external stream
尾喷管外流干扰试验

exhaust manifold　排气集气筒

exhaust nozzle test　尾喷管试验

exhaust pipe　排气管

extension pipe　延伸管

fan exhaust　风扇排气

fan thrust reverse　风扇函道反推装置

fuse (electric)　保险丝(电)

graphite lime grease　石墨钙基润滑脂

graphite lithium grease　石墨锂基润滑脂

heat shroud　隔热屏

hoop　箍带

horizontal axis line　水平轴线

hydraulic deploy hose　液压开伞软管

included angle　夹角

inner cone　内锥体

installation edge/mounting edge　安装边

installation seat/mounting base　安装座

integral thrust reverser　整体式反推装置

internal stream characteristic test of exhaust
nozzle　尾喷管内流试验

jet cone/jet bullet　喷管整流锥

jet exhaust excess impulse　排气富裕冲量

jet exhaust impulse　排气冲量

knurling/score coining　压花

matching of exhaust nozzle and aircraft　尾
喷管与飞机的匹配

matching of exhaust nozzle and engine　尾喷
管与发动机的匹配

neutral soapsuds　中性肥皂水

noise suppressor　消音匣

nozzle base drag　喷管的底阻

nozzle efficiency　喷管效率

nozzle expansion ratio　喷管膨胀比

nozzle thrust coefficient　喷管推力系数

nozzle thrust ratio　喷管推力比

oil vapor separator　油雾分离器

outlet guide vanes　出口导流(向)叶片

plug nozzle　塞式喷管

point welding　点焊

pressure regulating port　调压口

quick-release ring　快卸环

radial difference　径向偏差

residual fuel pipe　余油管

reverse thrust lever　反推操纵手柄

reverser actuator　反向器开伞

reverser deploy　反向器致动器

reverser throttle interlock　反向器风口内锁

reverser unlocked switch　反向器松锁开关

reverse-thrust nozzle　反推力喷管

rolled weld　滚焊

safety piece　保险片

sealing cavity　封严腔

semi-pipe　半管

shield 护罩

shroud 外罩

sound-suppression nozzle 消音喷管

stiffened edge 加强边

stop bushing 限动衬套

subsonic nozzle 亚音速喷管

supersonic nozzle 超音速喷管

supplementary air 补充空气

support piece 支撑件

support plate 支板

support plate seat 支板座

swiveling nozzle 转向喷管

synchronized actuator 同步致动器

synchronizing flexible shaft 同步软轴

T/R control position sensor 反推装置控制位置传感器

T/R feedback push-pull cable 反推装置回授推拉钢索

T/R indicating position sensor 反推装置指示位置传感器

tail cone 尾部锥体

tail extension pipe 尾喷管延伸管

tail nozzle 尾喷口

tailpipe 尾喷管

tailpipe nozzle 尾喷管

tailpipe thrust reverser 尾喷管反推装置

target-type thrust reverser 折流板式反推装置

tear 撕裂

tension ferrule 拉紧箍圈

tertiary-air door 第三股空气门

theoretical clearance 理论间隙

thermal load 热负荷

thermocouple 热电偶

thrust increase of exhaust nozzle 尾喷管的推力增益

thrust reverser control valve 反推装置控制活门

thrust reverser torque box 反推装置扭矩盒

thrust reversers（T/R） 反推装置

translating sleeve 反推装置移动套

turbine blowing shroud 涡轮吹风罩

turbine casing 涡轮机匣

turbine exhaust 涡轮排气

turbine exhaust cone（plug） 涡轮排气整流锥

turbine thrust reverser 内函道反推装置

two-stream nozzle 双流道(气动调节)喷管

vertical axis line 垂直轴线

warp 翘曲

OIL 滑油系统

absolute value circuit 绝对值电路

accessory casing oil filter 附件机匣滑油滤

accessory gear（drive）box 附件传动机匣

acid value 酸值

adapter 管接头/接管嘴

adjusting screw 调节螺钉

aero oil 航空滑油

air compressor 压气机

air inlet 进气口

air intake 进气道

air separator 空气分离器

air tightness test 气密试验

air-oil separator 油气分离器

air-sealing pressure 空气密封压力

allowable temperature 允许温度

aluminum alloy plate 铝合金板材

aluminum powder organic silicon heat-resistant paint 铝粉有机硅耐热漆

anti-negative G-load partition 抗负过载隔板

anti-siphon tube 防虹吸管

asbestos gasket 石棉垫

ash content 灰份

auto-regulating system 自动调节系统

auxiliary oil pump 辅助滑油泵

axial clearance 轴向间隙

axial load 轴向荷载

backup oil quantity 备用油量

ball-shaped pipe nipple 球形管接嘴

bi-directional series motor 双向串激式电动机

bite switch 机内测试开关

blade-filter 滤片

bonding jumper/bounding wire 搭接线

bonding strip 搭接片

boosting stage pump 增压级油泵

bow-shaped clamp 弓形夹

brass 黄铜

bridge arm resistance 桥臂电阻

bronze base 青铜座

buffer 缓冲器

built-in-test equipment 机内测试设备

bypass valve 旁通活门

cam 凸轮

capacity 容积

casing 机匣

center instrument panel 中央仪表板

centrifugal oil-mist separator 离心式油雾分离器

centrifugalization mechanism 离心分离原理

check ring 挡圈

check valve 单向活门

circuit breaker 自动保护开关

circuit ground check 地面检查线路

clamp locking wire 卡箍保险

clearance 间隙

clogging indicator 阻塞指示器

closed loop 封闭环路

combustion casing 燃烧室机匣

communicator circuit breaker board 通讯员自保开关板

complex hydrocarbon aviation lubricating oil 合成烃航空润滑油

concave pulley 凹轮

conducting piece 导电片

cone cylinder 锥形筒

connecting bolt 连接螺栓

connecting disc 结合盘

connecting pipe 连接导管

constant-speed drive oil 恒速传动装置滑油

cooler 冷却器

cooling oil 冷却滑油

corrosion preventive oil 防腐滑油

corrosion proofing layer 防腐层

cotter pin 开口销

cover plate 盖板

cup-shape cylinder 杯形筒

curve scale 弧线刻度

damping time 阻尼时间

DC bridge 直流电桥

DC motor 直流电动机

deep scrape 深擦痕

delay 迟滞

dent 压痕

depreservation 启封

destroying of pressure 破坏压力

dial 刻度盘

digital scale 数字刻度

dip stick 油量尺

disc valve 盘形活门

drain cock 放油开关

drain pipe 漏油管

drain plug/oil draining 放油塞

drain tube 排泄管

drive shaft 转动轴

driving box 传动盒

driving pull rod 传动拉杆

dual-coil logometer 双线圈电流比计

ejection radiation valve 引射散热活门

elastic washer 弹簧垫片

elbow outlet 弯管出口

electric mechanism 电动机构

electrical plug 电插头

electrical seal cavity 电气密封腔

element damage　元件损坏

engine airborne kit　发动机随机工具

engine auxiliary mounting bracket　发动机本体辅助安装支架

engine bracing　发动机撑杆

engine I capacitor　I 发电容器

engine I fuse　I 发熔断器

engine lifting service ladder　发动机升降工作梯

engine lower cover　发房下盖

engine lubrication　发动机润滑

engine nacelle　发动机短舱

engine oil　发动机润滑油

engine running-up　发动机试车

engine service ladder　发动机工作梯

engine shutdown　发动机停车

engine speed reducer　发动机减速器

engine warmup　发动机并暖机

exhaust device　排气装置

exhaust duct　排气风道

exhaust pipe　尾喷口

extension bar　加长杆

fabric hose/laminated fabric/rubber hose　夹布胶管

failure analysis　故障分析

fastening bolt　紧固螺栓

fastening screw　紧固螺杆

fault part　故障件

feathering pump　顺浆泵

feathering pump oil supply pipe joint　顺浆泵供油接头

feed pipe　供油管

feedback resistance　反馈电阻

filler cap　加油口盖

film hard core　膜片硬心

filter　油滤

filter cartridge　滤芯

filter cover/cap　油滤盖

filter screen　滤网

fine-pitch pipe line pressure annunciator　小距油路压力信号器

fixed pitch pipeline　定距油路

fixing nut　固定螺母

fixing strip　固定带

fixing support　固定支座

flash point　闪点

flat-head axle　平头轴

float-type oil indicator　浮子式油量表

freezing point　凝固点

fuel control unit　燃油调节器

full scale　刻度盘满量程

gasket/spacer　垫片

general distribution box　总配电盒

ground run test　地面试车

heat collector　集热器

heat exchanger　热交换器

hollow bolt　空心螺栓

honeycomb tube　蜂巢管

hose　软管

idling　慢车

impression mark　压痕

inertia displacement　惯性滑移量

inner oil vent pipe　内部通气管

insensitive zone　非敏感区域

insulation resistance　绝缘电阻

insulator　绝缘子

interchangeability　互换性

kinematic viscosity　运动黏度

lead sealing block　铅封块

left engine bracing　发动机左侧撑杆

left view　左侧视图

limit position　极限位置

limit switch/terminal switch　终点开关

link rod　连杆

lock valve　闭锁活门

loose weld　脱焊

low oil pressure light　低滑油压力指示灯

low oil pressure warning　低滑油压力告警

lower cover 下盖

lubrication unit 润滑装置

magnetic chip detector 磁碎屑探测器

magnetic plug 磁塞(吸除金属屑用)

main oil pump 主滑油泵

main oil pump boosting stage 主滑油泵增压级

manual regulating system 人工调节系统

mechanical impurity 机械杂质

mica polar plate 云母极板

mica sheet 云母片

microswitch 微动开关

mounting seat 安装座

mounting surface 安装面

multimeter 三用表

negative thrust auto-feathering annunciator
负拉力自动顺桨压力信号器

negative wire 负线

net weight 净重

neutral position 中立位置

nickel enamel wire 镍漆包线

oil accumulation 滑油堆积

oil bag 油兜

oil barrel 油桶

oil consumption 滑油消耗量

oil control system 滑油控制系统

oil deflector/oil retainer 挡油圈/导油器

oil distribution system 滑油分配系统

oil drain hose 放滑油软管

oil drain screw plug 放油螺塞

oil drip tray 接油盘

oil ejection radiation system 滑油引射散热系统

oil filter bypass light 油滤旁通指示灯

oil filter bypass warning 滑油油滤旁通告警

oil filter differential switch 滑油油滤差动开关

oil gauge sensor 油量表传感器

oil indication 滑油指示

oil inlet hole 进油孔

oil leakage 漏油

oil mist 油雾

oil passage orifice 通过滑油窗孔

oil plumbing 滑油管路

oil pressure 油压

oil pressure gauge 滑油压力表

oil pressure indicator 滑油压力指示器

oil pressure pump 滑油增压泵

oil pressure transmitter 滑油压力传感器

oil pump assembly 滑油泵组合件

oil quality 油质

oil quantity 油量

oil quantity gauge 滑油油量表

oil quantity transmitter 滑油油量传感器

oil radiating efficiency 滑油散热效率

oil radiating ejector 滑油散热引射器

oil radiator 滑油散热器

oil scavenge cavity 回油腔

oil scavenge tube 滑油回油泵

oil seepage 渗油

oil separator 滑油分离器

oil suction pipe 吸油管

oil sump 收油池

oil supply 滑油供给

oil supply duct 滑油供油管

oil supply pump pipe 供油泵导管

oil system 滑油系统

oil system indicating 滑油系统指示

oil tank antisiphon device 油箱防虹吸装置

oil tank inner casing 油箱内侧壳体

oil tank sediment 滑油箱沉淀物

oil tank/oil storage 滑油箱

oil temperature auto-regulating circuit 滑油温度自动调节电路

oil temperature control box 滑油温控盒

oil temperature indicator 滑油温度指示器

oil temperature regulating system 滑油温度

调节系统

oil temperature regulator　滑油温度调节器

oil trunk　滑油加油车

oil turn pipe adapter　回油管接头

oil vent tank　通气油箱

oil viscosity　滑油粘度

oil-in cavity　进油腔

oil-mist separator　油雾分离器

open circuit　开路

opening angle　打开角度

operating altitude　工作高度

operating medium　工作介质

operating pressure　工作压力

operating stroke　工作行程

operating temperature　工作温度

orifice　窗孔

outlet pressure　出口压力

parallel resistance　并联电阻

partition/bulkhead　隔板

performance　性能

pilot center instrument panel　驾驶员中央仪表板

pipe line　管路

piston ring　涨圈

pointer travel uniformity　指针行程均匀性

polarity comparator　极性比较器

position indicator　位置指示器

positive electricity　正电

potentiometer　电位计

power amplifier　功率放大器

power consumption　功率消耗

power voltage stabilizer　电源稳压器

power-on check　通电检查

prerequisite　先决条件

pressure annunciator　压力信号器

pressure filter　压力油滤

pressure relief valve　释压活门

pressure-regulating valve　调压活门

propeller feathering　螺旋桨顺桨

protective cover　保护罩

pulse drive　脉冲驱动

pulse-width modulator　脉宽调节器

quick-release clamp　快卸卡箍

radiating area　散热面积

radiating oil line　散热油路

radiating tube　散热管

radiating tube quantity　散热管总数

radiator　散热器

rated load　额定负载

rear bearing　后轴承

rear turbine-bearing oil sump　后轴承收油池

refueling port　加油口

regulating bearing　调节轴承

reinforced rod　加强杆

relay control circuit　继电器控制电路

remote filling port　外式注油口

remote overflow port　外式溢流口

repeat short-time duty　重复短时工作制

residual oil quantity　剩余油量

residual oil quantity signal　剩油信号

resistance box　电阻箱

restrictor　节流活门

rib-sealing cavity　肋形封严腔

rocker arm　摇臂

rotary joint　转动接头

rotary magnet　旋转磁铁

rubber asbestos gasket　橡胶石棉垫

rubber pad　胶垫

rubber pressing strip　橡胶压条

rubber strip　橡皮带

sawtooth generator　锯齿波发生器

scale beginning point　刻度始点

scavenge filter/oil return filter　回油滤

scavenge pump　回油泵

scavenge stage　回油级

screwdriver　螺丝刀

sealing gasket　封严垫片

sealing pad　封严垫

sealing ring　封严圈/密封圈

sensitive zone　敏感区

sensitivity　灵敏度

sensor casing　传感器壳体

series resistance　串联电阻

service life　使用寿命

short-circuit closed type　短路闭合式

shutter position indicator　风门位置指示器

single stage gear oil pump　单级齿轮式油泵

single stage high pressure gear pump　单级齿轮式高压油泵

soapsuds　肥皂水

specific location　指定位置

speed governor　调速器

speed reducer　减速器

speed reducer part　减速器零件

spline　花键

spongy washer　海绵垫圈

spring lock valve　弹簧闭锁活门

spring sheet　弹簧片

spring washer　弹性垫圈

stick-core　芯杆

stop ring　制动圈

storage system　贮存系统

stud　螺桩

supply pump　供油泵

synthetical magnetic field　合成磁场

temperature control box　温度控制盒

temperature indicator　温度指示器

temperature sensing element　感温元件

temperature sensing resistance　感温电阻

terminal box　接线盒

thermal resistance　热敏电阻

threaded rod　螺杆

three-position switch　三位开关

throttle pull rod　油门拉杆

throttling washer　节流垫圈

tightening torque　拧紧力矩

torque measuring pump　转矩测量泵

total resistance　全电阻

turbine lubricating oil　涡轮润滑油

turbine-bearing cavity　涡轮轴承腔

turnbuckle rod　螺套拉杆

two-stage gear oil pump　两级齿轮式油泵

valve plate　活门板

variable resistance　可变电阻

vent pipe　通气管

vent pressure　通气压力

vent tube　通风管

venting system　通风系统

view A　A向

water-soluble acid and soda　水溶性酸和碱

welded seam　焊缝

wheel chock　轮挡

white silk cloth　白绸布

white smoke　白烟

windward area　迎风面积

wire frame　线框

wrench/spanner　扳手

SPECIAL ELECTRONIC EQUIPMENT
特种电子设备

air data computer　大气数据计算机

air identification interrogation signal　对空识别询问信号

airborne transponder　机载应答机

altitude interface signal　高度接口信号

antenna feeder　天馈系统

antenna selection switch　天线选择开关

brightness adjustment knob　亮度调节按钮

confidential computer module　保密计算机模块

confidentiality operating mode/security operating mode　保密工作模式

crosslinking relationship　交联关系

cursor　光标

decoding　译码

distance measuring equipment　测距器

electromagnetic compatibility　电磁兼容性

field tester　外场检查仪

fuse　保险管

identification of friend or foe system　敌我识别系统

identity identification　身份识别

initial position　初始位置

interfacingrelationship　接口关系

key　密钥

key loader　密钥加载器

knurled nut　滚花螺母

level　电平

locking signal　闭锁信号

M1-mode code display　M1 模式代码显示

main control processing module　主控处理模块

mainframe　主机

military aircraft security identification mode　军机安全识别模式

mode selection knob　模式选择按钮

nixie tube　数码管

operating mode indication　工作模式指示

overcurrent　过电流

overvoltage　过电压

power filtering box　电源滤波盒

power supply network　供电电网

power-off state　关机状态

power-on test　加电检测

quick-release screw　快卸螺钉

radio frequency（RF）signal　射频信号

radio frequency switch module　射频开关模块

receiver　接收机

response light　应答灯

response mode switchover knob　应答模式切换开关

response probability　应答概率数值

response signal　应答信号

sea level　海平面

self-test processing module　自检处理模块

serial port　串口

short circuit　短路

side-lobe pulse　旁瓣脉冲

signal crosslinking box　信号交联盒

signal processing module　信号处理模块

special identification button　特殊识别开关

special position replay　特殊位置应答

standby interface　待机界面

sub-unit　分机

support cable　配套电缆

TACAN　塔康

target code　目标代码

toggle switch　拨码开关

touch screen　触摸式显示屏

transceiver　收发机

transceiver panel　收发机面板

trans-illuminated panel　导光板

transmitted power　发射功率

transmitting unit　发射单元

troubleshooting　故障处理

wrapping paper　包装纸

FREIGHT SYSTEM　货运系统

actual weight　实际重量

adjustable hook　调节钩

adjustable mechanism　调节机构

aerial delivery　空运

aerovan　运货飞机

after cargo compartment　后货仓

after locking device　后向锁装置

air cargo　空运的货物

air cargo system　空运系统

air express　空运包裹

air freight　航空货运

air transportability　航空运输性

aircraft container　飞机集装箱

aircraft on-board weight　飞机载货重量

aircraft pallet　飞机集装板

aircraft pallet net　飞机集装板网

aircraftunit load device　飞机集装单元

airdrop　空投空降

airdrop platform　空投空降平台

airdrop system　空投空降系统

airdrop weight　空投空降重量

airworthiness　适航性/飞行性能

all-freighter aircraft　全货运型飞机

allowable load　许用载荷

auxiliary wheel　辅助轮

axle load　轴载荷

balance system　平衡系统

ball transfer panel　滚珠传输板

ball transfer unit　滚珠传输装置

barrier net　（货物）阻拦网

bill of lading　提(货)单

brake shoe　制动器

bulk cargo　散装货物

bulk cargo door　散装货舱门

bulk compartment　散装货舱

bulkhead　隔板

cargo　货物

cargo aircraft　货机

cargo aperture　货物开度

cargo compartment　货舱

cargo conversion　货运改型

cargo conveyance system　货物传输系统

cargo envelope　货物包线

cargo handling system　货物装卸系统

cargo hold　货舱

cargo loading char　货物装载图表

cargo parachute　投物伞

cargo restrain net　货物限制网

cargo tiedown track　货物系留滑轨

cargo track　货运地轨

caster assembly　角轮装置

caster tray　角轮托盘

ceiling panel　天花板

center guide/restraint　中央导轨/限动器

center of gravity　重心

center damper　中央阻尼器

certificated ULD　合格审定的集装单元

class I aircraft loading and restraint system
飞机装载 I 级限动系统

class II aircraft loading and restraint system
飞机装载 II 级限动系统

clearance limit　余隙限度/净空(界)限

combi　客货两用运输机

crash condition　冲击条件

crew member　机组成员

date of delivery　交货日期

deck angle　地板角

decompression panel　减压板

deffector rail　导向滑轨

deflection　挠度

delivery order　交货单/提货单

delivery receipt　运货回单

delivery time　交货时间

door barrier　门栅栏/门保护网

door sill　门槛

door sill latch　门槛锁

door sill restraint　门槛限动装置

double pallet lock　双向锁扣

drainage　排水设备

dry　保持干燥

elevator　升降机

elevator-type loader　升降型装载机

emergency access　应急通道

end stop　末端止动器

end stop assembly　末端止动装置

external dimension　外部尺寸

external volume　外部体积

extraction parachute　牵引伞

fitting　连接件

fixed end stop　固定端止动器

fixed guide　固定导轨

flame resistant　耐火性

floor fitting　地板接头

floor loading　地板载荷

folding power drive unit　可收缩动力驱动
装置

folding side restraint　可收缩侧限动装置

fore locking device　前向锁装置

forklift capability　叉举能力

forklift pocket　叉槽

freight　运费

full size container　全尺寸集装箱

full swiveling caster　万向脚轮

general cargo　普通货物

general list　总清单

general purpose freight　通用集装箱

G-force　过载

gravity drop　重力空投

gross for net　以毛作净

gross rigged weight　集装总重量

gross weight　总重

ground equipment　地面设备

ground handling equipment　地面装卸设备

guide rail（assy）　导向滑轨（组件）

half size container　半尺寸集装箱

handle with care　小心装卸

haul　起吊点

hazardous cargo　危险货物

heated container　加热集装箱

height overall　停机高度

high velocity airdrop　高速空投

hiosting gear　提升绞车

igloo　圆拱结构

impact load　冲击载荷

inflammable　易燃性

inside diameter　内径

inside dimension　内部尺寸

insulated container　绝热集装箱

inter modal　联运

interchangeability　互换性

internal tiedown　箱内系留

jack　千斤顶

label　标签

lashing ring　系留环

lateral guide　横向导轨

length of fuselage　机身长度

limit load　使用载荷（限制载荷）

load factor　载荷系数

load manifest　（运输机）货物清单

loader　装载机/装货机

loading　载荷/载重

loading system　载荷系统

low velocity airdrop　低速空投

low velocity platform airdrop　低速平台空投

lower cargo compartment　下货舱

lower deck　下货舱

lower deck container　下舱集装箱

main cargo compartment　主货舱

main deck　主货舱

maintainability　维修性

manual loading　人工装载

manual unloading　人工卸载

marking　标记

maximum capacity　最大载货能力

maximum gross weight　最大总重量

maximum payload　最大商载重

mid cargo compartment　中间货舱

minimum clearance　最小净空间

minimum gross weight　最小总重量

named cargo types container　货物命名的集
装箱

net mesh　系留网格

net weight　净重

nominal dimension　公称尺寸

norm　标准

omni-directional caster　万向角轮

out board guide/restraint rail　外侧导向限动
滑轨

outside diameter　外径

outside dimension　外部尺寸

outsized cargo　特大尺寸货物

over charge　超载

packing list　装箱细目表

pallet　集装板

pallet attachment fitting　集装板连接件

pallet lock　集装板锁

pallet net　集装板网

passenger/cargo version　客货混装型

payload　载重/货载质量

platform　平台

platform load　平面载荷

port of destination　目的港

port of embarkation　发航港

power drive unit　动力驱动装置

powered loading　动力装载

powered unloading　动力卸装载

quick-release attachment　快卸装置

ramp　货桥

rear cargo compartment　后货桥

refrigerated container　冷藏集装箱

releasing mechanism　释放机构

reliability　可靠性

restraint system　限动系统

retractable end restraint　可收缩末端限动装置

retractable guide restraint　可收缩导向限动装置

retractable guide roller　可收缩导向滚棒

rigger load　集装载荷

roller track　滚道地轨

roller tray assembly　滚棒托盘装置

rollout stop　延伸止动器/门槛限动装置

seat track　座椅滑轨

separation net　隔离网

shipping mark　发货标记

shipping weight　出运重量

side guide　侧导轨

side lock　侧面锁

side wall panel　侧壁板

sill lock　门槛锁

sill roller　门槛滚棒

special cargo　特殊货物

specification　规范

spreader　机械分流器

stacking　堆码

standard　标准

tank container　罐式集装箱

tare weight　自重/空载质量

technical certificate　技术保证书

technical manual　技术手册

technical specification　技术规范

tensioning mechanism　拉紧机构

terminal　场站

test condition　试验条件

test date　试验数据

test loading　试验载荷

test result　试验结果

thermal barrier　保温栅栏

thermal container　保温集装箱

tiedown chain assembly　系留链条

tiedown fitting　系留接头

tiedown hook　系留钩

tiedown net　系留网

tiedown point　系留点

tiedown provision　系留装置

tiedown ring　系留环

tiedown rope　系留钢索

tiedown webbing　系留带

tire footprint loading　轮胎压面载荷

tow bar　牵引杆

towing facility　牵引设备

towing tractor　牵引车

transition rail　传输导轨

transport　运输机

transportable　运输性

transporter　运输车/运载工具

treadway　踏板

ultimate loading　设计载荷(极限载荷)

ultimate strength 极限强度

uncertificated ULD 未经合格审定的集装单元

validation loading 有效载荷

vertical restraint device 垂直限动装置

volume of cargo compartment 货舱容积

waterline 水准/平线

webbing strip 系留/编织带

wheel load 机轮载荷

wheeled equipment 轮式设备

yield strength 屈服强度

FLIGHT TEST
飞行试验

abort altitude 紧急中断飞行高度

absolute altitude 绝对高度

accelerate-stop distance 加速-停止距离

accelerating-run method 加速法

acceptance flight test 验收试飞

advise established inbound on ILS 向台建立盲降报告

aerobatic flight 花样飞行

aerobatics flight test 特技飞行试验

aerodrome weather minimum 机场气象最低标准

air tactics trainer 空中战术训练器

airplane level measurement 飞机水平测量

airplane weighing 飞机称重

airspeed 空速

air-tight test/air-tightness test 气密试验

altitude 高度

approach altitude 进场高度

approach facilities 进近设备

approaching lights in sight 看见进近灯

average airspeed 平均空速

back to the center line 回到中心线

base leg 四边

base speed 第四边飞行速度

basic airspeed 修正(基本)表速

belly landing 机腹着陆

best cost cruising speed 最经济巡航速度

bird-impact test 鸟撞试验

block speed 轮挡速度

bore sight harmonization 冷校靶

bumping test 连续冲击试验/撞击试验

cabin altitude 座舱高度

cabin crew training equipment 座舱乘员训练设备

cabin service trainer 座舱服务训练器

calibrated airspeed (CAS) 校正空速

call on downwind 三边呼叫

call procedure turn completed 转弯改出呼叫

call starting procedure turn 开始程序转弯呼叫

caution wake turbulence 注意尾流

ceiling altitude/peak altitude 最大飞行高度/升限

check again on short final 短五边再检查

check climb 测实用升限

check flight for range 航程检验飞行

clear to land 可以降落

cleared for a practices ILS approach 允许练习盲降进近

cleared for an ILS approach 允许盲降进近

cleared for simulating let-down 允许做模拟(练习)穿云

cleared for standard instrument approach 允许标准仪表进近

cleared for touch and go landing 允许连续起飞

cleared for visual approach 允许能见进近

cleared of all clouds 完全出云

cleared to join leftdownwind leg 允许加入左三边

cleared to join right traffic pattern (circuit) 允许加入右航线

cleared to land 可以降落

cleared to make a tear drop let-down　可以做修正角穿云

cleared to make full stop landing　允许做全停降落

clearway　净空道

climb to ceiling　上升到实用升限

cockpit familiarization trainer　熟悉座舱训练器

cockpit procedure trainer（CPT）　座舱程序训练器

cockpit system simulators（CSS）　座舱系统模拟器

cold front　冷锋

combat simulator　空战模拟器

composite trainer　复合训练器

compressibility correction　压缩修正量

computing system　计算系统

contact speed/touchdown speed　着陆速度

continue holding　继续等待

continuous climbing method　连续爬升法

coordinated sideslips method　协调侧滑法

corrected altitude　修正高度

correction to the left　向左修正

critical engine　临界发动机

critical speed　临界速度

cross wind　侧风

cross wind leg　二边

cross wind landing　侧风着陆

cruise airspeed　巡航空速

cruise status　巡航状态

cruise-in altitude　返航高度

cruise-out altitude　出航高度

cruising altitude　巡航高度

cutoff altitude　停车高度

datum airspeed　基准空速

decision altitude/decision height　决断高度

delivery flight test　交付(出厂)试飞

density airspeed　密度空速

density altitude　密度高度

descend down　下降到

design airspeed　设计空速

determination of climbing performance　爬升性能的测定

determination of horizontal accelerating-decelerating performance　水平加减速性能测定

determination of maneuvering characteristic　机动特性测定

determination of maximum speed　最大速度的测定

determination of range and endurance　航程和航时的测定

determination of the takeoff and landing performance　起飞着陆性能的测定

determination of turning performance　盘旋性能测定

development flight test　调整试飞

differential-correction method　微分修正量法

dim light　调暗灯光

distress landing　带故障着陆

diving speed　俯冲速度

diving test　俯冲试验

downwind leg　三边

downwind landing　顺风着陆

downwind speed　第三边飞行速度

drift out to the right　偏右

drip-proof test　淋雨试验

dry air flight test　干空气飞行试验

duration life test/durability test　寿命试验

egress trainer　救生训练器

embedded training　嵌入训练

emergency landing　紧急着陆

emergency spin recovery device　应急改出尾旋装置

endurance on board　机上续航时间

endurance test　持久/续航试验

engine-speed method　转速法

enter the holding pattern 进入等待航线

environmental measurements of flight vibration 飞行振动环境测量

environmental test 环境试验

equivalent airspeed（EAS） 当量空速

equivalent-altitude method 等量高度法

established on the ILS 盲降报告

evaluation flight test 鉴定试飞

expect approach clearance 预计许可进近

expect ILS approach 预计盲降进近

expect long final approach 预计长五边进近

expect straight-in approach 预计直接进近

experimental and research airplane 试验研究机

extend downwind leg 延长三边

extend the traffic pattern 延长航线

extension andretraction test 收放试验

extreme airspeed 极限空速

ferrying speed 转场飞行速度

final 五边

final leg speed 第五边飞行速度

flare speed （着陆前)拉平速度

flight airspeed 飞行空速

flight altitude 飞行高度

flight determination of buffet boundary 抖振边界的飞行测定

flight envelop 飞行包线

flight flutter test 飞行颤振试验

flight idling 空中慢车

flight load measurement 飞行载荷测定

flight platform 飞行转台

flight simulator 飞行模拟器

flight stress measurement 飞行应力测定

flight test mission sheet 试飞任务单

flight test modifications 飞行试验改装

flight test of stability and control 稳定性操纵性试飞

flight test program 飞行试验大纲

flight test/trial-flight 试飞

flying qualities of airplanes 飞机的飞行品质

FO beacon FO 导航台

forced landing 被迫着陆

frequency 频率

front 锋面

full flight simulator 全任务模拟器

full stop landing 全停着陆

full-load test 满负荷试验

GCA approach 地面控制进近

general purpose trainer 通用训练器

general trainer 一般训练器

geometrical altitude 几何高度

glide speed 下滑速度

ground idling 地面慢车

ground simulation test of primary control system 主操纵系统地面模拟试验

ground speed 地速

ground-controlled landing 地面控制着陆

guided landing 地面引导着陆

gust 阵风

have the ground in sight 能见地面

head wind 逆风

headwind landing/upwind landing 逆风着陆

high intensity light 高强度灯

high pressure 高气压

high temperature test 高温试验

high-speed landing/fast landing 大速度着陆

hypersonic speed 高超音速

ignition test 点火试验

ILS approach 盲降进近

indicated airspeed（IAS） 指示空速(表速)

indicated altitude 指示高度

in-flight simulator 空中飞行模拟器

initial approach 起始进近

instructor station 教员台

instrument flight trainer 仪表飞行训练器

intercept the localizer 切到航道

in-trim landing 配平着陆

iron bird 铁鸟

keep/maintain VMC approach　保持能见进近

keep sufficient spacing/distance separation　保持足够间隔距离

lag correction　延迟修正量

landing configuration　着陆状态

landing lane　着陆航道

landing run　着陆滑跑(距离)

left traffic pattern/circuit　左起落航线

level of flying quality　飞行品质等级

level-out altitude　(着陆前)拉平高度

lift-off speed　离地速度

loading spectrum　载荷谱

lock-releasing test/unlocking test　开锁试验

long final (short final)　长五边(短五边)

longitudinal static stability coefficient　纵向静稳定性系数

lose time　消磨时间

low cycle fatigue test　低循环疲劳试验

low pressure　低气压

low temperature test　低温试验

low-air-pressure test　低气压试验

maintenance simulator　维修模拟器

make a left short circle　向左飞一小圈

make a long approach　做大航线进近

make a normal approach　做正常航线进近

make a short approach　做小航线进近

make a short traffic pattern　做一个小航线

make a trial on ILS approach　试做盲降进近

malfunction training　故障训练

maneuver load　机动载荷

maximum cruise speed　最大巡航速度

maximum operating altitude　最大使用高度

maximum operating limit speed　最大使用限制速度

maximum permissible operating speed　最大允许飞行速度

maximum threshold speed　跑道入口最大速度

measurement of flight speed　飞行速度的测量

measurement of level flight performance　平飞性能测定

measurement of Mach number　飞行马赫数的测量

measurement of position error in airspeed system　空速系统位置误差的测定

measurement of powered descent performance　动力下降性能的测定

measurement of pressure altitude　气压高度的测量

measurement of primary control system characteristic　主操纵系统性能测定

measurement of vertical climbing performance　垂直爬升性能测定

meteorological information　气象情报

meteorological satellite　气象卫星

method of turns　盘旋法

minimum altitude　最低高度

minimum control speed　最小操纵速度

minimum control speed in air　空中最小可操纵速度

minimum control speed on ground　地面最小可操纵速度

minimum decision altitude　最低决断高度

minimum demonstrated threshold speed　跑道入口最小验证速度

minimum night altitude　夜航安全高度

minimum threshold speed　跑道入口最小速度

missed-approach altitude　复飞高度

mission trainer　任务训练器

motion system　运动系统

natural icing flight test　自然结冰飞行试验

NDB approach　导航台进近

negative/not approved for straight-in approach　不同意长五边进近

operation weather limits　飞行最低气象条件

operational flight test　使用试飞

operational flight trainer　操作飞行训练器

overhead the runway　在跑道上空

overload test　超负荷试验

overspeed test　超速试验

parachute jumps flight　跳伞飞行

part-task trainer　部分任务训练器

pitching test　俯仰试验

position error　位置误差

power-off speed　停车飞行速度

present position　现在位置

press-break test　压爆试验

pressure altitude　气压高度

pressure test　压力试验

pressure-drop test　压降试验

pull back the throttle　收油门

pull up and go around　拉升/复飞

pulse-ailerons method　脉冲操纵副翼法

pulse-elevator method　脉冲操纵升降舵法

pulse-rudder method　脉冲操纵方向舵法

R/W visual range　跑道目视距离

radar position　雷达位置

radio and radar facilities　无线电和雷达设备

randomly-checking flight test　典型试飞

reciprocal traffic　相对活动

recovery altitude　改出高度

rectangular let down　方块(盒)穿云

reduce speed　减速

refueling altitude　空中加油高度

reject take-off speed　中断起飞速度

relative airspeed　相对空速

relative altitude　相对高度

remain on heading　保持航向

remaining fuel on board　剩余油量

report exact passing altitude　报告精确的通过高度

report field in sight　看到机场报告

report in sight　看到跑道报告

report turning in bound　向台报告

report visual contact　能见地面报告

request approach sequence　请问进近次序

request landing sequence　请问着陆次序

request let-down sequence　请问穿云次序

research flight test　研究性试飞

residual oscillation　剩余振荡

rich running　富油运转

right traffic pattern/circuit　右起落航线

roger　明白

rolling rate oscillation　滚转速振荡

rolling takeoff method　滑跑起飞法

runway slope　跑道坡度

runway visual range（RVR）　跑道能见距离

salt spray test/salt fog test　盐雾试验

scenario training　场景训练

search light for landing　探照灯降落

second-segment climb　起飞第二阶段上升

segment climb　爬升阶段

shimmy measures　前轮摆振测量

shock measurement　冲击测量

shorten the pattern　航线减小

simulation icing flight test　模拟结冰飞行试验

situation display　态势显示

slant visibility　斜视能见度

slightly off to the right of the center line　稍偏在中心线右面

slightly off track to the left side　稍微偏在航线左面

sonic speed　音速

specification of airplane flying qualities　飞机的飞行品质规范

specification of the flight test mission　飞行试验任务书

speed on go-around　复飞速度

stall flight test　失速飞行试验

stall warning buffeting　失速警告抖振

stalling speed　失速速度

stationary front　静止锋

steady level straight flight method　定常水平直线飞行法

steady pull-up method　稳定拉起法

steady straight flight method　定常直线飞行法

steady turn method　稳定转弯法

steer to the left　向左操纵

step-ailerons method　阶跃操纵副翼法

step-rudder method　阶跃操纵方向舵法

stick-fixed maneuver margin　握杆机动余量

stick-free maneuver margin　松杆机动余量

stick-free static stability margin　松杆静稳定性余量

stop way　停机地带

straight-in approach　直接进近/进场

strength test　强度试验

subsonic speed　亚音速

sunlight radiation test　太阳辐射试验

supersonic speed　超音速

surg test　喘振试验

tactics trainer　战术训练器

tail wind　顺风

takeoff airspeed　起飞空速

takeoff decision speed　起飞决断速度

takeoff ground speed　起飞滑跑速度

takeoff roll　起飞滑跑

takeoff run　起飞滑跑（距离）

takeoff safety speed　安全起飞速度

takeoff status　起飞状态

taxiing lane　滑行道

technical specifications for modifications　试飞改装技术条件

temperature altitude　温度高度

temperature impacting test　温度冲击试验

temperature-altitude test　温度高度试验

terrain model　地形模型

tethered hovering method　系留悬停法

thunderstorm　雷暴

touch and go landing　连续起落

tower　塔台

trainee console　学员控制台

trainee station　学员工作台

trainer　训练器

training flight　飞行训练

transonic speed　跨音速

trial speed　试飞速度

trim tab method　调整片法

true airspeed　真空速

true cruise airspeed　巡航真空速

turn up the approach light　调亮进近灯

turning base（turn to base）　三转弯（转向四边）

turning final（turn to final）　四转弯（转向五边）

turning speed　转弯速度

unable to approve straight-in approach　不同意直接进近

unpaved-runway landing　土跑道着陆

up wind leg　一边

vertical climbing method　垂直爬升法

visibility　能见度

visual approach　能见进近

visual landing/contact landing　目视着陆

visual simulation　视觉模拟

visual system　视景系统

warm front　暖锋

watch out the altitude　注意高度

weak running　贫油运转

weapon tactics trainer　武器战术训练器

weather map/weather chart　天气图

weather radar　气象雷达

weather system　天气系统

wind shear　风切变

zero-flight time simulator　零飞行小时模拟器

缩略语

A

A	accepted	接收(电子飞行飞行系统显示符号)
A	airborn	空中的,飞行的,机载的
A	ampere	安培
A	anode	阳极
A	area	面积
A	apect ratio	展弦比
A	atomic	原子的
A	attack	攻击机,强击机(美国军用飞机机种代码)
A	aileron	副翼
A/A	airdrome to airdrome	机场间的
A/A	air-to-air	空对空,空空
A/C	aircraft	飞机
A/D	aerodrome	机场
A/D	alarm and display	告警并显示
A/D	analog to digital	模拟-数字转换
A/DF	assembly/disassembly facility	装配/拆卸设备
A/F	airfield	机场
A/F	airframe	机体
A/G	air-to-ground	空对地,空地
A/G/A	air-to-ground-to-air	空-地-空
A/N/B SK	A/N/B selection knob	A/N/B 选择旋钮
A/P	airplane	飞机
A/P	airport	机场,航空站,航空港
A/P	autopilot	自动驾驶(仪)
A/S	air-to-surface	空对面,空面
A/U	air to underwater	空对水下
A/V	audio/video	声频/视频
AA	air abort	中断飞行
AA	absolute altitude	绝对高度
AA	acquisition aiding	帮助截获
AA	adaptive array	自调谐天线阵,自适应阵列

AA	air almanac	航空历
AA	air assault	空中突击,空降突击
AA	airborne alert	空中警戒
AA	alarm assignment	警报任务
AA	artificial antenna	仿真天线
AA	automatic approach	自动进场,自动进近
AA	aviation annex	航空附件
AA	advisor yarea	咨询区
AA	alternate airfield	备用机场
AA	autopilot actuator	自动驾驶仪舵机
AAA	airport advisory area	机场咨询区
AAADB	autopilot audio alarm device box	驾驶仪语音报警装置箱
AAB	adaptive angle bias	自适应角偏差
AAB	autopilot aonnection box	驾驶仪交联接线盒
AAC	aircraft airworthiness certificate	飞机适航证
AAC	aviation advisory commission	航空咨询委员会
AACBS	aircraft anti-collision beacon system	飞机防撞信标系统
AACS	aileron active control system	副翼主动控制系统
AACS	active attitude control system	主动姿态控制系统
AAD	antiaircraft defence	防空
AAD	assigned altitude deviation	指定高度偏差
AAD	average absolute deviation	平均绝对偏差
AADA	antiaircraft defended area	防控地区
AADC	advanced avionics digital computer	先进航空电子数字计算机
AADHS	advanced avionics data handling system	先进航空电子设备数据处理系统
AADS	advanced air data system	先进大气数据系统
AADS	advanced air defense system	先进防空系统
AADS	automatic aircraft diagnostic system	飞机自动诊断系统
AAES	advanced aircraft electrical system	先进飞机电气系统
AAEV	automatic air exhaust valve	自动排气阀
AAFCS	advanced automatic flight control system	先进飞行自动控制系统
AAFIS	advanced avionics fault isolation system	先进航空电子设备故障隔离系统
AAH	automatic altitude hold	自动保持高度
AAH	automatic attitude hold	自动姿态保持
AAI	air-to-air identification	空对空目标识别
AAI	air-to-air interrogator	空对空询问机
AAI	all attitude indication	全[飞行]姿态显示
AAI	angle of approach indication	进场[进近]角指示
AAI	angle of approach indicator	进场[进近]角指示器

AAIFFS	air-to-air identification friend or foe system	空对空敌我识别系统
AAIM	aircraft autonomous integrity monitor	飞机自主完好性监控
AAL	above airfield level	离机场海平面的(高度)
AAL	aircraft approach light	飞机进场[进近]灯光
AAL	aircraft approach limitations	飞机进场[进近]限制
AAL	angle of approach light	进场[进近]下滑角指示灯
AAL	antiaircraft light	防空探照灯
AALS	advanced approach and landing system	先进进场[进近]与着陆系统
AALT	automatic azimuth laying theodolite	自动方位确定经纬仪
AAN	advanced alternation notice	提前更改通知书
AAR	aircraft accident report	飞机事故报告
AAR	air-to-air refueling	空中加油
AAR	area of air refueling	空中加油区
AAR	automatic air refueling	自动空中加油
AARA	air-to-air refueling area	空中加油区
AARB	advanced aerial refueling boots	先进空中加油(伸缩)套管
AARPLS	advanced airborne radio position location system	先进机载无线电定位系统
AARS	attitude and azimuth reference system	姿态与方位基准系统,姿态与方位参考系统
AARS	attitude altitude retention system	姿态高度保持系统
AARS	automatic altitude reporting system	高度自动报告系统
AAS	advanced administration system	先进管理系统
AAS	advanced antenna system	先进天线系统
AAS	airborne antenna system	机载天线系统
AASL	antiaircraft search light	防空(高射)探照灯
AASMA	altitude and speed measuring assembly	高度速度测量组件
AASW	airborne antisubmarine warfare	空中反潜战
AAT	accelerated aging test	加速老化试验
AATMS	advanced air traffic management system	先进空中交通管理系统
AATS	advanced automatic test system	先进自动测试系统
AATS	advanced automatic training system	先进自动化培训系统
AATS	alerting automatic telling status	自动报警装置
AATS	alternate aircraft takeoff system	飞机备用起飞系统
AATS	automatic altitude trim system	自动高度配平系统
AAU	absolute alignment update	绝对对准适时修正
AAVCS	airborne automatic voice communication(s) system	机载自动语音通信系统

AAVCS	automatic aircraft vectoring control system	飞机自动引导控制系统
AAVS	automatic aircraft vectoring system	飞机自动引导系统
AB	afterbuner	加力燃烧室,复燃加力燃烧室,复燃室,补燃室
AB	address bus	地址总线
AB	air bag	空气囊
AB	air blast	空气射流,空中冲击
AB	airborne	空中的,飞行的,机载的
AB	aerodrome beacon	机场信标台
ABA	airborne alert	飞机警报,空中警戒
ABC	airborne control	空中控制
ABCC	automatic brightness contrast control	自动亮度对比(反差)调整
ABCN	airdrome beacon	机场信标
ABID	aircraft based infrared detector	机载红外探测器
ABILA	airborne instrument landing approach	机载仪表进场[进近]着陆
ABM	aileron booster mark	副翼助力器标记
ABS	antilock brake system	防抱死制动装置
ABS	antiskid brake system	防滑刹车系统
ABSV	air bleed safety valve	放气安全阀
ABT	afterburning turbojet	复燃加力涡轮喷气(发动机)
ABTS	airborne transponder subsystem	机载发射机应答器子系统
ABU	audio backup unit	声频(响)备用设备
AC	acceptance and checkout	接收与检查
AC	approach control	进近管制
AC	air cover	空中掩护
AC	(aeronautical) approach chart	航行进场[进近]图
AC	absolute ceiling	绝对升限
AC	advisory circular	咨询通报
AC	aerial combat	空战
AC	aerodrome control	机场指挥,机场调度
AC	airframe change	机体更改
AC	aligned continuous	(空中受油探管的)连续对准
AC	application channel	应用通道
AC	alternating current	交流电
AC	aileron clutch	副翼离合器
AC	all clear	解除警报
AC	autopilot console	驾驶仪操纵台
AC	approach control	进场[进近]控制(系统)
ACA	airborne control approach	空中控制进场[进近]

ACA	attitude control assembly	姿态控制装置
ACAS	automatic central alarm system	中央自动报警系统
ACAS	airborne collision avoidance system	机载自动防撞系统
ACB	audio control box	声频控制盒
ACB	automatic call back	自动叫回
ACC	air command center	空军指挥[控制]中心
ACC	air control center	空中管制中心
ACC	area control code	区域管制代码
ACC	area coverage change	(雷达)探测地区改变
ACC	area [aerodrome] control center	区域[机场]管制中心
ACC	automatic carrier control	载波自动控制
ACC	automatic combustion control	自动燃烧控制
ACC	avionics control computer	航空电子控制计算机
ACCR	alternate current cleaner receptacle	交流吸尘器插座
ACCS	automatic checkout and control system	自动检测与控制系统
ACCU	accumulator	蓄电池
ACCU	accuracy	精度
ACD	active control device	主动控制装置
ACD	aircraft certification directorate	飞机适航审定中心
ACD	advanced control device	先进控制装置
ACD	aircraft configuration data	飞机构型数据
ACD	aircraft depot	机库
ACD	alternating current dump	交流电源切断
ACD	antenna control display	天线控制显示(器)
ACD	automatic chart display	自动航图显示,自动地图显示器
ACD	automatic closing device	自动开关装置
ACDB	alternating current distribution board	交流配电盘
ACDS	advanced command data system	先进指挥数据系统
ACDS	alarm communication(s) and display system	警报传递与显示系统
ACDU	automatic chart display unit	自动航图显示装置
ACE	air conditioning equipment	空气调节设备
ACE	aircraft condition evaluation	飞机状况评估
ACE	atmospheric control experimentation	大气控制实验
ACE	attitude control electronics	姿态控制电子设备
ACE	automatic checkout equipment	自动检测装置
ACF	advanced communication function	高级通讯功能
ACFT	aircraft flying training	飞行器飞行训练
ACG	alternating current generator	交流发电机

ACGCMI	AC generator current mutual inductor	交流发电机电流互感器
ACGF	alternating current generator failure	交流发电机故障
ACH	autopilot control handle	驾驶仪操纵手柄
ACI	acoustic comfort index	容许噪声指数
ACI	allocated configuration identification	配置构型识别
ACI	attitude control indicator	姿态控制指示器
ACI	aircraft identification	飞机识别
ACIDS	aircraft integrated data system	飞机集成数据系统
ACIM	aircraft component intensive management	飞机部件集中管理
ACIP	avionics communication(s) and information processing	航空电子通信与信息处理
ACIS	avionics central information system	航空电子中央信息系统
ACIU	aircraft interface unit	飞机接口单元
ACJB	alternating current junction board	交流接线板
ACJB	auto control junction box	自动控制接线盒
ACL	aircraft control link	飞机控制数据链
ACL	anti-collision light	防撞灯
ACLGDCO	auto-closing landing gear door circuit off	断开自动关闭起落架舱门
ACLICS	airborne communication location identification and collection system	机载通信定位识别与采集系统
ACLS	airborne command and launch subsystem	机载控制与发射子系统
ACLS	automated control and landing system	自动控制与着陆系统
ACME	attitude control and maneuvering electronics	姿态控制与机动飞行电子设备
ACMF	airplane condition monitoring function	飞机状态监控功能
ACMS	aircraft condition monitoring system	飞机状态监控系统
ACN	advance change notice	提前更改通知
ACN	aircraft classification number	飞机分类号
ACN	automatic celestial navigation	自动天体导航
ACNSS	advanced communication navigation surveillance system	先进通信导航监视系统
ACO	assembly and checkout	装配与检测
ACOG	aircraft on ground	停飞待用飞机
ACP	auxiliary control panel	辅助控制仪表板
ACP	azimuth change pulse	方位变化脉冲
ACP	audio control panel	音频控制板
ACP	AHRS control panel	AHRS 操作板
ACPA	adaptive-controlled phased array	自适应控制相控阵(天线)
ACPS	attitude control propulsion system	姿态控制推进系统

ACPS	auxiliary control propulsion system	辅助控制推进系统
ACR	airfield control radar	机场控制雷达,机场引导雷达
ACR	approach control radar	进场[进近]管制雷达
ACRA	airflow control regulator and actuator	气流控制调节器与作动筒
ACRJB	assistant control relay junction box	辅助操纵继电器接线盒
ACRS	automatic communication(s) and recording system	自动通讯设备与记录系统
ACS	active control system	主动控制系统
ACS	air conditioning system	空调系统
ACS	air cycle system	空气循环系统
ACS	aircraft call signal	飞机呼叫信号
ACS	airborne communication system	机载通信系统
ACS	aircraft control system	飞机控制系统
ACS	all channel signaling	全信道信令
ACS	attitude control system	姿态控制系统
ACS	audio communication system	声频通信系统
ACS	automatic checkout system	自动检测系统
ACS	automatic coding system	自动编码系统
ACS	automatic control system	自动控制系统
ACS	auxiliary cooling system	辅助冷却系统
ACS	azimuth control system	方位(角)控制系统
ACSEP	aircraft certification system evaluation program	航空器审定系统评审大纲
ACT	advanced composites technology	先进复合材料技术
ACT	advanced control technology	先进控制技术
ACT	air combat turnaround	空战再次出动准备
ACT	airborne crew trainer	(英国空军)空勤组教练机
ACT	aircrew coordination training	机组协调训练
ACT	analogical circuit technique	模拟电路技术
ACT	actuator	舵机
ACTPS	application control and teleprocessing system	应用控制与远程(信息)处理
ACU	autopilot control unit	自动驾驶控制单元
ACU	avionics control unit	航空电子控制装置
ACW	airborne collision warning	空中碰撞告警
ACW	aircraft control and warning	飞机控制与告警(系统)
ACWS	aircraft control and warning site	飞机飞行控制与警报站
ACWS	aircraft control and warning service	飞机控制与告警勤务
ACWS	aircraft control and warning system	飞机控制与告警系统

AD	air defence	防空
AD	attention display	提醒显示,特殊情况显示
AD	aerodynamic damping	气动阻尼
AD	air distance	空中飞行距离,无风距离
AD	airworthiness directive	适航指令
AD	attitude determination	姿态确定
AD	autopilot disconnect	自动驾驶仪断开
AD	autopilot disconnection	驾驶仪速断
ADA	automatic data acquisition	自动数据采集
ADAC	automatic data acquisition center	自动数据采集中心
ADAM	automatic distance and angle measurement	自动测距和测角
ADAMS	aeromagnetic data automatic mapping system	航磁数据自动绘图系统
ADAMS	airborne data analysis/monitor system	机载数据分析/监控系统
ADAP	aerodynamic data analysis program	气动数据分析计划
ADAS	airborne dynamic alignment system	机载动态校准系统
ADAS	automatic data acquisition system	自动数据采集系统
ADC	air data computer	大气数据计算机
ADC	autopilot directing console	驾驶指引操纵台
ADCC	air data computer calibration	大气机校准
ADCLS	advanced data collection/location system	先进数据收集与定位系统
ADDAR	automatic digital data acquisition and recording	自动数字数据采集与记录
ADDAS	airborne digital data acquisition system	机载数字数据采集系统
ADDAS	automatic digital data assembly system	自动数字数据汇编系统
ADDER	automatic digital data error recorder	自动数字数据误差记录器
ADDI	automated digital data interchange	自动化数字数据交换
ADDPB	automatic diluter-demand pressure breathing	自动调节氧气浓度的增压供氧
ADDS	advanced data display system	先进数据显示系统
ADDS	airborne decoy dispensing system	机载假目标投放系统
ADDS	application development data system	应用发展数据系统
ADDS	applied digital data system	应用数字数据系统
ADDS	automatic data digitizing system	自动数据数字化系统
ADDS	automatic data distribution system	数据自动分配系统
ADEMS	airborne display electrical management system	机载显示器电气管理系统
ADEP	airport of departure	起飞机场
ADES	automatic data encoding system	自动数据编码系统

ADES	automatic data entry system	自动数据输入系统
ADESS	automatic data editing and switching system	自动数据编辑与转接系统
ADEU	automatic data entry unit	自动数据输入设备
ADF	automatic direction finder	自动定向仪,无线电罗盘
ADFE	automatic direction finding equipment	自动测向装置
ADFRC	automatic direction finder remote-controlled	自动遥控测向仪
AADG	aircraft accessory drive gear	飞机附件传动机匣
ADI	attitude director indicator	姿态航向指示仪,指引地平仪,垂直位置指示器
ADIL	air defense identification line	防空识别线
ADIRS	air data inertial reference system	大气数据惯性基准系统
ADIRU	air dataand inertial reference unit	大气数据与惯性基准装置
ADIS	airborne digital instrumentation system	机载数字仪表系统
ADIS	automatic data interchange system	自动数据互换系统
ADIT	analog-digital integrating translator	模拟-数字积分变换器
ADIZ	air defence identification zone	防空识别区
ADL	automatic data link	自动数据链
ADL	automatic data logger	自动数据记录仪
ADM	air data module	大气数据模块
ADP	advanced development prototype	先期发展(性)样机
ADP	aerodynamic design point	气动设计点
ADPG	atmospheric dynamic payload group	大气动力有效载荷组
ADR	accident data recorder	事故数据记录仪
ADR	air defence radar	防空雷达
ADRFS	advanced dynamic RF simulator	先进动态射频模拟器
ADRN	advanced document revision notice	临时更改单
ADRS	airborne digital recording system	机载数字记录系统
ADS	active denial system	主动拒绝系统
ADS	advanced debugging system	先进程序调试系统
ADS	advanced delivery system	先进投放系统
ADS	aerial delivery system	空投系统
ADS	air data sensor	大气数据传感器
ADS	air data system	大气数据系统
ADS	airborne display system	机载显示系统
ADS	angular displacement sensor	角位移传感器
ADS	attitude determination system	姿态确定系统
ADS	audio distribution system	声频分配系统

ADS	automatic dependent surveillance	自动相关监视
ADS	autopilot disengage switch	自动驾驶仪断开开关
ADSI	analog display service interface	模拟显示服务接口
ADSP	advanced digital signal processor	先进数字信号处理机
ADSS	aircraft damage sensing system	飞机损伤传感系统
ADU	air data unit	大气数据装置
ADU	auxiliary display unit	辅助显示装置
ADU	avionics display unit	航空电子设备显示组件
AE	armament and electronics	军械与电子设备
AE	azimuth and elevation	方位-仰角,方位-高度
AE	audio equipment	声频设备
AE	autopilot engage	驾驶仪工作
AEA	air electronic attack	空中电子攻击
AEA	airborne electronic attack	机载电子攻击
AEB	avionics equipment bay	航空电子设备舱
AECM	active electronic countermeasures	有源电子对抗
AECS	adaptive engine control system	自适应发动机控制系统
AECS	advanced environmental control system	先进环境控制系统
AEI	azimuth error indicator	方位角误差指示器
AEPS	aircraft electric power system	飞机电源系统
AER	air expansion ratio	空气膨胀比
AERIS	airborne electronic ranging instrumentation system	机载电子测距仪设备系统
AERIS	automatic electronic ranging information system	自动电子测距信息系统
AET	average execution time	平均执行时间
AET	aviation electronic technology	航空电子技术
AEW	air electronic warfare	空中电子战
AEW	airborne early warning	空中预警
AEWA	airborne early warning aircraft	空中预警飞机
AEWC	airborne early warning and control	空中预警与控制
AEWR	airborne early warning radar	机载预警雷达
AEWS	advanced early warning system	先进预警系统
AEWS	airborned electronic warfare system	机载电子战系统
AF	air force	空军
AF	air formation	飞机编队,飞行编队
AF	airframe	机身,机体
AF	airway facilities	航路设施
AF	audio frequency	声频

AF	automatic following	自动跟随
AF	alternate field	备降机场
AF	autopilot failure	自动驾驶仪故障
AF	area forecast	区域(天气)预报
AF	attitude failure	姿态故障
AF	auto fly	自动飞行
AF	air facility	航空设施
AFC	audio frequency change	声频变换
AFC	automatic feedback control	(随控布局飞机的)自动反馈操纵
AFC	automatic fight control	飞行自动控制
AFC	automatic following control	自动跟随控制
AFC	automatic frequency control	自动频率控制
AFC	auto-feathering check	自动顺桨检查
AFCAS	advanced flight control actuation system	先进的飞行控制传动系统
AFCAS	automatic flight control augmentation system	自动飞行控制增稳系统
AFCE	automatic flight control equipment	自动飞行控制设备,自动驾驶仪
AFCR	air flow control regulator	空气流量控制调节器
AFCS	aircraft fire control system	飞机火控系统
AFCS	automatic fire control system	自动火控系统
AFCS	automatic flight control system	自动飞行控制系统
AFCS	automatic fuel control system	自动燃油控制系统
AFCS	auxiliary flight control system	辅助飞行控制系统
AFCS	avionics flight control system	航空电子飞行控制系统
AFCSS	air force communication(s) support system	空军通信保障系统
AFD	airfield data	机场数据
AFD	amplitude frequency distribution	振幅频率分布
AFDAS	aircraft fatigue data analysis system	飞机疲劳数据分析系统
AFDB	auxiliary fuel distributing box	辅助燃油配电盒
AFDMS	airborne flight deflection measurement system	机载飞行偏转角测量系统
AFDS	autopilot and flight director system	自动驾驶仪与飞行指引仪系统
AFF	automatic fault finding	自动故障探测
AFGS	automatic flight guidance system	自动飞行制导系统
AFIS	airborne flight information system	机载飞行信息系统
AFIS	aircraft fault identification system	飞机故障识别系统
AFIS	automated flight inspection system	自动飞行检查系统
AFL	actual flight level	实际飞行高度

AFLS	approach flash lighting system	进场(进近)照明系统
AFM	aircraft flight manual	飞机飞行手册
AFMS	auxiliary fuel management system	辅助燃油管理系统
AFP	approach flight phase	进场[进近]飞行阶段
AFPS	adjustable frequency power supply	可调频电源
AFR	acceptable failure rate	容许故障率,合格故障率
AFR	automatic field recognition	自动现场识别
AFR	auto-feathering relay	自动顺桨继电器
AFRC	area frequency response characteristic	区域频率响应特性
AFS	automatic flight system	自动飞行系统
AFS	azimuth follow up system	方位(角)跟随系统
AFS	antenna-feeder system	天馈系统
AFSC	automatic fuel shutoff controller	自动切油控制器
AFSM	anti fire switch mark	防火开关标记
AG	air-to-ground	空对地
AG	arresting gear (hook)	着陆拦阻装置(钩)
AGACS	automatic ground-to-air communication (s) system	地对空自动通信系统
AGASS	automated geomagnetic airborne survey system	机载自动地磁测量系统
AGC	air-to-ground communication	空地通信(设备)
AGC	arresting gear control	着陆拦阻装置控制
AGCA	automatic ground controlled approach	地面控制自动进场[进近](着陆)
AGCAS	automatic ground collision avoidance system	自动地面防撞系统
AGCL	automatic ground controlled landing	自动引导着陆
AGCS	advanced guidance and control system	先进制导与控制系统
AGE	aircraft ground equipment	飞机地面设备
AGILS	airborne general illumination light set	机载通用照明设备
AGL	above ground level	离地高度
AGL	aeronautical ground light	地面航行灯
AGL	air-ground liaison	空地联络
AGR	air-to-ground ranging (radar mode)	空对地测距(雷达模式)
AGS	aircraft general specification	飞机通用规范
AH	after hatch	后舱口
AH	artificial horizon	人工地平线
AH	attitude hold	姿态保持
AHARS	airborne heading-attitude reference system	机载姿态与航向参考系统
AHF	auxiliary hydraulic off	备助液压未通

AHIS	attitudeheading indicator system	姿态航向指示系统
AHMA	attitude heading measuring assembly	航姿测量组件
AHRS	attitude heading reference system	航向姿态参考系统
AHRSB	attitude heading reference system battery	航向姿态参考系统蓄电瓶
AHS	antenna homing system	天线寻的系统
AHSR	air height surveillance radar	飞行高度监视雷达
AHSR	airborne height surveillance radar	机载高度监视雷达
AHT	aero-hydrodynamic tunnel	气动力-流体动力风洞
AHT	average handling time	平均处理时间
AHTR	auto horizontal tail retrimming	水平尾翼自动调整配平
AI	alteration and inspection	改变与检查
AI	aircraft icing	飞机结冰
AI	affordability initiative	经济可承受性计划
AI	aircraft identification	飞机识别
AI	airspeed indicator	空速指示器
AI	anti-icing	防冰
AI	accident investigation	事故调查
AIBIT	automatic initiated built-in test	自动启动自检测
AIC	aircraft in commission	现役飞机/执行任务的飞机
AICS	air intake control system	进气口控制系统
AICS	aircraft identification and control system	飞机识别与控制系统
AID	aircraft identification determination	飞机识别判断
AIDAS	advanced instrumentation and data analysis system	先进仪表测量与数据分析系统
AIDRA	automated integrated data recording system	自动综合数据记录系统
AIDS	airborne integrated data system	机载综合数据系统
AIDS	aircraft integrated data system	飞机综合数据系统
AIF	anti-icing fuse	防冰熔断器
AIG	all inertial guidance	全惯性制导
AIG	augmented inertial guidance	增强惯性制导
AIGH	air inlet guide heating	进气导向器加温
AIH	anti icing heating	防冰加温
AILAS	airborne instrument landing approach system	机载仪表着陆进场[进近]系统
AILAS	automatic instrument landing approach system	自动仪表着陆进场[进近]系统
AILAS	automatic instrument low approach system	自动仪表低空进场[进近]系统

AILS	advanced instrument landing system	先进仪表着陆系统
AILS	advanced integrated landing system	先进综合着陆系统
AIMS	aircraft identification monitoring system	飞机识别监控系统
AIMS	aircraft information management system	飞机信息管理系统
AIMS	attitude indicator measurement system	姿态指示器测量系统
AINS	advanced inertial navigation system	先进惯导系统
AINS	airborne inertial navigation system	机载惯导系统
AINS	area inertial navigation system	区域惯导系统
AIS	anti-icing switch	防冰开关
AJBM	aircraft jack base mark	飞机千斤顶座标记
AL	approach and landing (chart)	进场[进近]与着陆(图)
AL	approach lighting	进场[进近]照明
ALC	automatic landing control	自动着陆控制
ALCH	approach light contact height	进场[进近]指示灯光目视高度
ALCS	active lift control system	升力主动控制系统
ALF	auxiliary landing field	辅助着陆场
ALG	advanced landing ground	前方降落场
ALG	autonomous landing guidance	自主式着陆引导
ALMM	aileron lock mechansim mark	副翼锁机构标记
ALS	aircraft landing system	飞机着陆系统
ALS	alternate landing site	备用着陆场
ALS	approach lighting system	进场[进近]灯光系统
ALS	azimuth laying set	方位瞄准装置,方位角瞄准器
ALT	altitude	高度
ALT	altitude hold	高度保持
ALT	autopilot light test	驾驶仪灯测试
ALTR	approach/landing thrust reverse	进场[进近]/着陆反推装置
AM	amplitude modulation	调幅
AM	awaiting maintenance	等待维修
AMC	aerodynamic mean chord	平均空气动力弦
AMI	airspeed Mach indicator	空速马赫数指示器
AMM	aircraft maintenance manual	飞机维修手册
AMMM	aviation, manning, maintenance and material	航空,人员配备,维修和材料
AMP	avionics master plan	(美国空军)航空电子主计划
AMPG	air miles per gallon	飞行英里每加仑
AMPP	air miles per pound	飞行英里每磅
AMPT	air miles per tone	飞行英里每吨
AMS	air material specification	航空材料规范

AMSEC	analytical method for system evaluation and control	系统评定与控制分析方法
AMSL	above mean sea level	海拔
AN	area navigation	区域导航
AND	aircraft nose down	飞机俯冲
ANDAS	automatic navigation and data acquisition system	自动导航与数据采集系统
ANDS	automatic navigation differential station	自动导航差分台
ANL	aircraft nose left	机头向左
ANMPG	air nautical miles per gallon	航空海里每加仑
ANMS	aircraft navigation and management system	飞机导航与管理系统
ANR	aircraft nose right	机头向右
ANRZ	alternating non return-to-zero	交替非归零制
ANS	aviation navigation system	航空导航系统
ANS	autonomous navigation system	自主导航系统
ANT	antenna	天线
ANU	aircraft nose up	机头上仰
ANVC	active noise and vibration control	主动噪声与振动控制
ANW	actual net weight	实际净重
AO	awaiting overhaul	等待大修
AO	autopilot off	自动驾驶仪断开
AO	autopilot on	自动驾驶仪接通
AOA	angle of attack	迎角,攻角
AOA	at or above	在…及以上
AOAC	angle of attack calculator	迎角计算器
AOAH	angle of attack hold(ing)	(代号)迎角保持
AOAH	angle of attack heating	攻角加温
AOAI	angle of attack indicator	迎角指示器
AOAIB	angle of attack indicator box	迎角指示器盒
AOAS	angle of attack sensor	迎角传感器
AOAS	angle of attack system	迎角系统
AOASZPR	angle of attack sensor zero position regulator	攻角传感器零位调节器
AOAT	angle of attack transmitter	迎角传感器
AOAT	allowed off-aircraft time	容许离机时间,容许离开飞机时间
AOAT	angle of attack true	实际迎角
AOAWS	angle of attack warning system	迎角告警系统
AOD	airport of destination	目的地机场

AOG	aircraft on ground (awaiting parts)	停在地面(等配件)的飞机
AOG	acceleration of gravity	重力加速度
AOG	aircraft on ground	停飞待用飞机
AOL	achieved overhaul life	大修寿命,大修周期
AOS	air oil separator	气油分离器
AP	airframe and powerplant	机体与动力装置
AP	absolute pressure	绝对压力
AP	approach pattern	进场［进近］航线图
AP	autopilot panel	自动驾驶仪操纵板
AP	altitude preselect	高度预选
AP	awaiting parts	等待零备件
APA	airport (or airfield) pressure altitude	机场压力高度
APAR	active phased array radar	主动相控阵雷达,有源相控阵雷达
APAR	automatic precision approach radar	自动精确测量进场［进近］雷达
APARS	automatic programming and recording system	自动程序设计与记录系统
APAS	automated pilot advisory system	自动驾驶咨询系统
APC	autopilot controller	自动驾驶仪控制器
APCS	approach path control system	进近航路控制系统
APMCU	autopilot monitor and control unit	自动驾驶仪监控与控制装置
APP	auxiliary powerplant	辅助动力装置
APPI	autopilot positioning indicator	自动驾驶仪定位指示器
APPR	approach	进场
APPS	auxiliary payload power system	辅助有效载荷动力系统,有效载荷辅助电源系统
APR	auxiliary pump room	辅助泵室
APS	antenna pointing system	天线定向系统
APSF	amperes per square foot	安倍每平方英尺
APSI	amperes per square inch	安培每平方英寸
APT	airport traffic (control) tower	机场调度塔台
APU	auxiliary power unit	辅助电源设备,辅助动力装置
APW	automatic pitch warning	自动俯仰告警
APWS	aircraft proximity warning system	飞行器接近告警系统
AQS	advanced quality system	先进质量体系
AR	assembly and repair	装配与修理
AR	alarm release	报警解除
AR	alternate route	备用航线,备用路线
AR	autopilot right	自动驾驶仪正常
AR	air [aerial, airborne] refueling	空中加油

AR	air resupply	空中补给
AR	airborne receiver	机载接收机
AR	aspect ratio	展弦比
AR	auxiliary relay	辅助继电器
ARFP	aerial recovery flight phase	空中改出飞行阶段
ARFP	ACS right forward pitch	姿态控制系统右前向俯仰
ARI	airborne radio instrument	机载无线电仪表
ARI	azimuth range indicator	方位距离指示器
ARM	availability, reliability and maintainability	可用性、可靠性和维修性
ARM	arm	准备
ARN	aeronautical radio navigation	航空无线电导航
ARNG	auto ranging	自动测距
ARO	aircraft requiring overhaul	需要大修的飞机
ARR	air refueling receiver	空中加油受油机
ARR	airborne radio relay	空中转播无线电中继通信,空中无线电中继
ARS	aerial reconnaissance and security	空中侦察与安全
ARS	attitude reference system	姿态参考系统,姿态基准系统
ARS	automatic recovery system	自动回收系统
ARS	auto-relight system	自动重新点火系统
ARTC	air route traffic control	航路交通管制
ARTS	automatic radar tracking system	自动雷达跟踪系统
ARU	attitude reference unit	姿态参考装置,姿态基准装置
ARWS	advanced radar warning system	先进雷达告警系统
AS	altitude select	高度选择
AS	air speed	空速
AS	alarm surveillance	告警监测
AS	autonomous system	自主系统
AS	avionics system	航空电子系统
AS	antisatellite system	反卫星系统
AS	alternation switch	交替开关
AS	attitude switching	姿态转换
ASA	azimuth, speed and altitude	方位、速度和高度
ASAP	aircraft synthesis analysis program	飞机综合分析计划
ASAS	alternate stability augmentation system	交替增稳系统
ASB	advanced system bus	先进系统总线
ASC	anti skid control	防滑操纵
ASC	autopilot self-check	自动驾驶仪自检测
ASC	aerodynamic surface control	空气动力舵面操纵

ASC	avionics system control	航空电子系统控制
ASCP	automatic system checkout program	自动系统检查程序
ASCS	area surveillance control system	地区监视控制系统
ASCS	automatic stabilization and control system	自动稳定与控制系统
ASCT	air system configuration table	航空系统配置表
ASCU	alarm system control unit	警报系统控制装置
ASCU	analog signal converter unit	模拟信号转换装置
ASD	anti skid detection	防滑探测
ASD	accelerate-stop distance	中断起飞距离,加速停止距离
ASD	air situation display	空中情况显示器
ASD	automated structure design	自动化结构设计
ASD	automatic synchronized discriminator	自动同步鉴别器
ASD	average sortie duration	平均架次出动持续时间
ASDA	accelerate-stop distance available	可用中断起飞距离
ASDE	antenna slave data equipment	天线从属数据设备
ASDR	airport surface detection radar	机场地面探测雷达
ASDS	aircraft sound description system	飞机噪声描述系统
ASE	actual steering error	实际操纵误差
ASE	aero servo elasticity	气动伺服弹性
ASE	air surveillance evaluation	空中侦察评估
ASE	airborne search equipment	机载搜索设备
ASE	airborne support environment	机载保障环境
ASE	airborne support equipment	机载保障设备
ASE	allowable steering error	容许操纵误差,容许驾驶误差
ASEW	airborneand surface early warning	空中和地面预警(系统)
ASF	accelerometerscale factor	加速表标度因数
ASI	airspeed indicator	空速表
ASL	above sea level	海拔
ASL	average service life	平均使用寿命
ASM	available seat-miles	可用座英里
ASP	automatic switching panel	自动开关板
ASPJ	advanced self-protection jamming	先进自保护干扰
ASPJ	airborne self-protection jamming	机载自保护干扰
ASPS	advanced self-protecting subsystem	先进自保护子系统
ASR	airport surveillance radar	机场监视雷达
ASR	access service request	访问服务请求
ASR	advance special receiver	先进专用接收机
ASR	air surveillance radar	对空监视雷达
ASR	airborne search radar	机载搜索雷达

ASR	airfield surveillance radar	机场[进近]监视雷达
ASR	airport surveillance radar	机场[航空港]监视雷达
ASR	altimeter setting region	高度表调整范围
ASR	approach surveillance radar	进场[进近]监视雷达
ASR	assigned signal recognition	指定信号识别
ASR	automated speech recognition	自动语音识别
ASR	aviation safety regulation	航空安全条例
ASRB	auxiliary start relay box	辅助起动继电器盒
ASRS	automatic storage/ retrieval system	自动存储与检索系统
ASS	aircraft security system	飞机警戒系统,飞机安全系统
ASS	attitude-sensing system	姿态敏感系统,姿态探测系统
AST	automatic scan tracking	自动扫描跟踪
ASTA	automatic system trouble analysis	自动系统故障分析
ASTAS	antiradar surveillance and target acquisition system	反雷达监视与目标截获系统
ASTC	airport surface traffic control	机场地面交通管制
ASTOL	alternate STOL	备份短距起落飞机
ASV	air solenoid valve	空气螺管(控制)阀门,空气电磁阀门
ASV	automatic shuttle valve	自动往复阀
AT	action time	作战时间,作用时间
AT	air temperature	大气温度
AT	angle tracker	角跟踪仪
AT	auto-throttle	自动油门
AT	actual temperature	实际温度
AT	autopilot test	自动驾驶仪测试
AT	ambient temperature	周围介质温度,环境温度
AT	ampere-turns	安倍匝数
AT	approach time	进场[进近]时间,接近时间
AT	autothrottle	自动油门
AT	autotransformer	自耦变压器
AT	atmospheric temperature	大气温度
ATA	actual time of arrival	实际到达时间
ATA	automated target acquisition	自动目标截获
ATA	automatic threshold adjust	自动阈值调整
ATA	automatic track(ing) acquisition	自动跟踪截获
ATA	automatic trouble analysis	自动故障分析
ATA	actual time of arrival	实际抵达时间
ATAMS	automated tracking and monitoring system	自动跟踪与监控系统
ATAR	acquisition tracking and recognition	截获跟踪与识别

ATB	actual time of block	实际停机时间
ATBM	average time between maintenance	平均维修间隔时间
ATC	actual time of completion	实际完成时间
ATC	approved type certificate	型号合格证,机型批准书
ATC	automatic threat countering	自动威胁对抗
ATC	automatic train control	自动序列控制系统
ATC	automatic tuning control	自动调谐控制
ATC	average total cost	平均总成本
ATC	air traffic control	空中交通管制
ATCAS	air traffic control automatic system	自动空中交通管制系统,空中交通管制自动化系统
ATCBGS	air traffic control beacon ground station	空中交通管制信标地面站
ATCBI	air traffic control beacon interrogator	空中交通管制信标询问器
ATCCC	air traffic control coordination center	空中交通管制协调中心
ATCRB	air traffic control radar beacon	空中交通管制雷达信标
ATCRBS	air traffic control radar beacon system	空中交通管制雷达信标系统
ATCRU	air traffic control radar unit	空中交通管制雷达设备
ATCS	active thermal control system	有源[主动]热控制系统
ATCSE	air traffic control system and equipment	空中交通管制系统与设备
ATCSS	air traffic control signaling system	空中交通管制信号系统
ATCT	air traffic control tower	空中交通管制塔台,机场指挥塔台
ATCU	attitude and translation control unit	姿态与平移控制装置
ATD	actual time of departure	实际离场时间
ATD	advanced technologies demonstration	先期技术演示验证
ATD	average temperature difference	平均温度差
ATDESA	automatic three dimensional electronics scanning array	自动三维电子扫描天线阵
ATDPS	airborne tactical data processing system	机载战术数据处理系统
ATDR	aeronautical technical directive requirement	航空技术指示性要求
ATDS	airborne tactical data system	机载战术数据系统
ATE	airborne test equipment	机载测试设备
ATE	along track error	距离偏差,纵向偏差
ATEM	aircraft test equipment modification	飞机试验设备改进
ATER	automatic testing, evaluation and reporting	自动测试、鉴定与报告
ATEWS	advanced tactical electronic warfare system	先进战术电子战系统
ATEX	automatic test and extraction	自动测试与提取
ATF	air track file	空中跟踪文件

ATF	air traffic flow	空中交通流量
ATF	automatic terrain following	自动地形跟随(雷达)
ATF	automatic transmission fluid	自动传输流体
ATF	aviation turbine fuel	航空燃气轮机燃油
ATFCB	auxiliary tank fire control box	副油箱火控盒
ATFM	air traffic flow management	空中交通流量管理
ATFR	automatic terrain following radar	自动地形跟随雷达
ATFS	automatic terrain following system	自动地形跟随系统
ATH	above the horizon	在水平线之上
ATH	automatic target handoff	目标自动交接
ATHS	airborne target handover system	空中目标交接系统
ATHS	automatic target handoff system	目标自动交接系统
ATI	actual time of interception	实际拦截时间
ATI	automatic track initiation	自动跟踪开始
ATIC	avionics test and integration complex	航空电子测试和综合设施
ATIGS	advanced tactical inertial guidance system	先进战术惯性制导系统
ATIMS	airborne turret infrared measurement system	机载回转红外测量系统
ATICM	advanced threat infrared countermeasure	先进威胁红外对抗
ATIS	airborne test instrumentation system	机载测试仪表系统
ATJC	annular turbojet combustor	环形涡轮喷气发动机燃烧室
ATJS	airborne tactical jamming system	机载战术干扰系统
ATK	aviation turbine kerosene	航空燃气轮机煤油
ATL	actual time of landing	实际着陆时间
ATL	actual total loss	实际总耗损
ATM	air traffic management	空中交通管理
ATM	air transport movement	空中运输动态
ATM	air turbine motor	空气涡轮发动机
ATMP	air target material program	空中目标材料计划
ATMP	air traffic material program	空中交通材料计划
ATMS	advanced terminal managementsystem	先进终端管理系统
ATMS	air traffic management system	空中交通管理系统
ATMSR	aircraft type model series report	飞机型号模型系列报告
ATO	actual time over	实际飞临目标上空时间
ATO	after takeoff	起飞后
ATO	air task order	空中任务分配指令
ATO	aircraft transfer order	飞机转移命令
ATOC	average total operating cost	平均总工作成本
ATOG	allowable takeoff gross(weight)	容许起飞重量

ATP	acquisition, tracking and pointing	捕获、跟踪与瞄准
ATP	augmented thrust propulsion	加力推进
ATPCS	automatic takeoff power control system	自动起飞功率控制系统
ATPD	ambient temperature and pressure	干燥环境温度和压力
ATR	acceptance test requirement	验收试验要求
ATR	actual time of refueling	实际(空中)加油时间
ATR	actual time of return	实际返回时间
ATR	advanced tactical radar	先进战术雷达
ATR	attained turn rate	可达到的(最大)转弯角速度
ATR	auto transformer rectifier	自耦变压器整流器
ATRAN	automatic terrain recognition and naviga-tion (system)	自动地形识别与导航(系统)
ATRID	automatic terrain recognition and identifi-cation device	自动地形识别与鉴定装置
ATRS	automatic terrain recognition system	自动地形识别系统
ATS	acceptance test specification	验收试验规范
ATS	acquisition and tracking system	截获与跟踪系统
ATS	automatic throttle system	自动油门系统
ATS	automatic trim system	自动配平系统
ATS	automatic tuning system	自动调谐系统
ATSS	aircraft trouble shooting system	飞机故障判断系统
ATT	aileron trim tab	副翼调整片
ATT	attenuation	衰减
ATT	attitude	姿态
ATTEMM	aileron trim tab electrical mechanism mark	副翼调整片电动机构标记
AU	above and under	以上与以下
AUCS	advanced UHF communication system	先进超高频通信系统
AUJS	advanced universal jamming system	先进通用干扰系统
AUL	average useful life	平均有效寿命
AUTO	automatic	自动
AA	automatic adaptation	自适应
AV	assistant ventilation	辅助通风
AVASIS	abbreviated visual approach slope indicator system	简化目视进场[近进]下滑道指示器系统
AVCS	attitude, velocity and control subsystem	姿态、速度与控制子系统
AVCS	avionics	航空电子
AVDR	axial velocity density ratio	轴向速度与密度比
AVDT	angular variable differential transformer	角度可变压差式变压器
AVG	average	平均

AVI	angular velocity indicator	角速度指示器
AVI	avionic system	航空电子系统
AVI	avoid verbal information	避免口头通知
AVIC	Aviation Industry Corporation of China	中国航空工业集团公司
AVLF	airborne very low frequency	机载甚低频
AVO	avoid verbal orders	禁用口头命令
AVOI	airspace volume of interest	空域容量
AVR	automatic voice recognition	自动语音识别
AVR	automatic voice relay	自动语音中继
AVV	assistant ventilation valve	辅助通风活门
AW	above-water	在水面上
AW	airway weather	航线天气
AW	airway	航路
AW	all-weather	全天候
AWA	all-weather aircraft	全天候飞机
AWADS	all-weather aerial delivery system	全天候空投系统
AWAR	airborne weather avoidance radar	机载气象警戒雷达
AWARAU	all-weather automatic radio aids unit	全天候自动无线电导航设备
AWARE	advanced warning of active radar emissions	先进有源雷达(电磁波)发射告警(装置)
AWARE	automatic warning and recording equipment	自动告警与记录设备
AWARS	airborne weather and reconnaissance system	机载气象与侦察系统
AWAS	airborne windshear alert sensor	机载风切边报警传感器
AWAS	all-weather approach system	全天候进场[进近]系统
AWAVS	aviation wide-angle visual system	航空广视角系统
AWC	absolute worst case	绝对最坏情况
AWC	airworthiness certification	适航证
AWCS	airborne warning and control system	机载告警与控制系统
AWCS	airborne warning computer system	机载告警计算机系统
AWCS	automatic weapon control system	自动武器控制系统
AWDDS	aircraft weapon delivery system	机载武器投放系统
AWDS	all-weather delivery distribution system	全天候投送分发系统
AWF	airdrome weather forecast	机场天气预报
AWF	aviation weather facility	航空气象设备
AWFGS	all-weather flight guidance system	全天候飞行制导系统
AWGS	all-weather guidance system	全天候制导系统
AWGSA	airborne wave guide slotted array	机上波导式隙缝天线阵

AWH	audio warning horn	音响告警器
AWI	aircraft weight indicator	飞机重量指示器
AWLS	all-weather landing system	全天候着陆系统
AWM	average working man	平均工作人数
AWMPT	AHS wiring module power transformer	AHS 接线模块电源变压器
AWP	aircraft without parts	缺件飞机
AWR	airborne weather radar	机载气象雷达
AWS	abrupt wing stall	机翼突然失速
AWS	advanced warning system	先进告警系统
AWS	air ward system	空中监护系统
AWS	altitude warning system	高度告警系统
AWS	audible [audio] warning system	音响告警系统
AWS	automatic warning system	自动告警系统
AWS	automatic wing sweep	机翼自动后掠
AWSM	air weather service manual	航空气象勤务手册
AWSO	all-weather surface observation	全天候地面观测
AWSS	altitude warning signal system	高度告警信号系统
AWSTAS	all-weather sea target acquisition system	全天候海上目标截获系统
AWU	aural warning unit	音响告警装置
AWY	airway	航路
AYC	aerodynamic yaw coupling	气动力偏航耦合
AYCP	aerodynamic yaw coupling parameters	气动力偏航耦合参数
AYI	angle of yaw indicator	偏航角指示器
AZE	azimuth and elevation	方位-仰角,方位高度
AZS	automatic zero set	自动归零,自动零位调节装置

B

B	bar	巴(压强单位)
B	barn	靶恩(核子有效截面单位)
B	bel	贝尔(电平单位)
B	bit	二进制数(字),二进制码,位
B	blue	蓝色(国际民航组织代码)
B	bomber	轰炸机
B/D	binary to decimal	二十进制,二进制换为十进制
B/M	bill or material	材料单
BA	bleed air	引气,放气
B	bank angle	倾斜角,滚转角
BL	base line	基准线,基线
BA	basic airplane	基本型飞机

BA	beam approach	波束引导进场[进近]着陆
BA	blind approach	仪表进场[进近]
BABS	beam approach beacon system	进场[进近]波束信标系统
BABS	blind approach beacon system	仪表进场[进近]信标系统
BACS	bleed air control system	放气控制系统,引气控制系统
BAD	boom avoidance distance	防声爆距离
BAE	beacon antenna equipment	信标天线系统
BAF	baffle	隔板,折流板,缓冲板,挡板,节气门
BAI	bearing altitude indicator	方位-高度指示器
BAI	blind approach indicator	仪表进场[进近]指示器
BAIS	blind approach instrument system	仪表进场[进近]仪表系统
BALS	blind approach landing system	仪表进场[进近]着陆系统
BAM	basic aircraft maneuvers	飞机的基本机动飞行
BAP	beacon aircraft position	以信标确定的飞机位置
BARS	backup attitude reference system	备用姿态参考系统
BAS	basic airspeed	基本表速
BAS	basic angle system	基本角系统
BAS	blind approach system	仪表进场[进近]系统
BASH	bird/aircraft strike hazard	鸟撞事故
BAT	boom avoidance technique	防声爆技术
BATH	best available true heading	最佳可用真航向
BATM	bureau of air traffic management	空中交通管理局
BATT	battery	蓄电池
BAU	British absolute unit	英制绝对单位
BAU	British association unit	英制标准单位
BB	bus-bar	汇流条,母线,工艺导线
BBL	beacons and blind landing	信标与盲目着陆
BC	battery charge	蓄电池充电
BC	back channel	背航道
BC	back course	反航线
BCA	best cruising altitude	最佳巡航高度
BCAS	beacon-based collision avoidance system	信标防撞系统
BCB	battery contactor board	蓄电池接触板
BCI	bank and climb indicator	倾斜与爬高指示器
BCL	braked conventional landing	用刹车常规着陆
BD	bearing and distance	方位与距离
BD	backward diode	反向二极管
BD-2	big dipper-2	北斗-2
BDE	beam deflection error	波束偏转误差

BDH	bearing, distance, heading	方位,距离与航向
BDP	bottom dead point	下死点
BFA	before flight abort	中断飞行之前
BFC	braking force coefficient	制动摩擦因数
BFD	book for drawing	图册
BFDAS	basic flight data acquisition system	基本飞行数据采集系统
BFDGW	basic flight design gross weight	基本飞行设计总重
BFE	basic flight envelope	基本飞行包线
BFF	body-freedom flutter	机身自由度颤振
BFMDS	base flight management data system	基地飞行管理数据系统
BFO	bit-frequency oscillator	位-频振荡器,比特-频率振荡器
BG	blast gauge	风压计
BGC	beginning climb	开始爬高
BGD	beginning descent	开始下降
BGSIA	beginning standard instrument approach	开始标准仪表进场[进近]
BH	blast hole	吹气孔,鼓风孔,爆破眼,炮眼
BH	bulkhead	舱壁,隔框
BI	barometric input	场压装订
BI	battery inverter	蓄电池变流器
BIFF	battlefield identification friend or foe	战场敌我识别
BEIC	before encountering IFR conditions	遇到盲目飞行条件之前
BIT	built-in test	机内测试,内部测试,内装测试,机内自检测
BITE	built-in test equipment	内装测试设备,机内自检设备
BITFI	built-in test and fault isolation	自检测及故障隔离
BL	boundary light	(着陆跑道)界灯
BLDA	bottom line display area	底行显示区域
BLL	below lower limit	低于下限
BLOS	beyond line of sight	超视距
BLS	beacon landing system	着陆信标系统
BMCS	backup master control station	备用地面控制站
BMEP	brake mean effective pressure	平均有效制动压力
BOA	breakout altitude	(进场着陆时)转入目视飞行高度
BOA	broad ocean area	公海区,外洋区
BOAS	blade outer air seal	(涡轮)叶片外部气封
BOC	bandwidth of coupling	耦合带宽
BOC	basic operational capability	基本工作能力
BOL	beginning of life	寿命初期
BOM	basic operating monitor	基本运行(操作)监督程序

BOM	bill of material	材料清单
BOQ	base ordering quantity	基本订购量
BORIS	breath on recirculation ignition system	循环进气点火系统
BOS	back off system	补偿系统
BOS	backup oxygen system	备份氧气系统
BOS	beam omni select	波束全向选择
BOW	basic operational weight	基本使用重量
BPA	basic pressure altitude	修正压力高度,标准修正表高
BPC	back-pressure control	反压力控制
BPC	barometric pressure control	大气压力控制
BPCU	bus power control unit	总线功率控制装置
BPD	boost pressure difference	增升压差
BPI	bits per inch	比特每英寸,二进位数每英寸
BPL	boot pressure low	气囊压力低
BPR	bypass ratio	涵道比
BPS	balanced pressure system	平衡压力系统
BPS	beacon processing system	信标处理系统
BPS	bits per second	位每秒,比特每秒
BPSC	bearing per standard compass	标准罗盘所示方向
BPT	base point	基点
BR	bit rate	位速率
BR	brake	刹车
BRC	base recovery course	返回主航向,返航航向
BRCS	bearing and range computer system	方位与距离计算机系统
BRI	bearing and range indicator	方位与距离指示器
BRP	beacon ranging pulse	信标测距脉冲
BRS	best range speed	最佳航程速度
BS	band switch	波段开关
BS	brightness sensor	亮度传感器
BS	bus system	总线系统
BB	baseband	基本频带
BSC	binary symmetric channel	二进制对称信道
BSDH	bus shared-data highway	总线共享数据信息通道
BSGS	bootstrap gyro system	自举陀螺系统
BSIA	beginning straight in approach	开始直飞进场[进近]
BSPA	brake system pressure accumulator	刹车系统蓄压瓶
BSRA	beginning standard range approach	开始标准距离进场[进近]
BT	bus tie	汇流条连接线
BTA	basic true altitude	基准绝对高度

BT 专用 E	brake thermal efficiency	制动热效率
BTG	beacon trigger generator	信标触发信号发生器
BTH	beyond the horizon	超视距
BTL	beacon tracking level	信标跟踪高度
BU	backup	备份
BUCS	backup control system	备用控制系统
BUR	bottom-up review	自下而上检查
BV	bleed valve	放气阀
BV	balance valve	平衡活门
BV	battery voltage	蓄电池电压
BV	back view	后视图
BV	balanced voltage	平衡电压
BV	ball valve	球阀
BV	breakdown voltage	击穿电压
BVD	beacon video digitizer	信标视频数字变换器
BVF	boundary vorticity flux	边界窝流量
BVI	blade vortex interaction	桨叶涡旋相互作用
BVP	beacon video processor	信标视频信号处理机
BVP	booster vacuum pump	助推器真空泵
BVPS	beacon video processing system	信标视频信号处理系统
BVR	beyond visual range	超视距,在视觉区以外
BVRI	beyond visual range identification	超视距识别
BW	backward wave	反射波,反向波
BW	bandwidth	(频)带宽
BW	beam width	波束宽度,天线方向图宽度
BWM	backward wave magnetron	回波磁控管
BWO	backward wave oscillator	回波振荡器
BWR	bandwidth ratio	带宽比
BZ	buzzer	蜂鸣器,蜂音器

C

C	capacitor	电容器
C	centigrade (in degrees)	摄氏度
C	clock	数显时钟
C	cockpit	驾驶舱
CA	certificate of airworthiness	(飞机的)适航证
CA	certification and accreditation	认证与认可
CA	compartment and access	舱与舱门
CA	calibrated altitude	校正高度

CA	certificated of airworthiness	适航证
CA	civil aviation	民用航空
CA	conversion angle	转换角,外形变换角
CA	course alignment	航线校准
CA	crab angle	偏航修正角
CA	copilot audio	右音频
CA	control actuator	舵机
CAA	collision avoidance aids	防撞辅助设备
CAAC	civil aviation advisory committee	民航咨询委员会
CAAS	computer aided approach spacing	计算机控制的进场[进近]间隔时间
CAAS	computer assisted approach sequencing	计算机辅助进场[进近]程序
CAATC	civil aeronautics administration type certificate	民航局飞机型号证书
CAB	common avionics baseline	通用航空电子设备基线
CAB	critical air blast	最低鼓风量,临界风量
CAB	cabin	座舱,客舱,机舱
CAC	centralised approach control	集中进场[进近]控制
CAC	changing to approach control	改为进场[进近]控制
CAC	constant altitude control	等高控制
CAC	contact approach control	目视进场[进近]控制
CAD	cushion augmentation device	增升装置,近地增升装置
CADL(S)	communication(s) and data link (system)	通讯与数据链(系统)
CADU	channel access data unit	信道访问数据装置
CAF	calibration flight	校正飞行
CAG	constant altitude glide	等高滑翔
CAGC	clutter automatic gain control	杂乱回波自动增益控制
CAGC	coded automatic gain control	编码自动增益控制
CAGE	common air ground environment system	通用地面防空警备系统
CAI	compression after impact	冲击后压缩强度
CAI	countermeasure airborne infrared	机载红外对抗
CAIE	control aileron in emergency	应急操纵副翼
CAIS	common airborne instrumentation system	通用机载仪表系统
CAL	calculated average life	平均计算寿命
CAL	calibration	标定,校准
CAM	cockpit angle measure	驾驶舱视角界限
CAMP	control and monitor panel	控制与监控板
CAN	correlation air navigation	相关空中导航
CANARI	communication and navigation airborne radio instrumentation	机载无线电通信与导航仪器

CANP	collision avoidance notification procedure	防撞程序说明书,防撞告知程序
CAOA	critical angle of attack	临界攻角
CAR	C-band angle receiver	C 波段角度数据接收机
CAR	civil aeronautics regulations	民用航空条例
CARA	combined-altitude radar altimeter	组合高度雷达高度表
CAS	calibrated airspeed	标定表速,修正空速
CAS	close air support	近距空中支援
CAS	cockpit avionics system	座舱航空电子系统
CAS	collision avoidance system	防撞系统
CAS	control augmentation system	控制增稳系统
CAS	calibrated air speed	修正表速
CASC	combined acceleration and speed control	加速度与速度综合控制
CASS	crab angle sensing system	偏航角传感系统
CASW	counter antisubmarine warfare	对抗反潜战
CAT	category	种类,范畴,类目,飞行阶段
CAT	clear air turbulence	晴空湍流
CAT	compressed air tunnel	压缩空气风洞
CATCV	cockpit area temperature control valve	驾驶舱区域温度控制活门
CAV	compensated actuating voltage	补偿作动电压
CAV	constant angular velocity	等角速度
CAWS	central aural warning system	中央音响告警系统
CAWS	common aviation weather subsystem	通用航空气象子系统
CB	circuit breaker	断路器,电路自动保险电门,断路开关
CB	communication bus	通信总线
CB	compass bearing	罗盘方位
CB	control box	控制盒
CB	control button	控制按钮
CB	contactor board	接触器板
CBAL	counterbalance	配重,平衡块,移轴补偿
CBF	compressor blade form	压气机叶片形式
CB	call back	回叫
CL	cable length	缆长,索长
CBP	control box panel	控制盒面板
CBR	continuous bit rate	连续比特率
CBT	C-band transponder	C 波段发射机应答器
CC	calibration and certification	校准与证明
CC	calibration and checkout	校准与检查
CC	command and control	指挥与控制

CC	communication(s) and control	通信与控制
CC	change of course	航向改变
CC	concentric cable	同轴电缆,同心电缆
CC	choke coil	扼流圈
CC	coefficient of correction	修正系数
CC	combustion chamber	燃烧室
CC	concentric cable	中轴电缆
CC	coupling condenser	耦合电容器
CC	critical condition(s)	临界条件,临界状态
CC	center console	中央操纵台
CC	changeover contactor	转换接触器
CC	coupling coefficient	耦合系数
CCALPB	contactor controlling anti-collision light power box	控制防撞灯电源盒的接触器
CCAR	China civil aviation regulations	中国民用航空规章(条例)
CCB	changeovercontrol box	转换控制盒
CCLG	cabin ceiling light group	座舱顶灯组
CCM	communication(s) countermeasures	通信对抗
CCM	cooperative countermeasures	协同对抗
CCM	counter-countermeasures	反对抗
CCO	crystal controlled oscillator	晶体控制振荡器
CCO	current controlled oscillator	电流控制振荡器
CCOC	combustion chamber outer casing	燃烧室外壳,燃烧室外套
CCP	critical compression pressure	临界压缩压力
CCP	center controller panel	集中控制板
CCPR	cruise compressor pressure ratio	巡航状态压缩机压力比
CCR	closed-circuit refueling	闭式加油
CCR	critical compression ratio	临界压缩比
CCSTP	cubic centimeters at standard temperature and pressure	在标准温度和压力下的立方厘米
CCW	counter-clockwise	逆时针,反时针方向(转动)
CD	controls and displays	控制器与显示器
CD	coder/decoder	编码器/译码器
CD	cabin door	舱门
CD	controlled diffusion	可控扩散度
CD	current density	电流密度
CD	coefficient of drag	阻力系数
CD	close door	关舱门
CD	coder-decoder	编码译码器

CD	current difference	电流差安(A)
CD	current distributor	分流器
CDA	concept demonstrator aircraft	概念验证机
CDA	continuous descent approach	连续下降进场[进近]
CDA	current data array	现行数据阵列(数组)
CDACU	controls and displays avionics control unit	控制与显示航空电子控制装置
CDAS	central data acquisition system	中央数据采集系统
CDC	course and distance computer	航线与距离计算机
CDG	capacitor diode gate	电容器二极管门
CDI	course deviation indicator	偏航指示器
CDMLS	commutated-Doppler microwave landing system	转换式多普勒微波着陆系统
CDS	course director system	航道指引系统
CDSR	controldisplay system -retrofit	控制显示系统-改装
CDT	control differential transmitter	控制差动变压器
CDU	cockpit display unit	座舱显示装置
CDU	control display unit	控制显示器
CDUO	coupling display unit-optics	光学耦合显示装置
CDW	charge density wave	电荷密度波
CE	circular error	径向偏差,圆误差
CE	clutter elimination	杂波消除
CE	charge exchange	电荷交换
CEA	circular error average	平均径向偏差,平均圆误差
CEA	combined electronics assembly	组合式电子设备
CEB	curve of equal bearings	等方位曲线
CEF	cost-effective flight	低成本高效飞行
CEFH	cumulative engine flight hours	累积发动机飞行小时
CEFR	contactor engaging feathering pump	接通顺桨泵的接触器
CEGS	capture effect glide slope	截获效应下滑斜度
CEI	critical engine inoperative	关键发动机不工作
CEIE	control elevator in emergency	应急操纵升降舵
CELSS	controlled ecological life support system	受控生态生命保障系统
CEPR	cabin emergency pressure release	座舱应急释压
CES	central electronics system	中央电子系统
CETC	cockpit electronic temperature controller	驾驶舱电子温度控制器
CEWI	combat electronic warfare intelligence	作战电子战情报
CF	flap control	襟翼控制
CF	chaff and flare	箔条与红外
CF	captive flight	载飞试验,系留飞行试验

CF	certainty factor	可信度因子,确信度
CF	characteristic frequency	特征频率
CF	check flight	技术检查飞行
CF	constant frequency	恒定频率
CFA	cleared for approach	为飞机进场清除障碍
CFC	cockpit flow control	驾驶舱流量控制
CFD	chaff/flare dispenser	箔条/曳光弹投放器,干扰投放机
CFD	cumulative frequency distribution	累积频率分布
CFDS	chaff/flare dispensing system	箔片/闪光弹投放系统
CFF	critical flicker frequency	临界闪烁频率
CFHS	communicator foot heating switch	通讯员脚加温开关
CFIT	controlled flight into terrain	可控飞行撞地
CFR	check feathering relay	检查顺桨继电器
CFWT	compressible flow wind tunnel	可压缩流风洞
CG	center of gravity	重心,重心位置
CGA	color graphic adaptor	彩色图形显示卡,彩色图形适配器
CGA	contrast gate amplifier	对比选通放大器
CGC	cruise guidance control	巡航制导控制
CGL	controlled ground landing	受控着陆
CGML	combined gust and maneuver load	突风与机动飞行合成载荷
CGS	combined guidance system	联合制导系统
CGSE	common ground support equipment	通用地面保障设备
CH	chaff/channel	箔条弹/波道
CH	control handle	操纵手柄
CHU	centigrade heat unit	摄氏热单位
CHUM	aeronautical chart updating manual	航空地图修订手册
CHV	channel valid	通道有效
CI	communication and instrumentation	通信与仪表设备
CIA	cockpit integration area	座舱综合区
CII	configuration identification index	构型识别标记
CIL	cockpit instruments and lighting	座舱仪表与照明
CIPR	continuous in-flight performance recorder	连续飞行性能记录器
CIPR	cubit inches per revolution	立方英寸每转
CIPS	cockpit instrument panel space	座舱仪表板空间
CIR	coherent imaging radar	相干成像雷达
CI	color infrared	彩色红外
CIR	completed inspection records	完整检验记录
CIT	compressor inlet temperature	压缩器进口温度
CL	centreline	中(心)线,(中)轴线

CL	checklist	（核对用的）清单
CLP	climb flight phase	爬高飞行阶段
CL	closed loop	闭环,闭合回路
CL	cockpit light	驾驶员照明灯
CL	control LG	操纵起落架
CLB	crash locator beacon	应急定位信标机
CLGCOG	cutoff landing gear control on the ground	地面断开起落架操纵
CLGD	closel anding gear door	关闭起落架舱门
CLR	clear	批准放飞,批准,允许,清除
CLS	contingency landing site	应急着陆场
CLV	ceiling limit value	升限值
CM	control and monitor	控制与监控
CM	calibrated Match	标准马赫数
CM	cargo module	货仓
CMA	contract maintenance activity	合同维修活动,保修活动
CMEPS	command module electrical power system	指挥舱电源系统
CMPM	cubic meter per minute	立方米每分钟
CMR	continuous maximum rating	最大连续功率
CMSR	commercial/military spares release	商用/军用备件支付
CMT	corrected mean temperature	平均修正温度
CN	communication and navigation	通信与导航
CN	control and navigation	控制与导航
CN	channel	频道,通道,波道
CN	channel normal	通道正常
CNEDIE	close nose emergency door in emergency	紧急关闭前应急门
CNEDN	close nose emergency door normally	正常关前应急门
CNI	communication navigation instrument	通信导航仪表
CNI	communication, navigation, identification	通信,导航,识别
CNL	circuit net loss	电路净耗损
CNL	constant net loss	恒定净耗损
CNL	corrected noise level	经修正的噪声级
CNR	carrier to noise ratio	载波噪声比
CNS	common nacelle system	共用短舱系统
CNS	communication, navigation and surveillance	通信、导航与监视
CO	cabin overpressure	座舱超压
CO	controllability and observability	可控性与可观察性
CO	compressor outlet	压气机出口
COA	change of assignment	任务更改

COF	centrifugal oil filter	离心(式)滑油滤
COFEC	cause of failure，effect and correction	失败原因、影响及纠正,失事原因、影响及改正
COFW	certification of flight worthiness	飞行适航性鉴定
COMM	communications	通信
COPE	controlled operating pressure engine	可控压比发动机
CPT	complete procedure turn	完全程序转弯(飞行)
CRA	coherent radar array	相干雷达阵
COT	checkout time	检查时间
COT	current operating time	现行使用时间
COLTS	compressor outlet temperature sensor	压气机出口温度传感器
COTS	container off loading and transfer system	集装箱卸货与传输系统
COW	clean operational weight	净飞行重量
CP	constant power	恒定功率,不变功率
CP	cabin pressure	座舱压力
CP	cabin pressurization	座舱增压
CP	central processor	中央处理机
CP	clock pulse	时钟脉冲
CP	coefficient of performance	性能系数
CP	constant potential	固定电位,恒定势
CP	constant pressure	恒压
CP	controllable pitch	可控桨距,可操纵桨距
CPA	closest point of approach	最近进场[进近]点
CPA	constant potential accelerator	静电加速器,等电势加速
CPA	critical path analysis	关键路径分析
CPC	cabin pressure controller	座舱压力控制器
CPCR	cutoff power circuit relay	断开供电电路继电器
CPD	contact potential difference	接触势差
CPGC	course per gyro compass	按陀螺罗经航向,按陀螺罗盘指示的航向
CPH	counts per hour	计算次数每小时
CPILS	correlation protected ILS	相关保护仪表着陆系统
CPM	cycles per minute	周每分钟
CPM	corrosion prevention manual	防腐手册
CPMV	cold plate manual valve	冷板手动阀
CPOPS	cold plate outlet pressure sensor	冷板出口压力传感器
CPOTS	cold plate outlet temperature sensor	冷板出口温度传感器
CPS	combined power supply	复合供电
CPS	cycles per second	周每秒

CPS	characters per second	字符数每秒
CPT	cockpit procedure trainer	飞机座舱程序训练机
CPTS	cockpit pipeline temperature sensor	驾驶舱管路温度传感器
CR	convoy and routing	护航与开辟航线
CR	combined receiver	组合接收机
CR	capacitance resistance	阻容
CR	countermeasures receiver	对抗接收机
CR	crash recorder	失事记录仪
CRECBPD	combination receiving equipment control box panel drawing	组合接收设备控制盒面板图
CRF	clutter reducing features	杂波减少特性
CRPM	compressor revolutions per minute	压缩器每分钟转数
CRR	calibration, repair and return	校正、修理与送回
CRR	change request reporting (system)	更改请求报告（系统）
CRSS	critical resolved shear stress	临界解析剪应力
CS	channel selection	波道选择
CS	communication(s) security	通信安全,通信保密
CS	cross section	横截[断]面
CS	current strength	电流强度
CS	control stick	驾驶杆
CSAS	command and stability augmentation system	控制增稳系统,指令增稳系统
CSB	code select button	代码选择旋钮
CSB	carrier and sideband	载波与边带
CSCF	constant speed constant frequency	恒速恒频
CSCFPS	constant-speed constant-frequency power system	恒速恒频交流电源系统
CSL	control surface lock	舵面锁
CSL	control surface lockout	舵锁定
CSLB	center signal light box	中央信号灯盒
CSLEM	control surface lockout electrical mechanism	舵面锁电动机构
CSLS	control surface lock signal	舵面锁信号
CSTOL	controlled short takeoff and landing	受控短距起落飞机
CSTS	cockpit skin temperature sensor	驾驶舱蒙皮温度传感器
CT	communication and tracking	通信与跟踪
CT	compatibility test	兼容性测试
CT	constant temperature	恒温
CT	current transformer	变流器,电流互感器

CTAM	climb to and maintain	爬升后保持高度
CTC	cabin temperature controller	座舱温度控制器
CTDG	command track and distance to go	给定航迹和待飞距离
CTE	coefficient of thermal expansion	热膨胀系数
CTR	critical temperature resistor	临界温度电阻器
CTS	cockpit temperature sensor	驾驶舱温度传感器
CTU	centigrade thermal unit	摄氏热量单位
CTW	conventional takeoff weight	常规起飞重量
CV	cutoff valve	断流活门
CV	cross valve	交叉活门
CV	velocity coefficient	流速系数
CV	calibrated velocity	校准速度
CV	constant value	恒定值
CV	constant velocity	等速
CV	continuous variation	连续变化
CVF	compressor vane form	压气机静子叶片成形
CVF	controlled visual flight	可控目视飞行
CVL	controlled-vortex lift	可控涡升力
CVR	cockpit voice recorder	座舱音频记录器
CVV	compressor variable vane	压气机可变静止叶片
CW	caution and warning	预防与告警
CW	control and warning	控制与告警
CW	carrier wave	载波
CW	clockwise	顺时针,顺时针(转动)的
CWAR	continuous wave acquisition radar	连续波截获雷达
CWC	cross wind component	侧风分量
CWEWR	centimeter wave early warning radar	厘米波预警雷达
CWF	cross wind force	侧向风力
CWL	control wheel lighting	驾驶盘照明
CWLB	center warning light box	中央告警灯盒
CWS	caution and warning system	预防与告警系统
CWS	control wheel steering	驾驶盘操纵
CWTR	climb well to right	爬升到右方
CWV	crest working voltage	峰值工作电压
CY	calendar year	历年

D

| D | diode | 二极管 |
| D | distance | 距离 |

D	diameter	直径
D	drag（force）	阻力
DA	decision altitude	决策高度
DA	drift angle	偏流角
DA	directional antenna	定向天线
DA	display area	显示区域
DA	departure airfield	出发机场
DABNS	discrete address beacon and navigation system	离散地址信标和导航系统
DAFC	departure airfield control	飞离机场控制
DALM	deflection angle limited mechanism	偏角限制机构
DALO	disconnect at lift off	起飞脱开
DAM	display ATC mode	显示 ATC 模式
DAMI	designated aircraft maintenance inspector	委任航空器维修检查员
DAR	designated airworthiness representative	委任适航代表
DARAC	damped aerodynamic righting attitude control	阻尼式气动力姿态校正控制
DARAS	direction and range acquisition system	方向与距离探测系统
DARE	Doppler and range evaluation	多普勒导航与航程估计
DARE	Doppler automatic reduction equipment	多普勒自动处理装置
DARE	DOVAP automatic reduction equipment	多普勒测速与定位自动换算装置
DARS	data acquisition and recording system	数据采集和记录系统
DARS	digital adaptive recording system	数字式自适应记录系统
DARS	digital attitude reference system	数字姿态基准系统
DARS	drogue air refueling system	锥套式空中加油系统
DAS	data acquisition system	数据采集系统
DAS	data analysis system	数据分析系统
DAS	digital avionics system	数字航空电子系统
DAS	door actuation system	舱门驱动系统
DAS	dynamo alert system	（直流）发电机警报系统
DASF	direct air support flight	直接空中支援飞行
DASR	daily aircraft status report	每日飞机状况报告
DASS	defence aids support system	防御手段保障系统
DATR	design approval test report	设计批准试验报告
DAU	data acquisition unit	数据采集单元
DAVI	dynamic antiresonant vibration isolator	动态反谐振缓冲器
DAWS	design of aircraft wing structure	机翼结构设计
Db	decibel	分贝
DB	distribution box	配电箱[盒],分线盒

DBI	differential bearing indicator	差动方位(角)指示器
DBL	distributing board lighting	配电盘照明
DBL	drawing breakdown list	图纸明细表
DBLGD	distributing box of landing gear door	起落架舱门配电盒
DBOLFP	distributing box of left fuel pump	左燃油泵配电盒
DBS	Doppler beam sharpening	多普勒波束锐化
DBTF	database terrain following	数据库地形跟随
DBTF	duct burning turbofan	外涵道燃烧加力涡扇发动机
DC	drift and correction	偏移修正,漂移校正,偏流校正
DC	direct current	直流电
DC	digital clock	数显时钟
DC	display controller	显示器控制盒
DC	dust cleaner	吸尘器
DC	discharge coefficient	放电系数
DC	drag coefficient	阻力系数
DCB	double cantilever beam	双悬臂梁
DCG	direct current generator	直流发电机
DCL	digital channel link	数字信道数据链
DCM	drip chamber mark	沉淀槽标记
DCP	data collection platforms	数据收集平台
DCP	digital communication protocol	数字通信协议
DCP	display controller panel	显示控制板
DCPS	direct current power system	直流电源系统
DCPU	display control power unit	显示控制电源组件
DCR	design characteristic review	设计(结构)特性检查
DCS	DC sensor	直流电流传感器
DCSA	direct current servo amplifier	直流伺服放大器
DCV	demonstrated crosswind velocity	指示侧风速度
DCV	directional control valve	定向控制阀
DD	departure delay	起飞延迟
DD	deflection down	下偏
DD	directional Doppler	方向性多普勒雷达
DDFS	direct digital frequency synthesizer	直接数字频率合成器
DDG	digital display generator	数字显示发生器
DDR	decoy discrimination radar	假目标识别雷达
DDR	determination of direction and range	定向与测距装置
DDT	durability and damage tolerance	耐久性和损伤容限
DEA	data exchange agreement	数据交换协议
DECM	defense electronic countermeasures	防御电子对抗

DECM	deception electronic countermeasures	欺骗电子对抗
DECR	digital electronic continue ranging	数字电子连续测距
DEFCS	digital electronic flight control system	数字式电子飞行控制系统
DER	declining error rate	衰减误差率,下倾误差率
DER	designated engineering representative	委任工程代表
DES	data encryption standard	数据加密标准
DEWIZ	distance early warning identification zone	远程预警识别区
DF	direction finder	测向器,无线电罗盘
DFA	delayed-flap approach	迟放襟翼进场[进近]
DFCF	double flap channels failure	襟翼双通道故障
DFCS	digital flight control system	数字式飞控系统
DFDR	digital flight data recorder	数字飞行数据记录仪
DFE	differential frictional effect	差动摩擦效应
DFE	directional frictional effect	定向摩擦效应,方向性摩擦效应
DFM	display FID mode	显示 FID 模式
DFP	deviant flight plan	偏离飞行计划
DFR	digital flight recorder	数字式飞行记录仪
DFR	dynamic flap restraint	动力襟翼限动器
DFS	discrete Fourier series	离散傅里叶级数
DFS	dynamic flight simulator	动力学飞行模拟器
DFT	discrete Fourier transform	离散傅里叶变换
DFT	distance from threshold	距跑道端头的距离
DFTI	distance-from-touchdown indicator	接地后滑跑距离指示器
DG	directional gyro	陀螺半罗盘,航向陀螺仪
DG	displacement gyro	位移陀螺
DG	directional gyro	航向陀螺
DGPS	differential global positioning system	差分式全球定位系统
DGRB	DC generator resistor box	直流发电机电阻盒
DGSI	drift and ground speed indicator	偏流地速指示器
DH	desired heading	规定航向
DH	decision height	决断高度
DHOO	daily hours of operation	每日运行小时数
DHOSVM	dismounting hydraulic oil system valve mark	拆卸液压油系统活门标记
DI	daily inspection	每日检查,日常检查
DIF	de-icing fuse	除冰熔断器
DIALS	digital integrated automatic landing system	数字式综合自动着陆系统
DIANS	Decca integrated airborne navigation sys-	台卡综合机载导航系统

猞猁

DIEL	Doppler inertial erection loops	多普勒惯性修正回路
DIG	differential laser gyro	差动激光陀螺
DIGC	Doppler inertial gyro compass	多普勒惯性陀螺罗盘
DILS	Doppler inertial LORAN system	多普勒惯性远程导航系统
DILS	Doppler instrument landing system	多普勒仪表着陆系统
DIM	dimmer	调光器,调光旋钮,亮度调节器
DINS	digital inertial navigation system	数字惯性导航系统
DL	down lock	下位锁
DLF	design load factor	设计载荷因数,设计过载
DLF	dynamic load factor	动载荷因数,动过载
DLL	design limit load	设计最大使用载荷
DLLF	design limit load factor	设计最大使用载荷因数
DLM	dead load moment	恒载力矩
DLME	direct lift and manoeuvre enhancement	直接升力和机动性增强
DLS	distance least squares	最小方差
DLS	DME landing system	测距仪着陆系统
DM	digital map	数字地图
DM	distributing module	配电模块
DMD	displacement measuring device	位移测量装置
DME	distance measuring equiment	测距设备
DMIR	designated manufacturing inspection representative	委任生产检查代表
DMS	deviation mean standard	平均标准偏差
DMS	digital map system	数字地图系统
DMTOGW	design mission takeoff gross weight	设计任务起飞总重
DN	drawing number	图号
DNC	day/night capability	昼夜能力
DNS	Doppler navigation system	多普勒导航系统
DOF	direction of flight	飞行方向
DOI	data of intercept	截获日期
DOL	difference of longitude	经度差
DOT	display operating transponder	显示正在工作的应答机
DPF	differential pressure feedback	差压反馈
DPI	dots per inch	点数每英寸
DPS	differential phase shift	差分相移
DPU	display processing unit	显示处理计算机
DR	distance ratio	距离比
DR	dead reckoning	推测领航

DR（C/L）	double receiving (communication/lifesaving)	双收
DRDB	dual-redundant data bus	双余度数据总线
DRS	distance ring selection (invalid)	距离环选择（无效）
DRVID	difference range versus integrated Doppler	不同距离与综合的多普勒雷达
DRVS	Doppler radar velocity sensor	多普勒雷达速度传感器
DRVS	Doppler radar velocity system	多普勒雷达测速系统
DRVS	Doppler radial velocity sensor	多普勒径向速度传感器
DRVS	Doppler radial velocity system	多普勒径向速度测量系统
DS	discrete signal	离散信号
DSAL	decelerated steed approach and landing	急剧减速进场[进近]着陆
DSATC	descend so as to cross	为了穿越而下降
DSATR	descend so as to reach	为了到达而下降
DSB	double sideband	双边带
DSBSC	double sideband suppressed carrier	双边带抑制载波
DSCT	double secondary current transformer	双次级电流变压器
DSF	design safety factor	设计安全系数
DSG	digital signal generator	数字信号发生器
DSMT	dual-speed magnetic transducer	双速磁传感器
DSPB	displacement signal processor box	位移信号调理器盒
DSPM	double strokes per minute	每分钟双行程
DSS	decision support system	决策支持系统
DSS	dynamic system synthesizer	动态系统合成器
DSSB	double single sideband	双单边带
DSTR	destructor	爆炸装置
DT	damage tolerance	损伤容限
DT	desired track	所需航迹,预定航线
DTC	data transfer card	数据传输卡
DTD	damage tolerant design	损伤容限设计
DTD	digital terrain data	数字地形数据
DTD	data transfer device	数据传输装置
DTDR	day to day repeatability	逐日重复性
DTG	distance to go	待飞距离
DTL	dead time log	失效时间记录薄
DTLCC	design to life cycle cost	（按）寿命周期费用设计
DTM	digital terrain model [matrices]	数字地形模型,数值地形
DTR	dive toss release	俯冲拉起投放
DTR	damage tolerance rating	损伤容限等级
DTVM	differential thermocouple voltmeter	差动热电偶电压表

DTVOR	Doppler terminal VOR	多普勒航站伏尔系统
DTWA	dual trailing wire antenna	双拖曳线天线
DU	deflection up	上偏
DU	display unit	显示装置
DUL	design ultimate load	设计极限载荷
DUVR	differential under voltage relay	差动低限继电器
DV	design validation	设计验证
DVARS	Doppler velocity altimeterradar set	多普勒速度-高度表雷达装置
DVM	digital voltmeter	数字式电压表
DVOR	Doppler VOR（system）	多普勒伏尔系统
DVHFRR	Doppler very high frequency radio range	多普勒甚高频无线电信标
DWN	down	下,放下
DZ	drop（pin）zone	空投区,空降区
DZA	drop zone area	空投区面积,空降区面积

E

E/E	electrical/electronic	电气/电子
E/R	extend/retract	收/放
EA	experimental aircraft	实验用飞机
EAB	engine air bleed	发动机供气
EAC	estimated approach control	估计进场[进近]控制
EAC	expected approach clearance	预期进场[进近]许可
EACS	electronic automatic chart system	自动电子地图显示系统
EAD	effective air distance	有效空中距离
EADI	electronic attitude director indicator	电子指引地平仪
EAF	effective attenuation factor	有效衰减系数
EAF	engine auto-feathering	发动机自动顺桨
EAFS	engien anti-fire switch	发动机防火开关
EAFSM	engine anti-fire switch mark	防火开关标记
EAGLE	elevation angle guidance landing equip-ment	仰角导引着陆装置
EAL	estimated average life	预计平均寿命
EAL	expected average life	预期平均寿命
EARS	emergency airborne reaction system	机载应急反应系统
EARS	environment analog recording system	环境模拟记录系统
EARTS	enroute automated radar tracking system	航路自动化雷达跟踪系统
EAS	equivalent air speed	等效空速,当量空速
EAS	electronic anti-skid	电子防滑
EAS	engine auto shutdown	发动机自动停车

EASASB	electric anti-skid and static brake	电子防滑和静刹车
EASCR	electronic anti-skid changeover relay	电子防滑转换继电器
EASCV	ECU air supply check valve	ECU 供气单向活门
EASS	engine automatic stop and start system	发动机自动停车与起动系统
EASS	electronically controlled automatic switching system	电子控制的自动转换系统
EASTS	ECU air supply temperature sensor	ECU 供气温度传感器
EAT	earliest arrival time	最早到达时间
EAT	electronic angle tracking	电子角度跟踪
EAT	estimated approach time	预计进场[进近]时间
EAT	expected approach time	预期进场[进近]时间
EAV	effective angular velocity	有效角速度
EB	essential bus	重要设备汇流条
EB	EICAS battery	EICAS 蓄电池
EB	emergency bus	应急汇流条
EBAO	engine bleed air overpressure	发动机引气超压
EBS	electronic braking system	电刹车系统
EBT	effective brightness temperature	有效亮度温度
EBW	effective bandwidth	有效带宽
EC	elasticity coefficient	弹性系数
EC	engineering change	工程更改
EC	ejection cooling	引射散热
EC	electromagnetic clutch	电磁离合器
EC	elevator clutch	升降舵离合器
EC	emergency control	应急操纵
EC	engine control	发动机控制
EC	equipment cabin	设备舱
EC	excitation compensation	励磁补偿
ECA	earth central angle	地心角
ECASS	electronic-controlled automatic switching system	电子控制的自动转换系统
ECATCV	equipment cabin area temperature control valve	设备舱区域温度控制活门
ECB	electronic control box	电子控制盒
ECC	embedded communication channel	嵌入式通信信道
ECD	eddy current damper	涡电流阻尼器
ECETC	equipment cabin electronic temperature controller	设备舱电子温度控制器
ECI	engineering change information	工程变更通知

ECLCB	equipment cabin lighting control board	设备舱照明控制板
ECLG	emergency control landing gear	应急控制起落架
ECLGD	emergency closing landing gear door	应急关闭起落架舱门
ECLGDHP	emergency closing landing gear door with hand pump	手摇泵应急关闭起落架舱门
ECLSS	environmental control and life support system	环境控制与生命保障系统
ECM	electronic countermeasures	电子对抗
ECM	engine condition monitoring	发动机状态监视[监控]
ECMS	ECM simulator	电子对抗模拟器
ECOS	electromagnetic compatibility operational system	电磁兼容性操作系统
ECPTS	equipment cabin pipeline temperature sensor	设备舱管路温度传感器
ECS	encryption control signal	加密控制信号
ECS	environmental control system	环控系统,环境控制系统
ECS	emergency control switch	应急操纵开关
ECSTS	equipment cabin skin temperature sensor	设备舱蒙皮温度传感器
ECT	estimated completion time	预计完成时间
ECTCV	equipment cabin temperature control valve	设备舱温控活门
ECU	electronic control unit	电子控制装置
ED	EICAS display	EICAS 显示器
ED	emergency distance	应急距离
ED	emergency door	应急舱门
ED	entry door	登机门
EDAC	error detection and correction	误差探测与校正
EDM	electronic distance measuring	电子测距
EDP	engine driven pump	发动机传动泵
EDP-I	EICAS display processor-I	EICAS 显示处理机-I
EDS	emergency detection system	应急故障探测系统
EDS	emergency disengaging start	紧急断开起动
EDSIR	engine disengaging starting interlock relay	发动机断开起动连锁继电器
EE	electrical and electronics	电气和电子
EE	emergency exit	紧急出口
EEB	external equipment bus	外部设备汇流条
EEC	engine electronic control	发动机电子控制
EED	emergency escape device	应急逃逸装置
EEE	electromagnetic environmental efforts	电磁环境效应
EEF	engine emergency feathering	发动机应急顺桨

EESS	emergency escape sequencing system	应急逃逸定序系统
EET	estimated elapsed time	预计经过时间,预计已飞时间,估计的航程时间
EETC	ECU electric temperature controller	ECU 电子温度控制器
EF	effective factor	效率因素
EF	engine feathering	发动机顺桨
EF	extending flap	放襟翼,打开襟翼
EFA	engine fire assembly	发动机灭火组件
EFAL	electronic flash approach lighting	电子闪光进场[进近]照明
EFATO	engine failure after takeoff	起飞后发动机停车
EFB	electronic flight bag	电子飞行包
EFCB	engine fire control box	发动机灭火控制盒
EFCS	electrical flight control system	电飞行控制系统
EFCS	emergency flight control system	应急飞行控制系统
EFCSA	emergency flight control system available	应急飞行控制系统可用
EFCSE	emergency flight control system engaged	应急飞行控制系统接通
EFCV	engine fuel crossfeed valve	发动机燃油连通开关
EFDS	electronic flight data system	电子飞行数据系统
EFFI	emergency full fuel indication	应急满油指示
EFFR	exit from feathering relay	退出顺桨继电器
EFG	electric field gradient	电场梯度
EFI	electronic flight instrument	电子飞行仪表
EFID	electronic flight instrument display	电子飞行仪表显示器
EFIS	electronic flight instruction	电子飞行指令
EFIS	electronic flight instrument system	电子飞行仪表系统
EFP	emergency flight phase	应急飞行阶段
EFPD	effective full power days	有效全功率天数
EFPH	equipment full power hours	设备全功率小时
EFR	equipment failure rate	设备故障率
EFR	enter feathering relay	进入顺桨继电器
EFS	electric field strength	电场强度
EFSR	engine fail signal relay	发动机故障信号继电器
EFSSS	engine failure sensing and shutdown system	发动机故障指示和停车系统
EFT	earliest finish time	最早完成时间
EFT	estimated flight time	预计飞行时间
EFT	engineering flight test	技术飞行试验,技术试飞
EFWHP	extending flap with hand pump	手摇泵打开襟翼
EG	emergency gyroscope	应急地平仪

EGC	electronic gyro compass	电子陀螺罗盘
EGCNS	electronic geographic coordinate navigation system	电子地理坐标导航系统
EGI	embedded GPS/ INS	嵌入式全球定位/惯性导航系统
EGPWS	enhanced ground proximity warning system	增强型近地告警系统
EGR	engine ground run	发动机地面试车
EGSE	electrical ground support equipment	电气地面保障设备
EGT	elapsed ground time	地面停留时间
EGT	exhaust gas temperature	排气温度
EHHM	engine heating hatch mark	发动机加温舱口标记
EHP	equivalent horsepower	当量马力
EHP	effective horsepower	有效马力
EHS	electrical hydraulic switch	电液开关
EHSI	electronic horizon situation indicator	电子航道罗盘
EHWTC	electric heating water tank control	电加热水箱控制
EIA	engine icing annunciator	发动机结冰信号器
EIC	engien inner cavity	发动机内腔
EICAS	engine indication and crew alert system	发动机指示和空勤告警系统 发动机(参数)指示与机组告警系统
EICFE	engine inner cavity fire extinguishing	发动机内腔灭火
EIEC	electromagnetic interference/electromagnetic compatibility	电磁干扰/电磁兼容
EIH	engine intake heating	发动机进气道加温
EII	engine intake icing	发动机进气道结冰
EIPR	ECU inlet pressure regulator	ECU 进口压力调节器
EIPS	engine ice protection system	发动机防冰系统
EIS	electronic instrument system	电子仪表系统
EJB	engine junction box	发动机接线盒
EK	engine kerosene	发动机煤油
EL	electronic logbook	电子飞行日志
EL	emergency light	应急灯
EL	engines lighting	发动机照明
ELAC	enroute low altitude chart	航线低空图
ELB	emergency locator beacon	应急定位信标
ELEV	elevator	升降舵
ELGE	emergency landing gear extension	起落架应急外伸
ELGH	emergency landing gear handle	应急起落架手柄
ELGIM	extend LG in emergency	起落架应急放下

ELGWHPE	extend LG with hand pump in emergency	手摇泵应急放下起落架
EI	electronic intelligence（system）	电子侦察（系统），电子情报（系统）
ELS	elastic limit under shear	抗剪弹性极限
ELS	emergency landing site	应急着陆场
ELT	emergency locator transmitter	应急定位器发射机
EM	electrical mechanism	电动机构
EM	escort mission	护航任务
EME	electromagnetic effects	电磁效应
EME	electromagnetic environment	电磁环境
EMER	emergency	应急
EMISM	electromagnetic interface safety margin	电磁干扰安全裕度
EMP	electromagnetic pulse	电磁脉冲
EMR	electromagnetic resonance	电磁共振
EMTBF	estimated mean time between failures	所估计的平均故障间隔时间
EMTF	equivalent mean time to failure	等效平均故障时间
EMTTF	estimated mean time to failure	所估计的平均故障时间
EMU	engine monitoring unit	发动机监控装置
EN	Euler number	欧拉数
ENG	engine	发动机
ENGA	engage	接通
ENR	equivalent noise ratio	等效噪声比
ENT	enter	确认
ENWLR	engine nacelle working light receptacle	发房工作灯插座
EOC	engaging onboard circuit	接通机上电路
EOCP	engine out of commission for parts	发动机因缺少零件停止使用
EOD	end of decent	下降结束
EOD	emergency open door	应急开舱门
EOD	effect of defects	缺陷效应
EOF	end of flight	飞行结束
EOGD	emergency opening lading gear door	应急打开起落架舱门
EOL	end of life	寿命末期
EOL	engine-off landing	发动机关机的着陆
EOP	engine operating point	发动机工作点
EOR	engine oil radiator	发动机滑油散热器
EOS	emergency oxygen system	应急氧气系统
EOVL	engine-out vertical landing	发动机停车垂直降落
EP	electric pump	电动泵
EP	emergency power	应急电源
EP	external power	外部电源

EPB	emergency power box	应急电源盒
EPC	flight path control [command]	飞行轨迹控制[指令]
EPCO	emergency power cutoff	紧急断电,紧急切断动力
EPCU	electrical power control unit	电源控制装置
EPDB	emergency power distributing box	应急电源配电盒
EPE	estimated position error	估计位置误差
EPERA	extractor parachute emergency release assembly	降落伞应急释放分离装置
EPG	emergency power generator	应急发电机
EPGS	electrical power generation system	发电系统
EPI	elevator (surface) position indicator	升降舵位置指示器
EPI	engine performance indicator	发动机性能指示器
EPIRB	emergency position indicator radio beacon	应急位置指示器无线电信标
EPL	engine power lever	油门杆
EPM	engine propeller maintenance	发动机螺旋桨维护
EPO	electric pump operating	电动泵工作
EPP	electric pump pressurization	电动泵增压
EPP	emergency power package	应急动力包,应急电源包
EPR	engaging parallel relay	接通并联工作继电器
EPR	engine pressure ratio	发动机压比
EPR	external power receptacle	外接电源插座
EPRL	engine pressure ratio limit	发动机压力比极限
EPS	emergency power supply	应急电源
EPS	electrical power system	电源系统
EPS	emergency power system	应急电源系统
EPS	engine preparation schedule	发动机准备程序
EPSS	engine pressure signal sensor	发动机压力信号传感器
EPU	emergency power unit	应急动力装置
EPUPS	emergency power unit pressure system	应急动力装置压力系统
ER	electronic reconnaissance	电子侦察
ERAC	ESM receiver access cover	ESM 接收机维护口盖
ERBM	electronic range/bearing marker	电子距离/方位标志
ERC	engine-related cause	与发动机有关的原因
ERC	extended runway centerline	延长的跑道中心线
ERF	extending/retracting flap	开关襟翼
ERGS	electronic route guidance system	电子线路导航系统
ERGS	enroute guidance system	中途导航系统
EHAC	enroute high altitude chart	航线高空图
ERI	electric revolution indicator	电动转数指示器

ERJM	engine rack joint mark	发动机架接头标记
ES	engine shutdown	发动机停车
ES	engine sensor	发动机传感器
ES	escape slide	紧急离机滑道[梯]
ESA	equalized sidelobe antenna	均衡旁瓣天线
ESAF	emergency shutdown and feathering	应急停车和应急顺桨
ESB	emergency switch board	应急配电盘
ESC	engine start-cranking	发动机起动-冷转
ESC	escape	返回,逃脱
ESC/A	essential system characteristics/ airborne	机载主要系统特性
ESCB	excitation source changerover box	激励源转换盒
ESCP	EFIS signal changeover processor	EFIS 信号转换处理机
ESCS	emergency SATCOM system	应急卫星通信系统
ESD	electrical schematic diagram	电气原理图
ESD	electrostatic discharge	静电放电
ESDR	engine shutdown relay	发动机停车继电器
ESG	electrostatically suspended gyro	静电悬浮陀螺
ESHP	equivalent shaft horsepower	当量轴马力
ESICC	engine start ignition coil contactor	发动机起动点火线圈接触器
ESIT	electrical system integrated test	电气系统综合测试
ESLR	electronically scanned laser radar	电子扫描激光雷达
ESM	electronic support measure	电子支援措施
ESM	electronic surveillance measures	电子监控措施
ESM	electronic surveillance and monitoring	电子监视监控
ESPM	engine start procedure mechanism	发动机起动程序机构
ESR	engine start relay	发动机起动继电器
ESR	enroute surveillance radar	航路监视雷达
ESRT	esscntial systems repair time	主要系统维修时间
ESS	electronic security system	电子保密系统,电子安全系统
ESS	equipment support structure	设备支承结构
ESSR	engine start select relay	发动机起动选择继电器
EST	elevation, slope, temperature	海拔、坡度、温度
ESTA	electronic-scanned TACAN antenna	电扫描塔康天线
ESTOL	extremely short takeoff and landing	极短距起飞与着陆
ESTT	elevator standby trim tab	升降舵备用调整片
ET	engine torque	发动机扭转力矩
ET	effective temperature	有效温度
ET	elapsed time	历程
ET	exhaust temperature	排气温度

ETA	effective turn angle	有效转折角,有效倾转角
ETA	estimated time of acquisition	预计截获时间
ETA	estimated time of arrival	预计到达时间
ETAWS	embedded terrain awareness warning system	内置式地形感知告警系统
ETCV	equipment temperature control valve	设备温控活门
ETD	estimated time of departure	预计出航时间
ETE	estimated time enroute	预计航线飞行时间
ETI	elapsed time indicator	经过时间指示器
ETO	estimated timeover	预计飞跃时间
ETOT	estimated time over target	预计飞跃目标时间
ETR	engine thrust request	发动机推力要求
ETR	estimated time for refueling	预计加油时间
ETR	estimated time of recovery	预计回收时间,预计恢复时间
ETR	estimated time of repairing	预计修理时间
ETR	estimated time of return	预计返回时刻,预计返航时间
ETR	expected time of response	预期响应时间
ETRB	estimated time to return to base	预计返场时间
ETS	estimated time of separation	预计分离时间,估计退役时间
ETTI	elevator trim tabneutral indication	升降舵调整片中立指示
EU	engine unfeathering	发动机回桨
EUDDL	equivalent uniformly distributed dead load	等效均匀分布静载荷
EUR	equipment's use rate	设备使用率
EV	electric valve	电动阀
EV	engine vibration	发动机振动
EVA	engines vibration amplifier	发动机振动放大器
EW	electric wiper	电动雨刷
EW/GCI	early warning and ground controlled intercept	预警与地面控制截击
EW/GWR	(aircraft) empty weight to gross weight ratio	(飞机)空重与总重比
EWAPAR	enlarged wide angle phase array radar	大空域相控阵雷达
EWB	emergency wheel brake	机轮应急刹车
EWC	early warning and control	预警与控制
EWM	EFIS wiring module	EFIS 接线模块
EWM	EICAS wiring module	EICAS 接线模块
EWOSG	excitation winding of starter generator	启动发电机激磁绕组
EWS	electronic warfare support	电子战支援
EWS	electronic warfare system	电子战系统

EWSM	early warning support measures	预警保障措施
EWSM	electronic warfare support measures	电子战支援措施
EWSM	electronic warfare surveillance measures	电子战监视措施
EWWS	electronic warfare warning system	电子战告警系统
EXC	excitation	励磁
EXT	extension	放下
EXTG	extinguishing	熄灭

F

FA	flap asymmetry	襟翼不对称
FA	final assembly	总装,最后装配
FA	flaperon angle	襟副翼偏转角
FA	frame alignment	帧定位
FA	fast alignment	快速对准
FAC	final approach course	最后进场[进近]航迹
FAC	full/ arc compass	全/半罗盘
FAD	fuselage automated design	机身自动化设计
FADD	fatigue and damage data	疲劳与损伤数据
FADEC	full authority digital engine/ electronic controller	全授权数字式发动机/电子控制器
FADS	fuselage analysis and design synthesis	机身分析与设计综合
FAE	final approach equipment	最后进场[进近]设备
FAF	final approach fix	第五边定位点,最后进近定位
FAF/L	final approach fix/left	最后进场[进近]定点左转弯
FAF/R	final approach fix/right	最后进场[进近]定点右转弯
FAIL	failure	故障
FAM	flight acceptance meeting	飞行验收会
FAN	fan	电风扇
FAOFM	flap actuator oil-filling mark	襟翼作动筒注轴标记
FAP	fleet average performance	机群平均性能
FAR	false alarm rate	虚警率
FAR	federal aviation regulations	联邦航空条例
FAR	fixed array radar	固定阵列天线雷达
FAS	final assemble schedule	总装进度计划
FAS	flight augmentation system	飞行增稳系统
FAS	follow-up alarm system	随动警报系统
FAS	final approach speed	(着陆)第五边速度
FASA	final approach spacing of aircraft	飞机最后进场[进近]间隔
FAT	flight acceptance test	飞行验收试验

FB	flap braking	襟翼制动
FB	feathering button	顺桨按钮
FB	fire button	灭火按钮
FBE	filter-band eliminater	带阻滤波器
FBM	fire bottle mark	灭火瓶标记
FBO	flights between overhauls	(两次)大修间飞行次数
FBP	filter-band pass	带通滤波器
FC	carrier frequency	载波频率
FC	flight chart	飞行图表
FC	FDR check	飞参地面检测
FC	flap control	襟翼操纵
FC	fuel consume	耗油
FC	fuel crossfeed	燃油连通
FC	fully closed	全关
FCB	flight control booster	飞控助力器
FCB	flight control box	飞行控制盒
FCB	fire check button	灭火检查按钮
FCB	fire control box	灭火控制盒
FCC	flight control clutch	飞控离合器
FCC	flight control computer	飞控计算机
FCD	flap control device	襟翼控制装置
FCE	flight control electronics	飞行控制电子设备
FCEUF	flap control electronices unit failure	襟翼操纵电子装置故障
FCF	flight check facility	飞行检查设备
FCH	flap control handle	襟翼操纵手柄
FCH	flight control hydraulic	飞控液压
FCH	flight controllers handbook	飞行控制员手册
FCLF	flap control lever failure	襟翼操纵机构故障
FCLG	flight control landing gear	飞控起落架舱
FCLGW	flight control LG well	飞控起落架舱
FCMDB	forward cabin master distributing box	前舱总配电盒
FCOM	flight crew operating manual	飞行人员操作规程
FCR	fire control radar	火控雷达
FCR	flight condition recognition	飞行条件识别
FCS	fire control system	火力控制系统,射击指挥系统
FCS	flight control system	飞行操纵系统,飞行控制系统
FCS	fire check switch	灭火检查开关
FCS	fire changeover switch	灭火转换开关
FCS	fuel consumption selection	耗油选择

FCSHOT	flight control system hydraulic oil tank	飞控系统油箱
FCSOR	flight control system oil return	飞控系统回油
FCSOS	flight control system oil suction	飞控系统吸油
FCSP	flight control system pressurization	飞控系统增压
FCSSV	flight control system sampling valve	飞控系统采样阀
FCTM	flight crew training manual	机组乘员训练手册
FCTR	flight control tank refueling	飞控油箱加油
FCU	flap control unit	襟翼操作装置
FCV	flow control valve	流量控制活门
FD	failure display	故障显示
FD	flight director	飞行指引
FD	frequency display	频率显示器
FD/FI	fault detection/fault isolation	故障诊断/故障隔离
FDAU	flight data acquisition unit	飞行数据采集器
FDCR	flight data check receptacle	飞行数据检测插座
FDGW	flight design gross weight	飞行设计总重
FDI	flight direction indicator	飞行指引仪
FDI	flight detection and isolation	故障检测与隔离
FDP	fluid drain port	放液口
FDPS	flight data processing system	飞行数据处理系统
FDR	flight data recorder	飞行数据记录仪
FDR	frequency diversity radar	频率分集雷达
FDRS	flight data recording system	飞行数据记录仪系统
FDV	flow distribution valve	流量分配活门
FDWM	flight data wiring module	飞行数据接线模块
FEA	failure effect analysis	故障影响分析
FEBL	fire extinguishing button light	灭火按钮灯
FEC	fire extinguishing check	灭火检查
FECR	fire extinguishing check relay	灭火检查继电器
FED	file encryption/ decryption	文件加密/解密
FEDO	front emergency door open	前应急门开
FEDO	forward emergency door opened	前应急门打开
FEE	fire extinguishing equipment	灭火设备
FEOC	flight engineer overhead console	机械师顶部操纵台
FES	fire extinguishing system	灭火系统
FF	first flight	首飞,首次飞行
FF	fuel flow	燃油流量
FFC	flight facilities check	飞行设备检查
FFC	fixed frequency communication	定呼,定频通信

FFP	flight fine pitch	飞行小距
FFP	fluid filling port	充液口
FFR	flight feasibility review	飞行可行性评审
FFRR	first flight readiness review	首次待飞状态评审,首飞就绪评审
FFSP	front fuel supply pump	前供油泵
FGV	field gradient voltage	场梯度电压
FHF	first horizontal flight	首次水平飞行
FI	fault isolation	故障隔离,差错隔离
FI	flight inspection	飞行检验,飞行检查
FI	flight instrument	飞行仪表
FIFO	fast-in fast-out	快进快出
FIFO	first-in first-out	先进先出
FF	flight forecast	飞行预报
FIH	flight information handbook	飞行情报手册
FILO	first-in last-out	先进后出
FILT	filter	过滤器,滤波器
FIM	fault isolation manual	故障隔离手册
FIR	fault isolation rate	故障隔离率
FIS	flight instrument system	飞行仪表系统
FL	flight level	飞行高度
FL	flight line	起机线,起飞线,航线
FL	flight	飞行
FLAP	flap	襟翼
FLC	full load current	满载电流
FLDMF	full load displacement margin factor	满载位移安全系数
FLFE	forced landingand fire extinguishing	迫降灭火
FLGL	forward LG lighting	前起落架舱照明
FLGWMF	full load gross weight margin factor	满载重量安全系数
FLL	forward landing light	前着陆灯
FLR	flight line reference	航线基准
FLR	forward looking radar	前视雷达
FLS	flashing light system	闪光灯系统
FLS	forward landing strip	前方简易机场
FLW	full load weight	满载重量
FM	flight manual	飞行手册
FM	frequency modulation	调频
FMA	flight mode annunciator	飞行状态信号牌,飞行状态指示器
FMC	fuel measuring computer	燃油测量计算机
FMCP	fuel measuring computer power	燃油测量计算机电源

FMD	force measuring device	力测量装置
FMDS	flight management data system	飞行管理数据系统
FMEP	friction mean effective pressure	平均有效摩擦压力
FMG	frequency modulation generator	调频信号发生器
FMR	field maintenance reliability	机场维修可靠性
FMRB	flight manual review board	飞行手册审查委员会
FMS	flight management system	飞行管理系统
FMTL	forward emergency top light	前舱应急顶灯
FN	flow normal	流量正常
FOB	fuel on board	（机上）剩余油量
FOC	fibre optic control	光纤控制
FOC	flares/off/chaff	电光弹/关闭/箔条
FOC	flight operations center	飞行控制中心
FOC	foreign-object check	外来物检查，多余物检查
FODB	flap output device	襟翼输出装置
FOD	fiber optic data bus	光纤数据总线
FOD	foreign-object damage	外来物损伤
FODB	fiber optic data bus	光纤数据总线
FOG	fiber optic guidance	纤维光学制导，光纤制导
FOG	fiber optic gyro	纤维光学陀螺，光纤陀螺
FOR	fuel oil remaining	剩余燃油
FOS	full operational status	完全工作状态，完全作战状态
FOT	frequency of optimum transmission	最佳发射频率
FOV	field of view	视场，视野，视界
FP	flight path	飞行轨迹，飞行路线，飞行计划
FP	flight position	飞行位置，准备飞行姿态
FP	flight progress	飞行进展情况，飞行进度
FP	fuel pressure	油压，燃料喷射压力
FP	flight plan	飞行计划
FP	fuel pressure	燃油压力
FPA	failure probability analysis	故障概率分析
FPA	flat plane antenna	平面天线
FPA	flight path accelerometer	飞行轨迹（航迹）加速度
FPA	flight path analysis	飞行轨迹（航迹）分析
FPA	flight path angle	飞行轨迹（航迹）倾角
FPA	flight plan approval	飞行计划批准
FPBL	flight parameter boundary limitation	飞行参数边界限制
FPC	flight path control [command]	飞行轨迹控制[指令]
FPF	fuel pump filter	油泵滤波器

FPG	force per g	单位过载的力
FPGL	flight plan gasoline load	飞行计划规定的汽油载荷
FPGS	flight parameter ground station	飞参地面设备
FPH	failure per hour	故障每小时
FPH	feet per hour	英尺每小时
FPI	flap position indicator	襟翼位置指示器
FPI	fuel pressure indicator	油压表,燃油压力指示表
FPL	fluctuating pressure level（s）	气压波动的高度（层）
FPL	friction pressure loss	摩擦压力损耗
FPL	full power level	全推力水平,全功率级
FPM	flight path marker	飞行轨迹标志
FPM	flight path miles	飞行轨迹英里数
FPM	feet per minute	英尺/分
FPO	feather pump operating	顺桨泵工作
FPP	fixed-pitch propeller	定距螺旋桨
FPPS	flight plan processing system	飞行计划处理系统
FPR	fan pressure ratio	风扇压比
FPRM	flight phase related mode	飞行阶段有关方式
FPS	fine pitch stop	小螺距限动钉
FPS	fire propection system	火焰探测系统
FPS	flight path stabilization	飞行轨迹稳定
FPS	foot per second	英尺/秒
FPU	failure protection unit	故障防护装置
FQ	flight quality	飞行品质
FQBV	fuel-quantity balance valve	油量平衡活门
FQC	fuel quantity computer	燃油油量计算机
FQI	fuel quantity indication	燃油量指示
FQI	fuel quantity indicatior	油量指示器
FQIS	fuel quantity indicator switch	油量指示器开关
FQMS	fuel quantity and measurement system	油量测量系统
FQP	flight qualities and performance	飞行品质与性能
FQSM	fuel-quantity sensor mark	油量传感器标记
FQTI	fuel quantity totalizer indicator	总油量指示器
FR	final run	进入第五边,进入攻击
FR	flight recorder	飞行记录仪
FR	flash ranging	闪光测距,光测
FR	flight recorder	飞行记录仪
FR	flight refueling	空中加油
FR	frequency range	频率范围,频段

FR	fuel remaining	剩油量
FR	fuel requested	需要油料,申请油料
FR	full range	满量程,满标度,全波段,全范围
FR	fuel residual	剩油
FRACA	failure reporting, analysis and corrective action	故障报告、分析与纠正措施
FRB	feathering relay box	顺桨继电器盒
FRB	fire relay box	灭火继电器盒
FRC	failure recurrence control	故障再现控制
FRCC	fuel refueling/consumption controller	加耗油控制器
FREDI	flight range and endurance data indicator	航程和续航数据指示器
FREM	flap retraction/extension mechanism	襟翼收放机构
FRH	flap-retraction height	收襟翼高度
FRL	fuselage reference line	机身基准线
FRM	failure-related mode	与故障有关的(工作)状态
FRM	fault reporting manual	故障报告手册
FRO	failure requiring overhaul	需要翻修故障
FRR	feathering ready relay	顺桨准备继电器
FRR	flight readiness review	待飞状态评审,飞行准备状态评审
FRS	fuel residual signal	剩油信号
FRT	fixed radar tracking	固定雷达跟踪
FRT	flight readiness test	飞行准备状态试验
FRU	field replaceable unit	现场可更换部件
FS	failure source	故障源
FS	flow sensor	流量传感器
FS	factor or safety	安全系数
FS	flight simulator	飞行模拟器
FS	flying safety	飞行安全
FS	front spar	前梁
FS	full scale	全尺寸
FS	fuselage station	机身站位,机身测量点
FS	force sensor	力传感器
FS	fuel system	燃油系统
FS	function switch	功能开关
FSB	flight standard board	飞行标准委员会
FSB	function select button	功能选择旋钮
FSD	full scale deflection	满刻度偏转
FSD	full scale development	全尺寸研制(阶段)
FSE	flight support equipment	飞行保障设备

FSED	full scale engineering development	全尺寸工程研制(阶段)
FSFT	full scale fatigue test	全尺寸疲劳试验,整机疲劳试验
FSL	full-stop landing	着陆至停放
FSLM	flap suspension lug mark	襟翼吊耳标记
FSOAFS	flow sensor of antenna feed system	天馈系统流量传感器
FSOC	fuel shutoff cock	燃油切断开关
FSOT	flow sensor of transmitter	发射机流量传感器
FSOV	fuel shutoff valve	燃油切断阀门,停车开关
FSPB	force signal processor box	力信号调理器盒
FSR	frequency scan radar	频率扫描雷达
FSS	flight support system	飞行支援系统
FST	free stream turbulence	自由流湍流度
FST	full scale tunnel	全尺寸风洞,实物风洞
FR	failure rate	故障率
FT	false target	假目标
FT	fight test	飞行试验
FT	flight time	飞行时间
FT	full throttle	全油门,最大油门
FT	feet	英尺
FTC	flight time constant	飞行时间不变
FTC	flying training course	飞行训练科目
FTD	flap transmission device	襟翼传动装置
FTD	flight time diagram	飞行时间图
FTD	flight test display	飞行试验显示器
FTH	full throttle horsepower	全油门功率,最大功率[马力]
FTI	fixed time interval	固定时间间隔
FTI	flight test instrument	飞行试验仪器
FTI	frequency time indicator	频率时间指示器
FTI	frequency time intensity	频率时间强度
FTIS	flight test instrumentation system	飞行试验仪表系统
FTIT	fan-turbine inlet temperature	风扇涡轮进口温度
FTO	flexible (derated) takeoff	降低功率起飞
FTO	flexible takeoff	弹性起飞
FTP	flight test procedure	飞行试验程序
FTP	flight test prototype	飞行试验样机
FTP	full throttle position	(发动机)节流阀全开位置,全油门位置
FTRPM	fuel tank refueling port mark	油箱加油口标记

FTDPM	fuel tank drain port mark	油箱放油口标记
FUF	favourable unbalanced field	有利的不平衡机场
FUSS	flaps-up safety speed	收襟翼安全速度
FVRTDVM	fuel vent residual tank drain valve mark	燃油通气余油箱放油开关标记
FVV	fan variable vane	可变风扇叶片
FW	fire warning	火警
FW	fire wall	防火墙,防火镉框
FW	full wave	全波
FW	full weight	全重
FWD	forward	前
FWH	forward windshield heating	前玻璃加温
FWS	flight warning system	飞行告警系统
FWWMR	fire, water, weather, mildew resistant	防火,防水,防风雨,防霉

G

G	gain	增益
GA	go around	复飞
GA	gust alleviation	突风缓和,突风减缓
GA	general average	一般平均
GACCC	ground air-conditioning cart connector	地面空调车接头
GACW	gust above constant wind	大于定常风的突风
GAD	general assembly drawing	总装图
GADL	ground-to-air data link	地对空数据链
GAINS	GPS, air data and inertial navigation system	全球定位系统、大气数据与惯性导航系统
GANS	gyro automatic navigation system	陀螺自动导航系统
GAR	ground abort rate	(起飞)地面中断率
GAR	ground avoidance radar	地形回避雷达
GACCSCV	ground air-conditioning cart system check valve	地面空调车系统单向活门
GAT	general air traffic	通用空中交通
GAT	general aviation terminal	通用航空终点站
GATO	gasoline at takeoff	起飞时汽油量
GAU	GPS antenna unit	全球定位系统天线装置
GAVRS	gyrocompassing altitude and velocity reference system	陀螺罗盘高度与速度参考系
GB	ground beacon	地面信标
GB	ground brush	接地电刷
GB	grounded base	接地基极

GBH	glide beam hold	侧滑波束控制
GBIT	ground built-in test	地面自检
GBJ	ground based jamming	路基干扰
GC	gyro compass	陀螺罗盘
GCA	ground-controlled approach	地面控制进场［进近］（系统）
GCAS	ground collision avoidance system	防撞地系统
GCAU	ground control approach unit	地面控制进场［进近］装置
GCG	gravity controlled gyro	重力控制陀螺
GCIR	ground current interlock relay	地面电流联锁继电器
GCJB	ground console junction board	地面操纵台接线板
GCL	ground control landing	地面控制着陆，地面指挥着陆
GCP	ground control point	地面控制点
GCS	ground control station(s)	地面控制站
GCS	gyrosyn compass system	陀螺同步罗盘系统
GCSS	global combat support system	全球作战支援系统
GCSS	global communication satellite system	全球通信卫星系统
GCTS	ground communication(s) tracking system	地面通信跟踪系统
GCU	generator control unit	发电机控制装置
GCWS	ground collision warning system	防撞地告警系统
GDAS	ground data acquisition system	地面数据采集系统
GDC	graphical display controller	图形显示控制器
GDC	gyro display coupler	陀螺显示耦合器
GDCS	ground distributed control system	地面分配控制系统
GDH	generalized data handler	综合数据处理机
GDMS	ground data management system	地面数据管理系统
GDOP	geodesic degradation of performance	短程性能下降
GDT	geometric dimensioning and tolerancing	几何尺寸与公差
GE	generator excitation	发电机激磁
GEF	ground equipment failure	地面设备故障
GEN	generator	发电机
GEON	gyro-erected optical navigation	陀螺仪稳定光学导航系统
GES	ground earth station	地面地球站
GET	ground elapsed time	地面经历时间
GETOL	ground effect takeoff and landing	利用地面效应起飞与降落，气垫短距起飞
GETS	generalized electronic trouble shooting	综合电子故障排除
GEV	ground effect vehicle	地效飞行器
GEWS	global early warning system	全球预警系统
GF	generator fail	发电机故障

GFP	generator feathering pump	发电机顺桨泵
GFP	ground fine pitch	地面小距
GFS	global forecasting system	全球天气预报系统
GFT	general flight test	一般飞行试验
GG	gas generator	燃气发生器,气体发生器
GG	gravity gradient	重力梯度
GH	gyro horizon	陀螺地平仪
GHARS	gyroscopic heading and attitude reference system	陀螺仪航向与姿态参考[基准]系统
GHDB	glass heating distributing box	玻璃加温配电盒
GHS	ground hydraulic system	地面液压系统,地面水力系统
GHV	gross heating value	总热值
GIA	gyro input axis	陀螺输入轴
GICS	ground instrumentation and communication system	地面仪表与通信系统
GIE	ground instrumentation equipment	地面仪表设备
GINS	GPS inertial navigation system	全球定位系统惯性导航系统
GIPS	ground information processing system	地面信息处理系统
GIPS	general image processing system	通用图像处理系统
GIRAS	geographic information retrieval and analysis system	地理信息检索与分析系统
GIRLS	generalized information retrieval and listing system	综合信息检索与编排系统
GIRLS	global interrogation, recording and location system	全球询问、记录与定位系统
GIS	general information system	通用信息系统
GIS	geographic indexing system	地理索引系统
GIS	geographic information system	地理信息系统
GIS	graphic input system	图形输入系统
GISP	general information system for planning	通用计划信息系统
GIS	ground identification of satellite	卫星地面识别
GLOW	gross lift off weight	起飞总重,起飞总重量
GLM	G load maximum	最大过载
GME	gyro motor excitation	陀螺电机励磁
GNCS	guidance, navigation and control subsystem	制导,导航与控制子系统
GND	ground	地面
GNPC	global navigation and planning chart	全球导航与计划图,全球领航与准备图

GNS	global navigation system	全球导航系统
GNSS	global navigation satellite system	全球导航卫星系统
GOC	generator output control	发电机输出控制
GOE	ground operational equipment	地面操作设备
GOF	glass optical fiber	玻璃光纤
GOF	gyro overhaul facilities	陀螺大修设施
GOI	ground object identification	地面目标识别
GOV	generator output voltage	发电机输出电压
GP	ground power	地面电源
GP	glide path	下滑轨迹,下滑道
GPCC	grams per cubic centimeter	克每立方厘米
GPLS	glide path landing system	下滑道着陆系统
GPO	ground power-on	接通地面电源
GPS	global positioning system	全球定位系统
GPS/INS	global positioning system/inertial navigation system	全球定位系统/惯性导航系统
GPSU	ground power supply unit	地面电源,机场动力供应单元
GPT	glide path tracking	下滑道跟踪
GPW	ground proximity warning	近地告警
GPWS	ground proximity warning system	近地告警系统
GRADS	ground radar aerial delivery system	地面雷达空投系统
GRG	ground-roll guidance	着陆滑跑引导
GRS	geographic reference system	地理参考[基准]系统
GS	ground speed	地速
GS	ground start	地面起动
GS	glideslope	下滑道
GS	ground speed	地速
GS	ground station	地面站
GS	ground supply	地面电源
GS/GTK	ground speed/ground track	对地速度/地面轨迹
GSC	generator start contactor	发电机起动接触器
GSE	ground support equipment	地面保障设备
GSI	glide slope indicator	下滑道偏离指针
GSR	ground surveillance radar	地面监视雷达
GSS	global surveillance system	全球监视系统
GSS	ground support system	地面保障系统
GSV	globe stop value	球形断流阀
GTC	gain time control	增益时间控制
GTC	gas turbine compressor	燃气涡轮压气机

GTC	gyro time constant	陀螺时间常数
GTC	gas turbine compressor	燃气涡轮压缩机
GTE	gas turbine engine	燃气涡轮发动机
GTE	ground test equipment	地面试验设备
GTE	ground transport equipment	地面运输设备
GTOL	ground takeoff and landing	地面起飞与着陆
GTOW	gross takeoff weight	起飞总重,起飞总重量
GTPU	gas turbine power unit	燃气涡轮动力装置
GV	gate valve	闸门阀,滑门阀
GV	ground visibility	地面能见度
GV	guide vane(s)	导流叶片
GW	gross weight	总重,毛重
GW/CG	gross weight and center of gravity	总重与重心
GYP	gyro yaw position	陀螺侧滑位置,陀螺偏航位置

H

HAA	height above airport	对机场的真高,在机场上空的高度
HAA	high angle of attack	大迎角
HAE	high altitude equipment	高空设备
HAF	high altitude facility	高空设备
HAFS	hydraulic anti-fire switch	液压防火开关
HAMOTS	high accuracy multiple objects tracking system	高精度多目标跟踪系统
HAMOTS	high altitude multiple objects tracking system	高空多目标跟踪系统
HARPI	height-azimuth-range-position indicator	高度、方位、距离与位置指示器
HARS	heading and attitude reference system	航向姿态参考[基准]系统
HAT	height above touchdown [threshold]	飞越接地区[机场入口]的高度
HAT	high-altitude test	高空试验
HATL	holding and approach-to-land	等待进场着陆
HATOL	horizontal-attitude takeoff and landing	水平姿态起飞与着陆
HAWC	homing and warning computer	寻的与告警计算机
HBPR	high bypass ratio	高涵道比,大涵道比
HCC	helix coaxial cable	螺线同轴电缆
HCC	honeycomb corrugated construction	蜂窝波纹结构
HCI	high current inductor	大电流感应器
HCL	horizontal center line	水平中心线
HCO	hand control on	手操纵接通
HCP	heading channel panel	航向航道控制板

HCV	hose connection valve	软管链接阀
HCV	hydraulic control valve	液压控制阀
HD	height difference	高度差
HD	high duty lubricating	高温高压润滑油
HDATZ	high density air traffic zone	高密度空中交通区
HDCU	hydraulic duplicated control unit	液压复合舵机,液压复合控制装置
HDDR	head-down display radar	下视雷达
HDDS	high density data system	高密度数据系统
HDG	heading	航向
HE	heat exchanger	热交换器
HED	human engineering data	人机工程数据
HF	high frequency	高频
HFAA	high frequency airborne antenna	机载高频天线
HFAV	heating flow adjusting valve	加温流量调节活门
HFCS	hydraulicand flight control sensor	液压及飞控传感器
HFD	high frequency destructor	短波毁钥
HFG	high frequency generator	高频发生器
HFO	high frequency oscillator	高频振荡器
HFR	height finder radar	测高雷达
HFRT	high frequency radio transmitter	高频无线电发射机
HFS	high fidelity simulator	高逼真度模拟器
HFSS	high frequency simulation system	高频仿真系统
HG	heading gyro	航向陀螺仪
HG	horizontal gyro	水平陀螺仪
HH	heading hold	航向保持
HHS	helmet and head set	飞行帽及头戴耳机
HIK	heading index knob	航向指标旋钮
HIPR	high internal phase ratio	高内相比
HIRAN	high precision range navigation	高精度近程导航(系统)
HIRF	high intensity radiated field	高辐射区
HIRL	high intensity runway light	高亮度跑道照明(灯)
HIS	high impedance surface	高阻抗表面
HIS	high intensity spectrometer	高强度光谱仪
HIT	high torque	高扭矩,大扭矩
HLO	heavy lift operability	大升力可操作性
HMI	human-machine interface	人机接口
HMU	hydromechanical unit	液压机械装置
HOAF	hydraulic oil anti-fire	液压油防火
HOFM	hydraulic oil filter mark	液压油滤标记

HP	heading preselect	航向预选
HP	high pressure	高压
HPA	high performance aircraft	高性能飞机
HPAR	high power acquisition radar	大功率目标探测雷达
HPC	hand-pedal control	手-脚操纵
HPCC	high pressure combustion chamber	高压燃烧室
HPDB	hydraulic pump distributing box	液压泵配电盒
HPEPHS	hand pump electric pump hydraulic system	手摇泵电动泵液压系统
HPF	high-pass filter	高通滤波器
HPF	high-pass filter	高通滤波器
HPFF	high pressure fuel filter	高压燃油过滤器
HPFP	high pressure fuel pump	高压燃油泵
HPG	high performance generator	高性能发生器,高性能振荡器
HPG	high power generator	大功率发电机
HPGC	heading per gyro compass	回转罗盘航向,陀螺罗盘航向
HPI	high performance intercept	高性能拦截,高性能截击
HPI	height position indicator	高度位置指示器
HPIR	high power illuminating radar	大功率照射雷达
HPNS	high performance navigation system	高性能导航系统
HPPAR	high performance precision approach radar	高性能精确进场[进近]雷达
HPPC	hydraulic pump power control	液压泵电源控制
HPR	high power radar	大功率雷达
HPRV	high pressure relief valve	高压安全阀门
HPS	horizon and pitch scale	地平与俯仰刻度
HPSC	heading per standard compass	按标准罗盘的航向(航行)
HPT	high-pressure test	高压试验
HR	heavy route	繁忙航线
HR	high resolution	高分辨率,高清晰度
HRB	horn release button	解除喇叭按钮
HRL	horizontal reference line	水平基准线
HS	horizontal stabilizer	水平安定面
HS	hydraulic system	液压系统,水力系统
HS	high speed	高速
HS	hydraulic system	液压系统
HSCB	high speed circuit breaker	高度电路断开器
HSD	hard surface dry	有硬路面的干(燥)跑道
HSI	horizontal situation indicator	水平位置指示器
HSI/CDI	horizontal situation indicator/course devia-	水平位置指示器/航向偏差指示器

	tion indicator	
HSI/EFIS	horizontal situation indicator/electronic flight instrument system	水平位置指示器/电子飞行仪表系统
HSJ	honeycomb sandwich joint	蜂窝夹层结构连接
HSK	heading select knob	航向选择按钮
HSLLADS	high speed low level aerial delivery system	低空高速投放系统
HSOTP	hydraulic system oil tank pressurization	液压系统油箱增压
HSPP	hydraulic system pipe placard	液压系统导管标牌
HT	horizontal tail	水平尾翼
HTA	high-time aircraft	(同型中)长服役期的飞机
HTG	heating	加温
HTIM	horizon tail ice melting	平尾融冰
HTO	horizontal takeoff	水平起飞
HTOL	horizontal takeoff/landing	水平起落
HTOVL	horizontal takeoff vertical landing	水平起飞垂直降落
HTSLM	horizontal tail suspension lug mark	平尾吊耳标记
HUD	head up display	平视显示,平视显示器
HUW	hours under way	航行时数
HVAC	heating，ventilation，air-conditioning	加热、通风、空调
HVRS	heading and vertical reference system	航向与垂直参考[基准]系统

I

I/O	input/output	输入/输出
I/Q	in-phase/quadrature	同相/正交
IA	inactive aircraft	非现役飞机,退役飞机
IA	initial approach	开始进场[进近]
IAAP	identification of aircraft aerodynamic parameter	飞机气动力参数识别
IABCS	integrated aircraft brake control system	飞机刹车综合控制系统
IACS	inertial attitude control system	惯性姿态控制系统
IACS	integrated avionics control system	综合航空电子控制系统
IAEDS	integrated advanced electronic display system	先进综合电子显示系统
IAF	initial approach fix	起始进场[进近]点
IALC	instrument approach and landing chart	仪表(指引)进场[进近]与着陆图
IAM	instrument approach minima	仪表进场[进近]最低气象条件
IAP	initial approach procedure	起始进场[进近]程序
IAP	integrated avionics package	综合航空电子系统包
IAS	indicated airspeed	表速,指示空速

IAS	integral airframe structures	整体机身结构
IAS	integrated avionics system	综合航空电子系统
IAS	indication air speed	指示空速
IAS	instrument approach system	仪表进场系统
IAT	indicated air temperature	指示气温
IB	identification beacon	(机场)识别信标
IB	identity [identification] beacon	识别信标(台)
IB	IDT button	IDT 按钮
IBIS	image based information system	图像信息系统
IBIT	initiated built-in test	启动自检测
IBLS	integrity beacon landing system	整体信标着陆系统
IBM	inertial biased mode	惯性偏差状态
IBRCC	illumination brightness regulation of copilot console	右操纵台照明亮度调节
IBRCC	illumination brightness regulation of center console	中央操纵台照明亮度调节
IBRPC	illumination brightness regulation of pilot console	左操纵台照明亮度调节
IC	inductance-capacitance	电感-电容
ICA	inertial cruise altitude	起始巡航高度
ICG	icing	结冰
ICI	initial capability inspection	起始能力检查
ICNI	integrated communication, navigation and identification	综合通信、导航与识别
ICS	intercom switch	机内通话开关
ICS	intercom system	机内通话系统
ICWAR	improved continuous wave acquisition radar	改进型连续波探测雷达
ID	identification	识别
ID	induced drag	诱导阻力
IDB	inverter distributing box	变流器配电盒
IDNE	inertial Doppler navigation equipment	惯性多普勒导航设备
IDP	integrated data processing	综合数据处理法
IDR	initial design review	初始设计评审
IDR	interim design review	中间设计评审,临时设计评审
IE	instrument equipment	仪表设备
IEL	internal/external lighting	机内外照明
IETM	interactive electronic technical manual	交互式电子技术手册
IEW	intelligence/electronic warfare	情报/电子战

IEWS	integrated electronic warfare system	综合电子战系统
IEWS	information and electronic warfare system	信息与电子战系统
IF	instrument flight	仪表飞行
IF	inverter failure	变流器故障
IFA	in-flight alignment	飞行校准
IFC	initial flight clearance	初始飞行许可
IFDL	in-flight data link	飞行数据链
IFE	in-flight emergency	飞行中的紧急情况
IFF	identification friend or foe	敌我识别(系统)
IFFCS	integrated fire/flight control system	综合火力/飞行控制系统
IFIP	integrated flight instrument panel	综合飞行仪表板
IFIS	integrated flight instrument system	综合飞行仪表系统
IFMA	in-flight mission abort	飞行中中断任务,中途停止执行任务
IFN	information	信息,情报
IFP	individual flight plan	单个飞行计划
IFP	initial flight path	起始飞行路线[航迹]
IFR	initial flight release	首次放飞
IFSD	in-flight shutdown	空中停车
IFT	in-flight test	飞行试验
IG	inverse gain	反增益
IGN	ignition	点火
IGV	inlet guide vane	进口导流叶片(用于压气机进口)
IH	inhibition height	(近地告警系统的)禁止高度
IH	initial heading	起始航向
IHS	infrared homing system	红外线自导系统
IIF	inserted in-flight	(数据、目标资料等)在飞行中插入
IIS	intake ice signal	进气道结冰信号
IL	instrument landing	仪表着陆
ILAS	instrument landing approach system	仪表着陆进场[进近]系统
ILEF	inboard leading edge flap	内侧前缘襟翼
ILLL	intense light of landing light	着陆灯强光
ILS	instrument landing system	仪表着陆系统
ILS	integrated logistics system	综合后勤系统
ILSTACAN	instrument landing system and tactical air navigation (system)	仪表着陆系统与战术空中导航(系统)
IM	inner marker	近距指点标
IM	identification mark	识别符,识别标志
IMC	instrument meteorological conditions	仪表飞行气象条件

IMI	interactive multimedia instruction	交互式多媒体教学
IMPS	integrated multipurpose system	多功能综合系统
IMS	integrated mission system	综合任务系统
IN	inertial navigation	惯性导航
IN	inertial navigator	惯性导航仪
INBD	inboard	机内的,舱内的,内侧的
INCAS	integrated navigation and collision avoidance system	导航防撞综合系统
INCR	increase	增加,增长,增大
IND	indicator	指示器
INE	inertial navigation equipment	惯性导航设备
INR	inertial navigation reliability	惯性导航可靠性
INS	inertial navigation system	惯性导航系统
INSP	inspection	检验
INT	interval	间隔
INTPH	interphone	机内通话装置
INVTR	inverter	变换器,变流器
IOC	input-output channel	输入-输出通道
IOSA	integrated optical spectrum analyzer	综合光谱分析仪
IP	intermediate processor	中间处理机
IPA	intermediate power amplifier	中间功率放大器
IPDS	integrated product delivery and support	综合产品交付与保障
IPECS	integrated powerand environmental control system	综合动力与环境控制系统
IPR	inches per revolution	英寸/转
IPS	instrument pointing system	仪器定向系统
IR	interchangeability and replaceability	互换性与可换性
IR	instrument rating	仪表飞行等级
IRCM	infrared countermeasures	红外对抗
IRCS	intrusion-resistant communication(s) system	抗干扰通信系统
IREW	integrated radar-electronic warfare	综合雷达电子战
IREWS	infrared early warning system	红外预警系统
IRF	instrument reliability factor	仪器可靠性系数
IRF	integrated radio frequency	综合射频
IRFIS	inertial reference flight inspection system	惯性参考[基准]飞行检验系统
IRPS	ice andrain protection system	防冰和防雨系统
IRS	inertial reference system	惯性参考[基准]系统
IRSS	inertial reference stabilization system	惯性参考[基准]稳定系统

IRVR	instrumented runway visual range	仪表跑道能见度
IRWR	IR warning receiver	红外告警接受机
IS	icing signal	结冰信号
IS	inspection stamp	检印
IS	instruction sensor	指令传感器
ISB	intermediate sideband	中边带
ISIS	integrated strike and interceptor system	综合攻击和截击系统
ISO	isolate	隔离
ISR	intelligence，surveillance，reconnaissance	情报、监视与侦查
ISS	interrogation sidelobe suppression	询问旁瓣抑制
IST	interim STOL transport	临时性的短距起落运输机
ITAC	integrated transmitter access cover	集成发射器口盖
ITO	instrument takeoff	仪表起飞，盲目起飞
ITR	integrated technology rotor	综合技术旋翼
ITT	interstage turbine temperature	涡轮级间温度
ITT	indicated-turbine temperature	显示涡轮温度
IVALA	integrated visual approach and landing aids	目视进场[进近]着陆综合导航系统
IVI	instant visibly index	顺时能见度指数
IVR	instrumented visual range	仪表可见距离
IVSI	instantaneous vertical speed indicator	瞬时垂直速度表
IW	integrated warning	集中告警
IW/EW	information warfare/electronic warfare	信息战与电子战
IWBS	integral weight and balance system	整体称重与平衡系统
IWC	integrated warning computer	集中告警计算机
IWIP	integrated warning instrument panel	集中告警仪表板

J

JAMR	jammer	干扰机
JB	junction board	接线板
JIT	just-in-time	适时制，零库存
JITT	just-in-time training	即时培训
JPALS	joint precision approach landing system	联合精确进场[进近]着陆系统
JSR	jammer saturation range	干扰机饱和距离
JSRR	jam/signal ratio required	所需干扰/信号比
JUNC	junction	连接，连接器
JW	jamming war	电子干扰战
JWS	jamming and warning system	干扰与告警系统

K

KBPS	k-bits per second	千位/秒
KC	key characteristic	关键特性
KEAS	knots equivalent airspeed	节当量空速,端速
KGF	kilogramme force	千克力
KGM	kilogram meter	千克米
KGPS	kilograms per second	千克/秒
KGS	knot ground speed	节地面速度
KIAS	knots indicated airspeed	以节表示的指示空速
KM	kilometre	千米
KM	kilomega	千兆
KPH	kilometers per hour	千米/小时
KPI	key performance index	关键性能指标
KPP	key performance parameter	关键性能参数
KPS	kilocycles per second	千周/秒,千赫
KPSC	kilogram per square centimeter	千克/平方厘米

L

L/RB	left/right bulge	左/右鼓包
L/RCF	left right cross-feed	左右连通
L/RMLG	left/right main landing gear	左/右主起落架
L/RNL	left/right navigation light	左右航行灯
LACA	low altitude control area	低空控制区
LACA	low altitude close air support	低空近距离空中支援
LACR	low altitude coverage radar	低空搜索雷达
LADF	lift augmented ducted fan	升力增强涵道风扇
LADS	low altitude detection system	低空探测系统
LAEC	low altitude enroute chart	低空航线飞行图
LAHSR	low altitude high speed route	低空高速航线
LAINS	low altitude inertial navigation system	低空惯性导航系统
LALD	low angle low drag	小角度低阻力
LANE	low altitude navigational equipment	低空导航设备
LAP	low altitude performance	低空性能
LAR	local acquisition radar	局部搜索雷达
LARA	low altitude radar altimeter	低空雷达高度表
LAS	local altimeter setting	本地高度表设定(值)
LASAP	land as soon as possible	立即着陆
LASR	low altitude surveillance radar	低空监视雷达,低空搜索雷达

LASS	low altitude surveillance system	低空监视系统
LAT	latitude	纬度，范围
LAW	low altitude waning	低高度告警
LB	left bus	左汇流条
LB	left bank	左坡度
LB	low band	低波段
LB	lighting button	照明测试按钮
LB/RBC	left bus and right bus connection	左右发电机连通
LBIA	low band integrated antenna	低波段综合天线
LBL	laminar boundary layer	层流边界层
LBPR	low bypass ratio	低涵道比
LC	life cycle	寿命周期
LC	lighting controller	照明控制器
LCB	lighting cortrol board	照明控制板
LCBOC	lighting control box of operating cabin	工作舱照明控制盒
LCS	landing control system	着陆控制系统
LCS	laser communication system	激光通信系统
LD	landing distance	着陆距离
LDA	landing distance available	可用着陆距离
LDG	landing	登陆，着陆
LDLLB	left danger level light box	左危险级灯盒
LDLWLB	left danger level warning light box	左危险级告警灯盒
LDN	long distance navigation	远距导航
LDR	landing distance required	所需着陆距离
LDR	laser Doppler radar	激光多普勒雷达
LDR	lift drag ratio	升阻比
LDV	liquid drain valve	放液阀
LE	left elevator	左升降舵
LE	leading edge	前缘，前沿，导边
LED	leading edge down	前缘(襟翼)放下
LED	light-emitting diode	发光二极管
LEF	leading edge flap	前缘襟翼
LEF	living equipment fuse	生活设备熔断器
LEFAS	leading edge flap actuation system	前缘襟翼作动系统
LEMAC	leading edge of mean aerodynamic chord	平均气动力弦前缘
LEMF	leading edge manoeuvre flap	前缘机动襟翼
LEPDB	left engine power distributing board	左发电源配电盘
LES	leading edge slat	前缘缝翼
LEU	leading edge up	前缘(襟翼)收起

LEVL	leading edge vortex lift	前缘涡升力
LF	lighting fuse	指示灯的熔断器
LFAR	low frequency acquisition and ranging	低频搜索与测距
LFAR	low frequency analysis and recording	低频频谱分析与记录
LFC	left flap control	左襟翼操纵
LFDB	left fuel distributing box	左燃油配电盒
LFF	load factor for flight	飞行载荷因数
LG	left glass	左玻璃
LG	landing gear	起落架
LG	landing ground	降落场
LG	laser gyro	激光陀螺
LGA	low gain antenna	低增益天线
LGC	landing gear control	起落架操纵
LGCDB	landing gear control distributing box	起落架操纵配电盒
LGCS	landing gear control and signal	起落架操纵和信号
LGD	landing gear down	放下起落架
LGDC	landing gear door closed	起落架舱门关闭
LGDO	landing gear door opened	起落架舱门打开
LGE	landing gear extension	起落架放下
LGES	landing gear extending speed	放起落架允许速度
LGL	landing gear light	着陆灯
LGL	landing gear lock	起落架锁
LGP	landing gear panel	起落架操纵板
LGR	landing gear retraction	起落架收起
LGS	landing gear signal	起落架信号
LGS	landing gear switch	起落架开关
LGT	landing gear tread	起落架(主)轮距
LGU	left gear uplock	左起落架上位锁
LGW	landing gross weight	着陆总重量
LGWB	landing gear wheelbase	起落架前后轮距
LGWOW	left gear weight on wheels	左起落架机轮重量
LH	left hydraulic	左液压
LH	lubrication hole	润滑油孔
LHO	left hydraulic off	左控液压未通
LHOQ	left hydraulic oil quantity	左液压油油量
LI	landing impact	着陆时撞击
LI	lane identification	航道识别号
LINS	laser inertial navigation system	激光惯性导航系统
LIS	LORAN inertial system	罗兰惯性导航系统

LK	lock	锁,锁定
LL	limit load	极限载荷
LL	lower limit	下限
LLA	latitude-longitude-altitude	纬度-经度-高度
LLB	landing light bright	着陆灯强光
LLBC	landing light bright contactor	着陆灯强光接触器
LLC	landing light control	着陆灯操纵
LLC	landing light control	着陆灯操纵
LLC	lift-lift/cruise	升力-升力/巡航
LLHE	liquid-liquid heat exchanger	液－液换热器
LLL	left landing light	左着落灯
LLWAS	low level windshear alert system	低空风切变告警系统
LMLF	limit manoeuvre load factor	限制机动飞机载荷因数
LMM	(compass) locator at middle marker	中距指点标无线电导航台,中距导航台
LMN	local Mach number	当地马赫数
LO	level off	改平
LO	lighting/off	照明/关
LOC	localizer	定位器,定位信标,航向信标
LOCATE	LORAN/Omega course and tracking equipment	罗兰/奥米伽航道与跟踪设备
LOCK	locked	锁定
LOF	line of force	力线
LON	longitude	经度,经线
LP	limit position	门开到位
LP	landing performance	着陆性能
LP	low pressure	低压
LPC	low pressure compressor	低压压缩机
LPDB	lighting power distributing box	照明电源配电盒
LPDCB	lighting power distributing control box	照明配电控制盒
LPF	low pass filter	低通滤波器
LPIB	low pressure in boot	气囊压力低
LPRS	low pressure rotor speed	低压转子速度
LR	load ratio	载重比,载荷比
LR	long radius	大活动半径
LR	long range	远距离,远程
LRA(S)	long range accuracy (system)	远程精确导航(系统),罗拉克导航(系统)
LRALS	long range approach and landing system	远距离进场[进近]与着陆系统

LRC	long range cruise	远程巡航
LRF	left recirculation fan	左回流风扇
LRFCB	left recirculation fan circuit breaker	左回流风扇断路器
LRN	long range navigation	远程导航
LRP	left refrigeration package	左制冷包
LRR	long range radar	远程雷达
LRRA	low range radio altimeter	低范围无线电高度表
LRU	line replaceable unit	外场可更换单元
LS	left system	左系统
LS	load spring	加载弹簧
LS	left side	左侧
LS	lifesaving	救生
LS	left system	左系统
LS	lighting switch	照明开关
LSB	lower sideband	低边带,下边带
LSCRED	left system close rear entry door	左系统关后登机门
LSE	lifesaving equipment	救生设备
LSJB	lifesaving jacket bag	救生衣袋
LSK	line select key	行选键
LSOR	left system oil return	左系统回油
LSOS	left system oil suction	左系统吸油
LSP	left system pressurization	左系统增压
LSPS	left starter power supply	左起动供电
LSR	lifesaving rope	救生绳
LSSAS	longitudinal static-stability augmentation system	纵向静安定度增强系统,纵向静增稳系统
LSSV	left system sample valve	左系统采样阀
LST	liquid storage tank	储液箱
LSTTS	liquid storage tank temperature sensor	储液箱温度传感器
LSWT	low speed wind tunnel	低速风洞
LT	lag time	滞后时间
LT	local time	当地时,地方时
LT	low temperature	低温
LT	light	灯
LT	light test	灯测试
LTG	lighting	照明
LTLV	low temperature limitation valve	低温限制活门
LTO	landing-takeoff	着陆-起飞
LTOT	latest time over target	最迟飞跃目标时间

LUF	lowest usable frequency	最低可用频率
LVDT	linear variable differential transformer	线性电压差动式传感器
LWSFS	left wing start fuel supply	左翼起动供油
LWTBF	left wing tip brake failure	左制动器故障
LYS	limiting yield stress	极限屈服应力
LZ	landing zone	降落区
LZT	local zone time	当地时,地方时

M

M	Mach	马赫数
M	meter	米
M	manual	人工,手动,手册,指南
MA	minor airfield	次要机场
MA	missed approach	进场[进近]失败
MA	mission abort	任务中断
MA	mission accomplished	任务完成
MAA	minimum approach altitude	最低进场[进近]高度
MAA	missed approach area	复飞层
MAA	monitoring angle of attack	监控迎角
MADS	meteorological airborne data system	航空气象数据系统
MADT	mean accumulated downtime	平均累积不工作时间,平均累计停用时间
MAE	mean absolute error	平均绝对差
MAG	maximum available gain	最大可用增益
MAG	magnetic	磁的,有磁性的
MAG	mobile arresting gear	活动式拦阻装置
MAI	Mach airspeed indicator	马赫数空速指示器
MAR	mission abort rate	(飞行的)任务中断率
MART	mean active repair time	平均有效修理时间
MASR	multiple antenna surveillance radar	多天线监视雷达
MAT	military aircraft type	军用飞机型号
MAX	maximum	最大,最大限度,极大
MBHO	main booster hydraulic off	主助液压未通
MBRW	maximum brake release weight	松开刹车时最大重量
MBS	marker beacon system	指点标系统
MC	magnetic clutch	电磁离合器
MC	manual control	手动控制
MCB	main contactor box	主接触器箱
MCBF	mean cycles between failures	平均故障间隔循环(次)数

MCL	maximum climb power	最大上升功率
MCL	master caution light	主告警灯
MCP	maximum cruise power	最大巡航功率
MCP	multifunction control panel	多功能控制板
MCR	maximum cruise rating	最大巡航速率
MCT	mean corrective time	平均修复时间
MD	manual deicing	手动除冰
MD	measure direction	测向
MD	message display	信息显示器
MDA	minimum decision altitude	最低决断高度
MDA	minimum descent altitude	最低下降高度
MDGW	maximum design gross weight	最大设计总量
MDGW	mission design gross weight	飞行总重量
MDH	maximum descent height	最大下降高度
MDT	mean downtime	平均空闲时间,平均停机时间
MEIA	minimum enroute IFR altitude	仪表飞行规则最低航线高度
MEL	master equipment list	主设备清单
MEL	minimum equipment list	最低设备清单
MEM	memory	存储器
MEOTBF	mean engine operating time between failures	发动机平均故障间隔时间
MET	mission elapsed time	任务耗用时间
METOP	maximum except takeoff power	除起飞外最大功率
MFC	main fuel control	主燃油控制
MFC	maximum fuel capacity	最大燃油容量
MFD	multi-function display	多功能显示器
MFED	maximum flat envelope delay	最大平包线延迟(时间)
MFHBCF	mean flight hours between critical failures	平均关键故障间隔飞行小时
MFHBF	mean flight hours between failures	平均故障间隔飞行小时
MFHBFA	mean flight hours between false alarms	平均虚警间隔飞行小时
MFHBMCF	mean flight hours between mission-critical failures	平均关键任务故障间隔飞行小时
MFHBME	mean flight hours between maintenance events	平均维修事件间隔飞行小时
MFHBOMF	mean flight hours between operational mission failures	平均作战任务故障间隔飞行小时
MFHBR	mean flight hours between removals	平均移动间隔飞行小时
MFHBUMA	mean flight hours between unscheduled maintenance action	平均非计划维修间隔飞行小时

MFL	maintenance fault list	维修故障表
MFL	minimum field length	最短机场长度
MFLJ	multi-function lifesaving jacket	多功能救生衣
MFOB	minimum fuel on board	机上最低限度油量
MFR	magnetic field error	磁差
MGTD	main (landing) gear touchdown	主起落架接地
MGTOW	maximum gross takeoff weight	最大起飞总重量
MGW	maximum gross weight	最大总重量
MH	magnetic heading	磁航向
MH	main hydraulic	主液压
MHA	minimum holding altitude	空中等待最低高度
MHBF	maintenance hours between flight	飞行间隔维修小时
MHO	main hydraulic off	主控液压未通
MHP	main hydraulic pump	主液压泵
MHRS	magnetic heading reference system	磁航向参考系统,磁航向基准系统
MHS	main hydraulic system	主液压系统
MHW	magnetic heading warning	磁航向告警
MHZ	Megahertz	兆赫
MIC	microphone	话筒,扩音器,麦克风
MID	middle	中间,中部
MIE	manoeuvre-induced error	机动(引起的)误差
MILS	microwave instrument landing system	微波仪表着陆系统
MILS	modular instrument landing system	模块式仪表着陆系统
MIN	minimum	最小
MIS	mission intercom system	任务机内通话系统
MITO	minimum interval takeoff	最短间隔(时间)起飞
ML	microwave landing	微波着陆
MLC	maximum lift coefficient	最大升力系数
MLDT	mean logistics delay time	平均后勤延误时间
MLFS	maximum level flight speed	最大平飞速度
MLG	main landing gear	主起落架
MLGH	main landing gear height	主起落架高度
MLGL	main landing gear lighting	主起舱照明
MLGLR	main landing gear lighting and receptacle	主起落架舱照明灯和插座
MLGW	main landing gear well	主起落架舱
MLS	microwave landing system	微波着陆系统
MM	mask mike	面罩话筒
MM	multiuser multitask	多用户多任务
MMALS	multimode approach and landing system	多模式进场[进近]与着陆系统

MMD	mean mission duration	平均任务持续时间
MMEL	master minimum equipment list	主最低设备总目表
MMH	maintenance man-hour(s)	维修工时
MMH/AC	maintenance man-hours/ aircraft	每架飞机的维修工时
MMH/FH	maintenance man-hours per flight hour	维修工时每飞行小时
MMH/OH	maintenance man-hours per operating hour	维修工时每工作小时
MMP(R)	mixture maximum power (rich)	富油最大功率
MMR	multi-mode receiver	多模式接收机
MMRHPFH	mean maintenance and repair hours per flight hour	平均维修小时每飞行小时
MN	Mach number	马赫数
MN	magnetic north	磁北
MNWP	maximum normal working pressure	最大额定工作压力
MOCA	minimum obstruction clearance altitude	最低净空高度,最低越障高度
MOMN	maximum operating Mach number	最大飞行马赫数
MOP	main oil pressure	主滑油压力
MOT	main oil temperature	主滑油温度
MP	main power	主电源
MP	manifold pressure	进气压力
MPA	maritime patrol aircraft	海上巡逻飞机
MPAR	modified precision approach radar	改进型精确进场[进近]雷达
MPLW	maximum permitted landing weight	最大允许着陆重量
MPMT	mean preventive maintenance time	平均预防性维修时间
MPS	minimum performance specifications	最低性能规范
MPS	minimum performance standards	最低性能标准
MPS	mission performance standard	任务性能标准
MPTO	maximum performance takeoff	最大性能起飞
MPW	maximum pavement width	最大路面宽度
MR	middle relay	中间继电器
MR	mooring ring	系留环
MR	main receiving	主收
MR	mixing room	混合室
MRB	marker radio beacon	无线电指向信标
MRB	maintenance review board	维修审查委员会
MRB	material review board	不合格品处理委员会
MRHFV	maintain runway heading for vector	保持跑道引导航向
MRIL	master repairable item list	主要可修复项目清单

MRO	maintenance，repair and operation	维修、修理与运行
MRO	maintenance，repair and overhaul	维修、修理与大修
MRT	mean repair time	平均修理时间
MRT	mean response time	平均响应时间
MRVAFS	manual regulating valve of antenna feed system	天馈系统手动调节阀
MRVOT	manual regulating valve of transmitter	发射机手动调节阀
MRW	maximum ramp weight	最大停机重量
MS	mechanical system	机械系统
MS	military standard	军用标准
MS	mode switching	模式转换
MS	margin of safety	安全余量,安全系数,安全界限
MSA	minimum safe altitude	最低安全高度
MSA	mission system avionics	任务系统航空电子设备
MSAW	minimum safe altitude warning	最低安全高度告警
MSB	micro switch bracket	微动开关支架
MSB	main switch board	总配电盘,总控制板
MSCM	metal-semiconductor-metal	金属-半导体-金属
MSD	minimum safe distance	最小安全距离
MSEP	maintenance standardization and evaluation program	维修标准化与鉴定大纲
MSG	maintenance steering group	维修指导小组
MSL	mean sea level	平均海平面
MSP	maintenance service plan	维修服务计划
MSP	maximum structural payload	最大结构有效载荷
MSPS	mission system power supply	任务系统供电
MSSDCB	mission support system display and control box	任务支持系统显控盒
MSTOLW	maximum short takeoff and landing weight	最大短距起落重量
MT	maximum torque	最大扭矩
MT	mean time	平均时间
MT	maximum thrust	最大推力
MTAT	mean turn around time	平均再次出动准备时间
MTBA	mean time between action	平均行动间隔时间
MTBA	mean time between alarm［warning］	平均告警间隔时间
MTBCD	mean time between confirmed defects	平均故障间隔时间
MTBCF	mean time between component failure	平均部件故障间隔时间
MTBCF	mean time between critical failures	平均严重故障间隔时间

MTBCM	mean time between corrective maintenance	平均恢复性维修间隔时间
MTBD	mean time between demands	平均需求间隔时间
MTBDD	mean time between defects downing	平均故障停用间隔时间
MTBDE	mean time between downing event	平均不能工作事件间隔时间
MTBDR	mean time between depot repair	平均基地修理间隔时间
MTBE	mean time between errors	平均错误间隔时间
MTBF	mean time between failures	平均故障间隔时间
MTBFRO	mean time between failures requiring overhaul	平均需要大修的故障间隔时间
MTBI	mean time between incidents	平均事故征候间隔时间
MTBIE	mean time between interrupt event	平均中断事件间隔时间
MTBM	mean time between maintenance	平均维修间隔时间
MTBMA	mean time between mission aborts	平均任务中止间隔时间
MTBMA	mean time between maintenance actions	平均维修活动间隔时间
MTBME	mean time between maintenance events	平均维修事件间隔时间
MTBO	mean time between overhauls	平均大修寿命,平均大修间隔时间
MTBPM	mean time between preventive maintenance	平均预防性维修间隔时间
MTBR	mean time between removals〔replacement〕	平均拆卸〔换件〕间隔时间
MTBR	mean time between repairs	平均修理间隔时间
MTBSF	mean time between significant failures	平均重大故障间隔时间
MTBUER	mean time between unscheduled engine removals	平均非预定发动机拆卸间隔时间
MTBUM	mean time between unscheduled maintenance	平均非预定维修间隔时间
MTBUR	mean time between unscheduled removals	平均非预定拆卸间隔时间
MTBW	mean time between warning	平均告警间隔时间
MTC	mean transinformation content	平均传送信息量
MTC	minimum terrain clearance	最小地形间距
MTCA	minimum terrain clearance altitude	离地最低(飞行)高度
MTCF	mean time to catastrophic failure	灾难性故障前平均时间
MTD	mean temperature difference	平均温度差
MTE	maximum tracking error	最大跟踪误差
MTGW	maximum taxi gross weight	最大滑行总重(量)
MTM	maximum takeoff mass	最大起飞质量
MTOGW	maximum takeoff gross weight	最大起飞总量
MTOW	maximum takeoff weight	最大起飞重量
MTTCO	mean time to checkout	平均检查时间

MTTD	mean time to detection	平均故障检查时间
MTTF	mean time to failure	平均故障前时间
MTTFF	mean time to first failure	平均首次故障时间
MTTFSF	mean time to first system failure	平均系统首次故障时间
MTTM	mean time to maintenance	平均维修时间
MTTR	maximum time to repair	最长修复时间
MTTR	mean time to repair	平均修复时间
MTTR	mean time to restore	平均恢复时间
MTTR (I)	mean time to repair (I)	(I)平均修复时间(中继级)
MTTR (O)	mean time to repair (O)	(O)平均修复时间(基层级)
MTTRF	mean time to restore functions	恢复功能用的平均时间
MTTRF	mission time to restore functions	恢复功能用的任务时间
MTTRS	mean time to restore system	平均系统恢复时间
MTTS	mean time to service	平均维护时间
MTTSF	mean time to system failure	平均系统故障时间
MTTT	mean time to test	平均试验时间
MTUR	mean time between unscheduled removals (or replacements)	平均非预定拆卸(或更换)间隔时间
MTWA	maximum takeoff weight authorized	批准的最大起飞重量
MU	measure unit	测量装置
MUA	maximum usable altitude	最大可用高度
MUT	mean up time	平均可使用时间,平均工作时间
MVS	minimum vector speed	最小矢量速度
MVS	minimum visible signal	最小可见信号
MVV	maximum ventilator volume	最大通风量
MW	main warning	主告警
MWTQ	maximum wing tank quantity	机翼油箱最大油量
MWV	maximum working voltage	最高工作电压
MZFW	maximum zero fuel weight	最大无燃油重量

N

N	north	北,北方,北部
N/S	normal/standby	正常/备用
NA	not applicable	不适用
NA	navigation aid	导航设备
NAB	normal acceleration bias	正常加速度偏差
NAGC	noise automatic gain control	噪声自动增益控制
NAM	normal air mode	正常空中工作状态
NAS	national aerospace standard	美国国家宇航标准

NASS	navigation satellite system	导航卫星系统
NATCAS	navigation，air traffic control and collision avoidance systems	导航、空中交通管制与防撞系统
NAV	navigation	导航
NAWTOL	night/all-weather take-off and landing	夜间/全天候起飞着陆
NC	no current	无电流
NC	normal control	正常操纵
NCA	normally control aileron	正常操纵副翼
NCBB	navigator circuit breaker box	领航员断路器箱
NCE	normally control elevator	正常操纵升降舵
NCP	navigation control panel	导航方式控制板
ND	nose down	俯冲、低头
ND	navigation display	导航显示
NDB	non-directional beacon	全向信标，导航台
NDR	navigation Doppler radar	导航多普勒雷达
NED	nose emergency door	前应急门
NEUT	neutral	中立
NFOV	narrow field of view	窄视场，小视场
NFP	net flight path	基本飞行轨迹[航迹]
NGF	navigation flare	航行灯
NGS	navigation glide slope	航行下滑坡度
NHARS	navigation，heading and attitude reference system	导航、航向与姿态参考[基准]系统
NI	neutral indication	中立指示
NL	navigation light	航行灯
NLC	night lighting condition	夜间照明条件
NLF	navigation light flasher	导航灯闪光装置
NLF	normal load factor	法向载荷因数
NLG	nose landing gear	前起落架
NLS	no-load start	空载启动
NLS	nonlinear system	非线性系统
NM	nautical mile	海里
NMAC	near mid-air collision	空中碰撞危险
NOE	nap of earth	掠地飞行，贴地飞行
NOLGD	neutral，open landing gear door	中立,开起落架舱门
NORM	normal	正常
NP	normal pressure	标准压力,正常压力
NP	nozzle position	喷管位置
NPI	nozzle position indicator	喷口位置指示器

NPR	noise power ratio	噪声功率比
NPR	nozzle pressure ratio	喷管压力比
NPT	normal pressure and temperature	常温常压,标准温度与压力
NRE	negative resistance effect	负电阻效应
NRE	negative resistance element	负电阻元件
NRP	normal rated power	标准额定功率
NRS	normal rated speed	标准额定速度
NRT	normal rated thrush	标称[额定]推力
NS	navigation system	导航系统
NSE	navigation system error	导航系统误差
NST	noise, spikes and transients	噪声、峰值与瞬值
NSV	negative sequence voltage	逆序电压
NSV	noise, shock and vibration	噪声、震动与振动
NTFC	negative thrust feathering check	负拉力顺桨检查
NTOL	normal takeoff and landing	正常起落
NTR	noise-temperature ratio	噪声温度比
NTS	negative torque signal	负转矩信号
NU	nose up	拉机头,抬头,机头上扬
NW	nose wheel	前轮
NWB	negative wire bus	负线汇流条
NWGM	negative wire ground module	负线接线模块
NWP	normal working pressure	正常工作压力
NWS	nose wheel steering	前轮转弯

O

O	outer	外部
OA	omnirange antenna	全向式无线电信标天线
OA	overhead approach	飞越机场进场[进近]
OAC	optimum approach course	最佳进场[进近]航线
OADS	omniderectional air data system	全向大气数据系统
OAI	omnidirectional airspeed indicator	全向空速指示器
OAT	operating ambient temperature	工作周围温度,工作环境温度
OBD	omnibearing distance	全向距离导航
OBI	omnibearing indicator	全向方位指示器
OBIFCO	on board in-flight check-out	机上飞行中检查
OBIT	operating built-in test	运行自检测
OBS	omnibearing selector	全方位选择器
OC	obstacle clearance	离障碍物高度
OC	on course	正航向,在航线上

OC	operating cabin	工作舱
OC	oven control	电烤箱控制
OCA	obstacle clearance altitude	越障高度
OCFI	oil cooling flap indicator	散热风门指示
OCH	obstacle clearance height	越障高度
OCL	optimum cruising level	最佳巡航高度
OCPTS	operating cabin pipeline temperature sensor	工作舱管路温度传感器
OCTS	operating cabin temperature sensor	工作舱温度传感器
OD	overall dimension	总尺寸
ODADS	omnidirectional air data system	全向大气数据系统
ODALS	omnidirectional approach lighting system	全向进场[进近]照明系统
ODR	omnidirectional radar	全向雷达
ODR	omnidirectional digital radar	全向数字式雷达
ODR	overland downlook radar	陆上下视雷达
ODRNS	omnidirectioanl radio navigation system	全向无线电导航系统
ODT	omnidirectional transmitter	全向发射机
OFC	optical fiber cable	光纤电缆
OFC	overfly correction	飞越校正
OFF	off	断开
OFI	omniflight indicator	全向飞行指示器
OFI	oxygen flow indicator	氧气示流器
OFP	operational flight package	作战飞行包
OFP	original flight plan	原始飞行计划
OFZ	obstacle-free zone	无障碍物区
OGE	out-of-ground effect	脱离地面效应,无地效
OGV	outlet guide vane	出口导向叶片
OGW	overload gross weight	超载总重,超载毛重
OH	overheat	过热
OHM	overhaul manual	大修手册
OHR	overhaul report	大修报告
OHR	over horizon radar	超视距雷达
OHS	overhaul specification	大修规范
OI	oxygen interface	氧气接口
OL	overall length	总长,全长
OL	over load	过载
OM	outer marker	远台指示
OM	operation mode	工作模式
OMI	operations and maintenance instruction	使用与维修说明书

ONS	Omega navigation system	奥米伽导航系统
OOOI	out-off-on-in	滑出-起飞-接地-停靠门位
OP	oil pressure	滑油压力
OP	operation procedure	操作程序
OP	over pressure	超压
OPDOT	optimum preliminary design of transport	运输机初步设计最佳化
OPER	operating	工作
OQ	oil quantity	滑油油量
OR	overhaul and repair	大修与修理
OR	oil ring	油环
OR	operating range	工作范围
OR	operational reliability	使用可靠性,操作可靠性
OR	oxygen regulator	氧气调节器
ORB	omnidirectronal radio beacon	全向无线电信标
ORB	omnidirectional radar beacon	全向雷达信标
ORM	overhaul and repair manual	大修与修理手册
ORZ	omnirange zone	全向无线电信标区
OS	oil suction	吸油
OS	oil system	滑油系统
OS	out of service	退出现役,非现役,不能使用,失效
OS	overspeed	超速
OS	oxygen supply	用氧
OSAA	oxygen supply altitude annunciator	用氧高度信号器
OSC	oscillator	振荡器
OSC	oscilloscope	示波器
OSG	over speed governor	过速控制器
OSI	oxygen supply interface	充氧接口
OSS	oxygen supply signal	用氧信号
OSSB	oxygen supply signal box	用氧信号盒
OT	overall test	全面测试
OT	over temperature	超温
OTIS	operation, transport, inspection, storage	操作,运输,检查,存储
OTM	operational technical manual	操作技术手册
OTSI	operating time since inspection	检修后运行时间
OVHD	overhead	顶部,在头顶上
OVP	over voltage protection	过压保护
OW	outboard wing	外翼
OWE	operating weight empty	空机使用重量,运转空重
OWE	optimum working efficiency	最佳工作效率

OWF	optimum working frequency	最佳工作频率
OWS	oil-water separator	油水分离器
OWSL	outboard wing suspension lug mark	外翼吊耳标记

P

P	pitch	俯仰
P	pitot	全压
P/B	power/brightness	电源/亮度
PA	power approach	带油门进场[进近]着陆
PA	power amplifier	功率放大器
PA	preamplifier	前置放大器
PAAR	precision approach airfield radar	机场精密进近雷达
PADS	primary attitude determination system	初始姿态确定系统
PAG	page	页面
PAI	precise angle indicator	精密角度指示器
PAI	pressure alarm indicator	压力告警指示器
PAI	propeller anti-icing	螺旋桨防冰
PALS	precision apparoah and landing system	精密进场[进近]和着陆系统
PAR	phased array radar	相控阵雷达
PATH	position attitude and true heading	位置姿态与真航向(系统)
PB	power button	电源按钮
PB	power from battery	蓄电瓶供电
PBDB	power from battery distributing box	蓄电池供电配电盒
PBDI	position，bearing and distance indicatior	位置、方位与距离指示器
PBR	propeller brake release	螺旋桨制动解除
PBRB	propeller brake relay box	螺旋桨制动继电器盒
PC	production certificate	生产许可证
PC	pilot console	驾驶员操纵台
PC	pressure controller	压力控制器
PC	pressure control	压力控制
PC	power controller	功率控制器
PCAS	pitch control augmentation system	俯仰控制增稳系统
PCS	pitch control system	俯仰控制系统
PCS	power conditioning system	功率调节系统
PCS	process control system	过程控制系统
PCU	pneumatic checkout unit	气动检测装置
PCU	propeller control unit	螺旋桨控制装置
PCU	power control unit	助力器
PD	potential difference	电位差,势差

PD	powered descent	带油门下降
PD	parasite drag	废阻力
PD	pulse Doppler	脉冲多普勒(雷达)
PD	pulse duration	脉冲持续时间,脉冲宽度
PD	propeller deicing	螺旋桨除冰
PDC	pre-departure clearance	起飞前放行许可
PDD	priority delivery date	优先交付日期
PDU	power distribution unit	配电装置
PE	position error	位置误差
PEC	pressure error correction	压力误差修正
PEC	propeller electronic control	螺旋桨电子控制
PFC	preflight check	飞行前检查
PFC	primary flight control	初级飞行控制
PFCES	primary flight control electronic system	主飞行控制电子系统
PFCS	primary flight control system	主飞行控制系统
PFCU	powered flight control unit	飞行操纵助力器
PFD	primary flight display	主飞行显示器
PFP	propeller feathering pump	螺旋桨顺桨泵
PFR	preflight review	起飞前检查
PFR	post flight report	飞行后报告
PFUO	pitch follow up operation	俯仰随动操作
PFUS	pitch follow up system	俯仰随动系统
PGPE	preflight ground pressurization equipment	飞行前地面增压设备
PH	pipeline heating	管路加温
PHAOA	positive high angel of attack	正大迎角
PHP	propeller horsepower	螺旋桨马力,推进器马力
PHPB	propeller heat power box	螺旋桨加温功率盒
PHPSCB	propeller heating power supply circuit breaker	螺旋桨加温供电断路器
PIA	pilot-interpreted approach	飞行员自主判断进场[进近]
PL	payload	有效载荷,净载荷,战斗部
PL	power lever	油门杆
PL	pipe label	导管标签字样
PLA	power lever angle	功率杆角度,油门杆角度
PLAOA	positive low angle of attack	正小迎角
PLAP	power lever angle prime	油门杆注油角度
PLER	power limiter execution relay	功率限制器执行继电器
PM	programmed mechanism	程序机构
PMA	parts manufacturer approval	零部件制造批准书

PMA	permanent magnet alternator	永磁交流发电机
PMAPA	probable missed approach per arrival	每次降落的复飞率
PMAT	portable multi-access terminal	多路存取
PMC	pitching moment coefficient	俯仰力矩系数
PMMP	permission mean maximum pressure	容许最大平均压力
PMRM	periodic maintenance requirements manual	定期维修需求手册
PN	part number	件号
PNI	pictorial navigation indicator	图形导航指示器
PNL	panel	板,仪表板
PNS	positioning and navigation system	定位导航系统
POP	pilot operation procedure	飞行员操作程序
POS	position	位置
POSS	power-off stall speed	无动力失速速度
POUP	post overhaul upkeep period	翻修后维护周期
POV	peak overshoot voltage	峰值上冲电压
PP	pressure pump	增压泵
PPBM	power plant buildup manual	动力装置手册
PPC	power parameter collector	电源参数采集器
PPI	power performance index	发送机性能指数
PPOPS	pressure pump outlet pressure sensor	增压泵出口压力传感器
PPSI	pounds per square inch	磅/英寸^2
PPSM	primary power supply module	主电源微型组件,主电源模块
PPUC	prompt position undating course	即时位置改航
PR	pitch rate	俯仰角速度
PR	planning reference	设计标准,设计参考资料
PR	pressure ratio	压力比,增压比
PR	pressure regulator	调压器,压力调节器
PR	propeller reversed	螺旋桨倒转
PRAM	producibility, reliability, availability, maintainability	生产性、可靠性、可用性与维修性
PRARS	pitch, roll, azimuth reference system	俯仰、横滚与方位参考系统
PRAWS	pitch/roll attitude warning system	俯仰/滚转姿态告警系统
PRESS	pressure	压力
PRESS	pressurization	增压
PRH	pressure reducer heating	减压器加温
PRI	primary	主要的,初步的,初级的
PRSV	pressure regulate safe valve	压力调节安全活门
PRTR	pressure reducer temperature relay	减压器温度继电器
PS	phase shift	周向移动,相移

PS	position sensor	位置传感器
PS	power system	电源系统
PS	pressure sensor	压力传感器
PS	propeller speed	螺旋桨速度
PS	preset	预置
PS	power source	电源
PSFCDB	power supply of forward cabin distributing box	前舱配电盒供电
PSH	preselected heading	航向预选
PSPFP	power supply of propeller feathering pump	螺旋桨顺泵供电
PSPSH	pitot static prssure sensor heating	全静压受感器加温
PSR	precision secondary radar	精密二次雷达
PSR	primary surveillance radar	一次监视雷达,主要雷达
PSRM	pressurization system regulator manifold	增压系统调节器歧管
PSSP	pitot static standby probe	全静压备用探头
PSSR	pitot static standby relay	全静压备用继电器
PST	propeller STOL transport	短距起落螺旋桨运输机
PT	power turbine	动力涡轮
PTC	provisional type certificate	临时型号合格证
PTGL	preicision touch-and-go landing	连续精确着陆起飞
PTO	power takeoff	功率输出,动力起飞
PTU	power transfer unit	动力传动装置
PU	power unit	动力装置
PUD	power unit deicing	动力装置除冰
PV	phase velocity	相位速度
PWI	preliminary warning instruction	预警指令
PWI	proximity warning indication	防撞告警指示
PWR	power	电源

Q

QAS	quality assurance system	质量保证体系
QD	quadrant depression	俯角
QE	quadrant elevation	仰角
QET	quick engine test	发动机快速试车
QM	quality management	质量管理
QRS/VTOL	quiet, reduced, short, vertical takeoff and landing	低噪声、降低速度、短距离/垂直起落
QWL	quick weight loss	快速失重

R

R	radio	电台
R	radius	半径
R	range	距离,航程,射程,靶场
R	rate	率,速率
R	ratio	比,比率
R	received	已收到
R	reconnaissance	侦察机
R	reliability	可靠性
R	resistance	阻力,抗阻
R	resistor	电阻器,电阻
R	right	右(跑道符号)
R	receive	接收
R	recirculation	回流,再循环
R	receptacle	插座
R	redundancy	余度
R	relay	继电器
R	reset	复位
R	resistor	电阻
R/C	radio/communication	无线电/通话
R/E	receiver/exciter	接收机/激励器
R/R	radar/ radio	雷达/无线电
R/T	receiver/ transmitter	接收/发送
RA	radar altimeter	雷达高度表,无线电高度表
RA	radio altimeter	无线电高度表
RA	radio altitude	无线电高度
RA	roll angle	横滚角
RA	radar altimeter	雷达高度计
RAA	radar approach aid	进场[进近]雷达导航设备
RAC	radar absorbing chaff	雷达吸波箔条
RAC	radar approach control	雷达进场[进近]控制
RACER	reliability, availability, compatibility, economy and reproducibility	可靠性、可用性、兼容性、经济性与再现性
RAE	range, azimuth and elevation	距离、方位与仰角,距离、方位与高度
RAF	resolution advisory fault	RA 故障
RAN	random area navigation	随机导航
RAP	right accumulator pressure	右蓄压瓶压力
RAP	radar absorbent [absorbing] paint	雷达吸收涂层

RAR	radio acoustic ranging	无线电声测距
RAS	rough air speed	颠簸气流空速
RAVDPI	radiator air vent door position indicator	散热风门指示器
RB	right bank	右倾斜
RB	right bus	右汇流条
RB	relative bearing	相对方位（角）
RBA	radar beacon antenna	雷达信标天线
RBG	rear boarding gate	后登机门操纵
RBS	radio beacon station	无线电信标台
RBSOR	rotary base system oil return	转台系统回油
RBSOS	rotary base system oil suction	转台系统吸油
RBSP	rotary base system pressurization	转台系统增压
RBSSV	rotary base system sample valve	转台系统采样阀
RC	radio compass	无线电罗盘
RC	rating code	等级代码
RC	resistance/capacitance	电阻/电容
RC	redundancy controller	余度控制器
RC	rest cabin	休息舱
RCA	reach cruising altitude	到达巡航高度
RCDC	radar course-directing control	雷达航向引导控制
RCL	recall	二次呼叫，收回，撤消
RCLB	release of control lock bell	舵警铃解除
RCP	radar control panel	雷达控制板
RCP	reversionary control panel	转换控制板
RCS	radio/ compass switch	电台/罗盘开关
RCTCV	rest cabin temperature control valve	休息舱温控活门
RCTS	rest cabin temperature sensor	休息舱温度传感器
RCU	remote control unit	遥控装置
RCW	redundancy controller work	余度控制器工作
RCWV	rated continuous working voltage	额定连续工作电压
RD	rate of descent	（垂直）下降速率
RDA	remote data access	远程数据访问
RDAL	rudder deflection angle limitation	方向舵偏限
RDALMF	rudder deflection angle limiter mechanism failure	方向舵偏角限制机构故障
RDCP	refuel defuel control panel	加油放油控制板
RDF	radar direction finder	雷达测向（仪）
RDF	radio distance finder	无线电测距（仪）
RDINS	radio-Doppler-inertial navigation system	无线电-多普勒-惯性组合导航系统

RDLLB	right danger level light box	右危险级灯盒
RDLWLB	right danger level warning light box	右危险级告警灯盒
RDR	radio detection and ranging	无线电探测与测距
RDS	rudder-disconnct speed	方向舵断开速度
RDY	ready	准备
RE	right elevator	右升降舵
RE	radio equipment	无线电设备
RE	release	解除,释放
REA	radar echoing area	雷达回波面积
REC	rear equipment cabin	后设备舱
REGAL	range and elevation guidance for approach and landing	进场[进近]着陆距离与仰角引导
REI	range elevation indicator	距离俯仰指示器
REPDB	right engine power distributing box	右发电源配电盒
RETR	retraction	收起
RF	recirculation fan	回流风扇
RF	retract flap	收襟翼
RFC	right flap control	右襟翼操纵
RFDB	right fuel distributing box	右燃油配电盒
RFDB	recirculation fan distribution box	回流风扇配电盒
RFL	rough field landing	不平坦机场着陆
RFM	reliability figure of merit	可靠性品质因数
RFSP	rear fuel supply pump	后供油泵
RFTO	ready for takeoff	准备起飞
RFU	ready for use	准备使用
RG	right glass	右玻璃
RGB	reduction gearbox	减速齿轮箱,减速器
RGBEPC	right generator bus emergency power contactor	右发汇流条应急供电接触器
RGC	recorder ground check	记录器地面检查
RGD	right gear downlock	右起落架下位锁
RGDC	right gear doors closed	右起落架舱门关闭
RGENV	right generator electric network voltage	右发电网电压
RGU	right gear uplock	右起落架上位锁
RGWOW	right gear weight on wheels	右起落架机轮即论重量
RH	right hydraulic	右液压
RH	relative height	相对高度
RH	relative humidity	相对湿度
RH	Rockwell hardness	洛氏硬度

RHI	range-height indicator	距离高度指示器
RHN	Rockwell hardness number	洛式硬度(指)数
RHO	right hydraulic off	右控液压未通
RHP	rated horsepower	额定马力
RHS	right hydraulic system	右液压系统
RHSP	right hydraulic system pressurization	右液压系统增压
RHW	radar homing and warning	雷达寻的与告警(系统)
RHWS	radar homing and warning system	雷达寻的与告警系统
RI	removal and installation	拆卸与安装
RI	radar interrogator	雷达询问机
RI	radio instrumentation	无线电仪表
RIA	rapid inertial alignment	快速惯性校准
RIA	radar inertial altimeter	雷达惯性高度仪
RIAL	runway identifiers and approach lighting	跑道标志灯与进场[进近]灯光设备
RIAP	revised instrument approach prodedure	修订的仪表进场[进近]程序
RIC	resistance-inductance-capacitance	电阻-电感-电容
RID	radio inertial guidance	无线电惯性制导
RIDP	radar-IFF data processor	雷达敌我识别数据处理机
RIDP	right ice detector probe	右结冰探测器
RIDS	radio information distribution system	无线电信息分配系统
RIM	radar intelligence map	雷达情报图
RINS	radio inertial navigation system	无线电惯性导航系统
RINS	redundant inertial navigation system	备份惯性导航系统
RIT	rotor inlet temperatrure	转子进口温度
RLD	reference landing distance	参考[基准]着陆距离
RLG	retracting landing gear	收起落架
RLL	right landing light	右着陆灯
RLY	relay	继电器
RMA	reliability, maintainability and availability	可靠性、维修性和可用性
RMI	radio magnetic indicator	无线电磁指示器
RMLG	right main landing gear	右主起落架
RMPS	releasing mid-pitch stop	解除中距限动
RMS	reliability, maintainability and support-ability	可靠性、维修性和保障性
RMT	reliability, maintainability and testability	可靠性、维修性和测试性
RMTS	reliability, maintainability and testability and supportability	可靠性、维修性和测试性与保障性
RNF	radio navigation facilities	无线电导航设施
RNG	range	量程

ROA	radius of action	活动半径,作用半径
ROC/A	rate-of-climb/altimeter	爬升速率/高度表
RP	refrigeration pack	制冷包
RP	reference point	参考[基准]点
RP	release point	投放点
RPS	residual pressure select	余压选择
RPS	range positioning system	区域定位系统
RPTC	refrigeration pack temperature controller	制冷包温度控制器
RPY	roll, pitch and yaw	滚转、俯仰与偏转[偏航]
RQL	rich-quench-lean	富油燃烧-快速燃烧-贫油燃烧
RR	radar ranging	雷达测距
RRB	radio range-beacon	无线电航向信标
RRF	right recirculation fan	右回流风扇
RRG	roll rate gyro	滚转角速度陀螺仪
RRP	right refrigeration package	右制冷包
RS	right system	右通用,右系统
RS	reference speed	特征速度,标准速度,稳定速度
RS	reference source	参考源
RS	radio station	无线电台
RS	range search	距离搜索
RS	rear spar	后梁
RS	rotational speed	转速
RS	right side	右侧
RSB	rectifier signal box	整流器信号盒
RSB	radio system bus	无线电系统汇流条
RSC	right start current	右启动电流
RSCB	refrigeration system control box	制冷系统控制盒
RSCRED	right system close rear entry door	右系统关后登机门
RSFS	right starter fuel supply	右起动供油
RSOR	right system oil return	右系统回油
RSOS	right system oil suction	右系统吸油
RSP	right system pressurization	右系统增压
RSSCU	rotative speed signal changeover unit	转速信号转换器
RSSV	right system sample valve	右系统采样阀
RT	real time	实时
RT	recovery time	恢复时间
RT	route	航线,航路,路线
RTA	receiver transmitter antenna	接收发射机天线
RTF	radar terrain following	雷达地形跟随

RTOW	regulated takeoff weight	规定起飞重量
RTT	rudder trim tab	方向舵调整片
RTTCB	rudder trim tab control box	方向舵调整片控制盒
RTTDALB	rudder trim tab deflection angle limiter box	方向舵调整片偏角限制控制盒
RUD	rudder	方向舵
RVR	runway visual range	跑道目视距离
RVTO	rolling vertical takeoff	滑跑垂直起飞
RVTOL	rolling vertical takeoff and landing	滑跑垂直起飞与着陆
RVV	runway visibility value	跑道能见度值
RWB	rear wheel brakes	后轮制动器,后轮刹车
RWBR	rudder warning bell release	舵警铃解除
RWTBF	right wing tip brake failure	右制动器故障
RWY	runway	跑道,滑道
RZ	relative azimuth	相对方位(角)

S

S	search	搜索
S	south	南,南方
S/DS	sand or dust storm	沙暴或尘暴
S/ND	signal-to-noise and distortion	信号-噪声与失真比
S/NM	signal-to-noise margin	信号噪声裕度
S/NR	signal to noise ratio	信号噪声比,信噪比
S/T	scanning/tuning	扫描/调谐
SA	safety altitude	安全高度
SA	servo amplifier	伺服放大器
SA	solenoid actuator	电磁制动器
SA	start annunciator	起动指示灯
SAC	system anti-collision	防撞系统
SALS	short approach-light system	近距进场[进近]照明系统
SAP	static air pressure	空气静压
SAR	system array radar	系统天线
SARAH	semiautomatic range, azimuth and height	半自动距离、方位与高度(探测系统)
SARH	semiactive radar homing	半主动雷达寻的,半主动雷达自导引
SAS	stability augmentation system	增稳系统
SAT	static air temperature	大气静温
SB	s-band	S频带,S波段
SB	sideband	边带,边频带

SB	sonic boom	(超声速)声爆
SB	select button	选择按钮
SB	start button	起动按钮
SB	static brake	静刹车
SBA	spot beam aerial [antenna]	点波速天线
SBAS	standard blind approachsystem	标准盲降进近系统
SBAS	standard beam approach system	标准波束[进近]系统
SBHO	standby booster hydraulic off	备助液压未通
SBILS	scanning beam instrument landing system	波束扫描仪表着陆系统
SCAT	speed control approach/takeoff	进场[进近]与起飞速度调节系统
SCB	signal changeover box	信号转换盒
SCCW	stabilizer control changeover switch	配平操纵转换开关
SCI	serial control interface	串行控制接口
SCU	steering controller unit	转向控制器装置
SCV	surge control valve	防喘阀
SD	scan duration	扫描持续时间
SD	shutdown	停车
SD	smoke detection	烟雾探测
SDA	sleeve dipole antenna	同轴偶极子天线
SDAR	sound detecting and ranging	声探测与测距
SE	starter excitation	起动激磁
SEL	select	选择
SF	signal flare	信号弹
SF	start failure	起动故障
SFA	special flight approval	特许飞行器批准书
SFC	side force coefficient	侧力系数
SFC	starter fail contactor	起动机故障接触器
SFCF	single flap channel failure	襟翼单通道故障
SFSR	start fail signal relay	起动故障信号继电器
SG	searching gain	搜索增益
SG	signal generator	信号发生器
SGL	static ground line	静电接地线
SGL	signal	信号
SH	Shore [scleroscope] hardness	肖氏硬度,回跳硬度
SHARS	strapdown heading attitude reference system	捷联航向姿态参考[基准]系统
SHP	shaft horsepower	轴马力
SHR	shear	剪切,剪切力,剪应力
SHT	static heating test	静压加温检测

SI	static inverter	静止变流器
SI	standby instrument	备份仪表
SIA	straight in approach	直接进场[进近]着陆
SIA	standard (instrument) arrival	标准仪表飞行进场
SIAL	standard instrument approach landing	标准仪表进场[进近]着陆
SIE	spark ignition engine	火花点火发动机
SIGN	strapdown inertial guidance and navigation	捷联式惯性制导与导航(系统)
SIR	search and interrogation radar	搜索与询问雷达
SIR	snow and ice on runway	跑道上积雪与结冰
SIS	stall-identification system	失速识别系统
SJ	spray jet	喷水器
SKP	skip	跳跃,跳动
SL	separating line	分隔线
SL	sea level	海平面
SL	service life	使用寿命
SL	short landing	短距着陆
SL	side light	舷灯,舷窗
SL	soft landing	软着陆
SLRP	signals of left/ right pressure	左/右压力信号
SLS	sea level speed	海平面(飞行)速度
SLS	sidelobe suppression	旁瓣抑制
SM	safety margin	安全裕度,安全系数
SM	small	小
SMC	standard mean chord	标准平均弦(长)
SMTCC	simulation, monitoring, training and command and control	模拟、监视、训练、指挥与控制
SN	serial number	序号
SN	system normal	系统正常
SO	system overload	系统过载,系统超载
SO	since overhaul	从大修后算起
SO	stabilization off	空稳关
SOA	spectrometric oil analysis	滑油光谱分析
SOA	speed of approach	进场[进近]速度
SOA	standard omnirange approach	标准全向导航进场[进近]
SOG	speed over ground	对地航速,地面速度
SOR	speed of rotation	转速
SOT	stator outlet temperature	静子出口温度
SOV	shut-off valve	断流阀,断流活门
SP	static pressure	静压

SP	start page	启动画面
SP	start power	起动电源
SPAR	super precision approach radar	高精密进场[进近]雷达
SPC	spare parts catalog	备件目录
SPKR	speaker	扬声器，喇叭
SPL	spare parts list	备件单
SPL	standard parts list	标准件表
SPLR	spoiler	阻流片，扰流板
SPS	serial-paralled storage	串行-并行存储器
SPS	stall protection system	失速预防系统
SPT	satellite positioning and tracking	卫星定位与跟踪
SPT	speed，position，track	速度、位置与跟踪
SQ	squelch	静噪
SR	specific range	给定航程，特定飞行距离
SR	squib releasing	爆炸帽释放
SR	short range	短程、近程、短距
SR	suppression ratio	抑制比
SR	start ready	起动准备好
SRB	start relay box	起动继电器盒
SRC	source	源，源极
SRH	search，recue，homing	搜索、营救与归航
SRH	semiactive radar homing	半主动雷达寻的
SRM	structure repair manual	结构修理手册
SRN	short-range navigation	近距导航
SRR	search and range radar	搜索与测距雷达
SRS	strapdown reference system	捷联式参考[基准]系统
SS	signal sideband	单边带
SS	shutdown sensor	刹车传感器
SS	signal source	信号源
SS	sequence switch	顺序开关、程序开关
SS	shear strength	抗剪强度
SS	start system	起动系统
SS	state selection	状态选择
SS	standby static	备用静压
SSALS	simplified short approach-light system	简化短距进场[进近]照明系统
SSCOW	standby stab change-over switch	备用配平转换开关
SSI	standby static inverter	备用静变交流机
SSM	single shot multivibrator	单冲多谐振荡器
SSM	single sideband modulation	单边带调制

SSM	system schematics manual	系统原理图手册
SSPAR	solid state phased array radar	固态相控阵雷达
SSSC	single sideband suppressed carrier	单边带抑制载波
ST	simple tone	单音
ST	shock tube	激波管
ST	shock tunnel	激波风洞
ST	start	启动
STB	simple tone button	单音按钮
STBY	standby	备用
STC	supplemental type certificate	补充型号合格证
STD	secheduled time of departure	计划离场[起飞]时间
STD	standard	标准
STO	short takeoff	短距起飞
STOAL	short takeoff and arrested landing	短距起飞拦阻着陆
STOGW	short takeoff gross weight	短距起飞总重量
STOL	short takeoff and landing	短距起落
STOL/VCD	short takeoff and landing/vertical climb and descent	短距起落/垂直升降
STOL/VTOL	short takeoff and landing/vertical takeoff and landing	短距起落/垂直起落
STOVL	short takeoff with vertical landing	短距起飞垂直降落
STRG	strong	强
STT	standby trim tab	备用调整片
STT	skid to turn	侧滑转弯
STTCMM	standby trim tab control mechanism mark	备用调整片操纵机构标记
SV	shunt valve	分路活门
SV	start voltage	启动电压
SV	static vent	静压通气孔
SVC	stop valve closed	断流活门关
SVCS	solenoid valve control switch	电磁阀操纵开关
SVD	safe vertical distance	垂直安全距离
SVFR	special VFR	特种目视飞机规则
SVL	slow vertical landing	低速垂直降落
SVTOL	short/vertical takeoff and landing	短距垂直起降
SW	safety wire	保险丝
SW	switch	开关
SW	stopway	应急停机地带,跑道保险道
SWH	side windshield heating	侧玻璃加温
SWI	stall warning indicator	失速告警器

SWI	standing wave indicator	驻波指示器
SWIL	smoke warning indication light	烟雾告警指示灯
SWSBIT	smoke warning system built-in test	烟雾告警自检
SWTG	switching	转换
SWW	severe weather warning	恶劣天气预报
SYNC	synchronize	同步

T

T	test	测试
T/M(H)	true/magnetic(heading)	真/磁航向
TA	true altitude	(航空)真高度,实际高度
TA	traffic advisory	交通通告
TA	turn around	往返飞行
TA	true azimuth	真方位角
TA	throttle angle	油门角度
TA	track angle	航迹角
TA	trim actuator	配平舵机
TA	trim amplifier	配平放大器
TAC	target acquisition capability	目标截获能力
TACAN	tactical air navigation	战术空中导航设备(简称"塔康")
TACAN/CB	TACAN control box	塔康控制盒
TACAN/PR	TACAN position resolver	塔康位置解算设备
TACB	TCAS/ATC control box	TCAS/ATC 控制盒
TAD	time available for delivery	有效投放时间
TAE	track angle error	航迹角误差
TAE	track angle error	航迹角误差
TAF	ture airspeed fail	真空速失效
TAI	tail anti-ice	尾翼防冰
TAIDB	tail anti-ice distributing box	尾翼防冰配电盒
TAN	tactical radar and navigation	战术雷达与导航
TAO	traffic advisory only	仅交通通告
TAR	terrain avoidance radar	地形回避雷达,地物防撞雷达
TARS	three-axis reference system	三轴坐标系
TAS	true air speed	真空速,实际空速
TAT	turn around time	再次出动准备时间,往返飞行时间
TAT	true air temperature	真实气温
TAT	total air temperature	总温
TATI	trim and tailplane incidence(indicator)	调整片与水平尾翼安装角指示器
TAV	temperature adjusting valve	调温活门

TAWS	terrain awareness warning system	地形提示与告警系统
TB	terminal board	接线板
TB	true bearing	真方位
TBA	total blade area	桨叶总面积
TBA	true bearing adapter	真方位测定仪
TBAI	tail boot anti-icing	尾翼气囊防冰
TBD	to be determined	待定
TBF	time between failures	故障间隔时间,失效间隔时间
TBF	time between faults	缺陷间隔时间
TBI	turn and bank indicator	转弯与倾斜指示仪
TBI	turn/bank indicator	转弯侧滑仪
TBL	turbulent boundary layer	湍流边界层
TBO	time between overhaul	大修间隔期
TBSS	time between scan start	扫描开始间隔时间
TBW	total bandwidth	全带宽,总带宽
TC	temperature coefficient	温度系数
TC	temperature controller	温度控制器
TC	terrain clearance	离地高度
TC	thermal control	热控制,热控,温控
TC	thermalcouple	热电偶
TC	type certificate	型号合格证
TC	time controller	时间控制器
TC	top console	顶控板
TC	thermal conductivity	导热率
TCA	time of closest approach	最近进场[进近]时间
TCA	track crossing angle	航迹交叉角
TCA	turbine cooling airflow	涡轮冷却空气流
TCAS	traffic alert and collision avoidance system	空中交通告警与防撞系统
TCAS/A	TCAS await	TCAS 待机
TCAS/T	TCAS transceiver	TCAS 收发机
TCB	type certification board	型号合格审定委员会
TCC	turbine case cooling	涡轮机匣冷却
TCGPR	to contactor for generator parallel run	接发电机并联工作的接触器
TCI	terrain clearance indicator	绝对高度指示器
TCL	throttle control level	油门控制杆
TCLPI	throttle-control level position indicator	油门杆位置指示器
TCM	terminate countermeasures	终止对抗
TCM	temperature control mode	温控模式
TCM	terrain contour-matching	地形匹配(导航技术)

TCP	transfer of control point	管制交接点
TCP	transmitter cold plate	发射机冷板
TCR	tracking，command and ranging	跟踪、指令与测距
TCR	temperature coefficient of resistance	电阻温度系数
TCS	tilt control switch	倾转操纵电门
TCS	temperature control system	温度调节系统,调温系统
TCU	temperature control unit	温度控制装置
TD	tail deicing	尾翼除冰
TD	time delay	时间延迟
TD	time difference	时差
TD	time domain	时域
TD	touchdown	着地,接地
TD	temporary duty	临时任务
TDC	top dead centre	上死点
TDC	type design change	型号设计更改
TDDA	time distance difference to arrival	到达时间距离差
TDF	traffic display fault	TD 故障
TDFG	three degree of freedom gyroscope	三自由度陀螺仪
TDFG	two degree of freedom gyroscope	二自由度陀螺仪
TDR	time delay relay	延时继电器
TDS	takeoff decision speed	起飞决断速度
TDS	takeoff safety speed	安全起飞速度
TDSS	time/distance/speed scale	时间/距离/速度刻度
TDU	time delay unit	延时装置
TDV	touchdown velocity	触地速度,着地速度
TE	tangent elevation	切向仰角,瞄准角
TE	turbine engine	涡轮发动机
TE	tensile strain	拉伸应变
TEC	turbine exhaust case	涡轮排气机匣
TEF	trailing edge flap	后缘襟翼
TEFS	trailing edge flap shroud	后缘襟翼罩
TEP	thermal equivalent power	热当量功率
TERP	turbine engine reliability programmes	涡轮发动机可靠性计划
TEST	test	测试
TET	tolerable exposure time	允许承受的时间
TEW	total equivalent weight	总当量
TF	terrain following	地形跟随
TF	thin film	薄膜
TF	time fuze	定时引信,时间引信

TF	turbine fuel	涡轮(发动机)燃料
TF	turbofan	涡轮风扇
TF	trim failure	配平故障,配平失效
TF/TA	terrain following/terrain avoidance	地形跟随/地形回避
TFB	tower fly-by	飞越塔台
TFC	total final consumption	最后总计消耗量
TFC	trim and flaps control	调整片和襟翼控制
TFCT	to flight control tank	至飞控油箱
TFD	total frictional drag	总摩擦阻力
TFG	thrust floated gyroscope	推力悬浮陀螺仪
TFM	tamed frequency modulation	平滑调频
TFOV	total field of view	总视场
TFOV	tracking field of view	跟踪视场
TFP	turbine fuel pump	涡轮燃油泵
TFP	total fuel quantity	燃油总量
TFP	through flight plan	直达飞行计划
TFR	terrain following radar	地形跟随雷达
TFR	total fuel remaining	总剩余燃油量
TFS	tensile field strength	拉伸屈服强度
TFSF	time to fisrt system failure	系统首次发生故障时间
TG	turbine generator	涡轮发电机
TGB	to generator bus	接发电机汇流条
TGSE	tactical ground support equipment	战术地面保障设备
TGSE	test ground support equipment	地面试验保障设备
TGT	turbine gas temperature	涡轮燃气温度
THAR	tyre height above runway	轮胎距跑道高度
THB	temperature-humidity-bias	温度-湿度-偏压
THP	to hydraulic pump	接液压泵
THROT	throttle	油门
THWS	tactical homing and warning system	战术寻的告警系统
TI	transponding indication	应答指示
TIA	type inspection authorization	型号检查核准书
TIAS	target identification and acquisition system	目标识别与截获系统
TIAS	true indicated airspeed	真实指示空速
TIG	tungsten inert gas (welding)	坞惰性气体保护焊
TIGER	terrifically insensitive to ground effect radar	对地效极不敏感的雷达
TINS	TERCOM and inertial navigation system	地形匹配与惯性导航系统
TIPLCB	trans-illuminated panel lighting	导光板照明控制盒

	control box	
TIPP	trans-illuminant panel power	导光板电源
TIR	type inspection report	型号检查报告
TIT	turbine inlet temperature	涡轮进口温度
TITE	TEWS intermediate test equipment	战术电子战系统中间测试设备
TITS	turbine inlet temperature sensor	涡轮进口温度传感器
TK	tank	油箱
TLA	throttle lever angle	油门杆角度
TLDA	top line display area	顶行显示区域
TLF	thrust for level flight	平飞推力
TLPI	throttle-lever position indicator	油门杆位置指示器
TM	tachometer	转速表,转速计
TM	terrain matching	地形匹配
TM	time modulation	时间调制
TMP	turnon manually pump	手动接通泵
TMU	timer monitor unit	定时监控装置
TMV	true mean value	真平均值
TN	tank number	油箱号
TN	true north	真北(非地磁北)
TNA	time of nearest approach	最近进场[进近]时间
TO	takeoff	起飞
TO/GA	takeoff/go around	起飞与复飞
TOA	time of arrival	到达时间
TOC	top of climb	爬升顶点
TOD	takeoff distance	起飞距离
TOD	time of departure	出航时间,起飞时间
TOD	top of decent	下降最低点
TODA	takeoff distance available	可用起飞距离
TODR	takeoff distance required	所需起飞距离
TOF	time of flight	飞行时间
TOFL	takeoff field length	起飞距离
TOFP	takeoff flight path	起飞飞行路线[航迹线]
TOG(W)	takeoff gross (weight)	起飞总重
TOHP	takeoff horsepower	起飞马力
TOI	time of ignition	点火时间
TOI	time of intercept	截获时间
TOJ	track on jamming	干扰跟踪
TOL	takeoff and landing	起落,起飞与着陆
TOLCAT	takeoff and landing clear-air turbulence	起落晴空湍流

TOLCAT	takeoff and landing critical atomosphere turbulence	起落时临界大气湍流
TOM	takeoff mass	起飞质量
TOP	takeoff power	起飞功率,起飞推力
TOP	torque oil pressure	扭矩油压
TOR	takeoff run	起飞滑跑(距离)
TORA	takeoff run available	起飞滑跑可用距离
TORR	takeoff run required	所需起飞滑跑距离
TORS	takeoff rotation speed	起飞滑跑抬前轮速度
TOSS	takeoff safety speed	起飞安全速度
TOT	turbine outlet temperature	涡轮出口温度
TOW	takeoff weight	起飞重量
TOW	time on wing	连续飞行使用时间
TOWA	terrain and obstacle warning and avoidance	地形与障碍告警与回避(系统)
TP	top priority	最高优先级
TP	total pressure	全压,总压
TPA	traffic pattern altitude	起落航线高度
TPACP	three-phase AC power	三相交流电源
TPH	temperature probe heating	温度探头加温
TPM	terrain profile matching	地形剖面匹配
TPWS	turbulence prediction and warning system	湍流预测与告警系统
TR	thrust reverser	反推力装置,推力反向器
TR	tracking radar	跟踪雷达
TR	transponder	转发器,应答机
TR	type rating	型号等级
TR	track required	应飞航迹[航迹角]
TR	throttle reset	油门复位
TR/TV	thrust reverser/ thrust vectoring	反推力装置/推力矢量
TRACK	track	导轨
TRR	total removal rate	总拆卸率
TRR	tyre [tire] rolling radius	轮胎滚动半径
TRT	takeoff rated thrust	起飞额定推力
TRT	turn round time	再次出动准备时间
TRU	true	真,真实的,正确的
TRVR	touchdown runway visual range	接地时跑道能见范围
TS	temperature selector	温度选择器
TS	temperature sensor	温度传感器
TS	throttle switch	风门开关
TS	time setup	时间设置

TS	time switching	时间切换
TS	trim system	配平系统
TS	tensile strength	抗拉［拉伸］强度
TS	thunderstorm	雷暴
TSDCG	test switch of DC generator	直流发电机测试开关
TSF	time to system failure	系统出现故障前的时间
TSI	true (air) speed indicator	真空速表
TSI	turn and slip indicator	转弯侧滑仪
TSM	time scheduled maintenance	定期维修
TSMO	time since major overhaul	大修后（使用）时间
TSN	time since new	从新的算起的使用时间,制造出厂后（使用）时间
TSO	technical standard order	技术标准规定
TSO	time sharing option	分时选择
TSO	time since overhaul	大修后（使用）时间
TSOA	technical standard order authorization	技术标准规定项目批准书
TSS	trouble shooting system	寻故（障）系统,故障搜索系统
TSSV	tapping self-sealing valve	放液自封活门
TST	trouble shooting time	故障检修时间
TT	trim tab (balancing linked tab)	调整片（随动补偿片）
TT	true track	真航迹角
TTAE	total time, airframe and engine	飞机（机体）与发动机总使用时间
TTCOS	trim tab change-over switch	调整片转换开关
TTF	time to failure	发生故障时间,用到失效的时间
TTFO	time to first overhaul	首次大修期限
TTGI	time to ground idling	至地面慢车时间
TTI	true tracking indicator	真航线指示器
TTO	trim for takeoff	起飞配平
TTOL	tangential takeoff and landing	切向起飞与着陆,滑跑起飞与着陆
TTR	target tracking radar	目标跟踪雷达
TTSCA	trim tab stab control actuator	调整片配平舵机
TTSMO	total time since major overhaul	大翻修后总（使用）时间
TTSN	total time since new	制造出厂后总（使用）时间
TURB	turbine	涡轮
TURB	turbulence	湍流
TV	thrust vectoring	推力矢量
TV	total volume	总容积
TVA	thrust vector alignment	推力矢量校准
TVC	thrust vector control	推力矢量控制

TVTO	true vertical takeoff	真垂直起飞(飞机)
TWC	two way channel	双向信道
TWR	thrust-weight ratio	推重比
TYP	type (of aircraft)	飞机型号
TYP	typical	典型

U

U	uncertainty	不可靠性,不精确性
U	utility	公用,效用;通用机
UC	universal code	通用代码
UFDR	universal flight data recorder	飞行数据记录仪
UFS	ultimate factor of safety	极限安全系数
UHF/HF	ultra high frequency/high frequency	特高频/高频
UL	ultimate load	最大负荷
ULF	ultimate load factor	极限载荷因数
ULR	utility light receptacle	工作灯插座
UNLK	unlocking	解锁
UOE	utilization of equipment	设备利用率,设备利用
UOS(SS)	unless otherwise shown [specified stated]	除非另有说明[规定]
UPS	universal power supply	通用电源
UREST	universal range, endurance, speed and time	通用航程、续航能力、速度与时间
USB	universal serial bus	通用串行总线
USB	upper sideband	上边带
USD	ultimate strength design	极限强度设计
UST	upper surface transition	上翼面转捩点
USTOL	ultrashort takeoff and landing	超短距起降
UT	universal time	世界时,格林威治时
UTS	ultimate tensile strength	极限抗拉强度
UTT	utility tactical transport	通用战术运输机
UV	under-voltage	欠压
UW	under wing	翼下

V

V	valve	活门
V	volatage	电压
V/AD	velocity/azimuth display	速度/方位显示器
V/CTOL	vertical/conventional takeoff and landing	垂直或常规起落
V/HUD	vertical/head up display	垂直/平视显示器

V/L	VOR/localizer	VOR/定位器,定位信标
V/STOL	vertical/short takeoff and landing	垂直/短距起落
VA	vertical acceleration	垂直加速度
VA	vertical navigation	垂直面导航
VA	volume adjustment	音量调整
VAD	vertical azimuth display	垂直方位显示器
VAH	velocity, acceleration, height	速度、加速度与高度
VARS	vertical and azimuth reference system	垂直与方位参考[基准]系统
VASI	visual approach slope indicator	目视进场[进近]下滑道指示器
VASIS	VASI system	目视进场[进近]下滑道指示器系统
VAVP	variable angle, variable pitch	可变角,可变螺距
VCD	variable capacitance diode	变容二极管
VCD	vertical climb and descend	垂直爬升与下降
VCI	visual course indicator	目视航向指示器
VCM	visual countermeasures	目视干扰,视觉干扰
VCW	velocity of crosswind	侧风速度
VD	void	砂眼,气泡,空隙
VDI	visual Doppler indicator	多普勒雷达目视显示器
VEN	variable exhaust nozzle	可调(截面)排气喷管
VFC	vortex flow control	涡流控制
VFHL	VF hot line	音频热线
VFPA	vertical flight path angle	垂直航迹角
VFR	visual flight rules	目视飞行规则
VFRA	visual flight rules approach	目视飞行规则进场[进近]
VG	vertical gyro(gyroscope)	垂直陀螺仪
VH	Vickers hardness	维氏硬度
VHF	very high frequency	甚高频
VHFO	very high frequency omnirange	甚高全向信标,伏尔
VHFOR	very high frequency omnidirectional range	甚高频全向(无线电)信标,伏尔系统
VIGV	variable incidence guide vane	变迎角导向叶片
VL	vertical lift	垂直升力
VM	voltage measure	电压测量
VOL	volume	音量
VOR	visual omni-range	目视飞行全向指标
VORL	VOR-localizer	甚高全向信标定位器
VORTALS	VHF omnirange takeoff, approach and landing system	甚高频全向信标起飞、进场[进近]与降落系统
VP	variable pitch	可变桨距
VP	ventilation pipe	通风管

VPI	valve position indicator	阀门位置指示器
VPL	volume of payload	有效载荷体积,满载容积
VPM	ventilation pipeline mark	通气管路标记
VPP	variable pitch propeller	可变桨距螺旋桨
VPS	variable parameter system	可变参数系统
VRC	vertical rate of climb	垂直爬升率
VRIL	visual rule instrument landing	目视规则仪表着陆
VRL	vertical reference line	垂直参考[基准]线
VRP	visible reference point	可视参考[基准]点
VS	ventilation system	通气系统
VS	vertical stablilizer	垂直安定面
VS	variable sweep	可变后掠(角)
VS	vertical speed	垂直速度
VS	touchdown speed	接地速度
VSBY	visibility	能见度,可见度,可见性
VSI	vertical speed indicator	垂直速度指示器
VSP	vertical speed preselect	垂直速度预选
VSW	variable sweep wing	变后掠翼
VSW	vertical speed and windshear	垂直速度与风切变
VV	vertical velocity	垂直速度
VVM	vacuum valve mark	真空活门标记

W

W	west	西,西方
W/E/F/T	wing/ engine (s)/ fuselage/ tail	机翼/发动机/机身/尾翼
W/T	weather/ terrain	气象/地形
WARN	warning	告警
WASR	wide area surveillance radar	广域监视雷达
WAT	weight/ altitude/ temperature	重量/高度/温度
WAT	wide-angle track	大角度跟踪
WB	wheel brake	机轮刹车
WBB	wideband bandwidth	宽带带宽
WBM	weight and balance manual	重量与平衡手册
WC	warning computer	告警计算机
WC	water closet	盥洗室
WCI	water circulation inlet	水循环入口
WCO	water circulation outlet	水循环出口
WD	web depth	梁腹高度,腹板深度
WD	wind direction	风向

WE	weight empty	空重,净重
WEA	weather	气象
WG	wave guide	波导管,导波器
WH	wing heating	机翼加温
WHL	wheels	机轮,车轮,轮子
WIF	weather instrument flight	气象仪表飞行
WII	wing ice inspection	机翼积冰检查
WING	wing	机翼
WL	warning light	告警灯
WL	water level [water line]	水线
WL	wave length	波长
WLDP	warning light display panel	告警灯显示板
WLE	wing leading edge	机翼前缘
WLFL	wet landing field length	湿着陆场长度
WM	weak mixture	贫(燃)油混合气
WM	weather message	天气通报
WOFF	weight of fuel flow	燃油流量重量
WOT	wide-open throttle	推大油门
WOW	weight on wheel	机轮着陆,轮上重量,轮载
WPD	wing profile drag	翼型阻力
WPT	waypoint	航路点
WR	weather radar	气象雷达
WR	wet runway	湿跑道
WRCBP	weather radar control box panel	气象雷达控制盒面板图
WRCP	weather radar control panel	气象雷达操作板
WRP	wing reference plane	机翼参考[基准]面
WS	water separator	水分离器
WS	windshield	风挡,风罩
WS	wind speed	风速
WSD	wind speed and direction	风速与风向
WSHS	windshield heating switch	风挡加温开关
WSI	weather severity index	天气严重性指标
WT	wind tunnel	风洞
WTAI	wing and tail anti-icing	机翼与尾翼防冰系统
WTDI	wing and tail deicing	机翼与尾翼除冰系统
WUTTO	warm up, taxi and takeoff	起动、滑行与起飞
WV	wind vector	风速矢量
WWC	wheel well control	轮舱门操纵

X

X-BSAR	X-band synthetic aperture radar	X 波段合成孔径雷达
X-BTA	X-band tracking aerial [antenna]	X 波段跟踪天线
X-BCT	X-band communication(s) transponder	X 波段通信应答器
XCTR	exciter	激励器,励磁器
XCVR	transceiver	收发机
XFMR	transformer	变压器,变换器,变量器
XMIT	transmit	传递
XMTR	transmitter	发射机
XPNDR	transponder	收发机
XT	cross talk	串音,通道间干扰
XTD	cross track distance	偏航距离

Y

Y	yaw	偏航
YA	yaw axis	偏航[转]轴
YBC	years between calibrations	校准间隔时间
YCP	yaw coupling parameter	偏航耦合参数
YD	yaw damper	偏航阻尼器
YDA	yaw damper actuator	偏航阻尼作动器
YM	yawing moment	偏航力矩
YMC	yawing moment coefficient	偏航力矩系数
YOS	year of service	服役数年
YP	yaw and pitch	偏航与俯仰
YP	yield point	屈服点
YP	yield pressure	屈服压力
YR	yard repair	入厂修理,厂修
YRG	yaw rate gyroscope	偏航率陀螺仪
YS	yield strength	屈服强度
YS	yield stress	屈服应力
YSE	yaw steering error	偏航操纵误差
YSF	yield safety factor	屈服安全因数
YSLF	yield strength load factor	屈服强度载荷因数
YTD	year to date	本年度到现在为止

Z

Z	Z-axis	Z 轴,垂直轴
Z	zone	区域,区

Z	azimuth	方位,方位角
ZAS	zero automatic set	自动归零,自动零位调节装置
ZFW	zero-fuel weight	无燃料时(飞机)重量
ZGE	zero gravity effect	失重效应,零重效应
ZL	zero line	零位线,基准线
ZL	zero lift	零升力
ZLA	zero lift angle	零升(力)角
ZLC	zero lift chord	零升力弦
ZLD	zero lift drag	零升阻力
ZROC	zero rate of climb	零爬升率
ZS	zero shift	零偏移
ZT	zone time	区时
ZTC	zero temperature coefficient	零温度系数
ZTDL	zero thrust descent and landing	零推力下降与着陆
ZZV	zero-zero visibility	能见度极差

随机资料清单

飞行手册　AIRCRAFT FLIGHT MANUAL

1	使用限制	Operational Limitation
2	应急程序	Emergency Procedures
3	正常程序	Normal Procedures
4	性能	Performance
5	飞机各系统设备	Aircraft System and Equipment
6	附录 A 仪表板和操纵台图	Appendix A Instrument Panels and Consoles
7	附录 B 有关常识介绍	Appendix B Common Knowledge Introduction
8	附录 C 各类云的特征及云中飞行条件	Appendix C Features of Various Clouds and Their Corresponding In-Cloud Flight Conditions
9	附录 D 飞机积冰强度等级	Appendix D Aircraft Icing Accretion Intensity Grade

主最低设备清单　MASTER MINIMUM EQUIPMENT LIST

环控系统　Air Conditioning

1	断流活门	Cutoff Valve
2	交叉活门	Cross Valve
3	分路活门	Shunt Valve
4	辅助通风活门	Auxiliary Vent Valve
5	座舱压力调节系统	Cabin Pressure Control System
6	制冷系统	Cooling System
7	温度控制系统	Temperature Control System

自动飞行系统　Auto Flight

1	自动驾驶仪	Autopilot
2	航空联系盒	Heading Liaison Box

通信系统　Communications

1	短波电台	HF Radio Set
2	超短波电台	VHF/UHF Radio Set
3	耳机	Headset

电源系统　Electrical Power

1	交流发电系统	AC Generating System
2	单相变流机	Single-Phase Inverter
3	交流电压表	AC Voltmeter
4	交流电流表	AC Ammeter
5	频率表	Frequency Meter
6	直流发电系统	DC Generating System
7	地面电源插座	Ground Power Socket
8	直流电压表	DC Voltmeter
9	直流电流表	DC Ammeter

设备　Equipment/Furnishings

1	货舱座椅	Cargo Cabin Seat

防火系统　Fire Protection

1	辅助动力装置火警探测和灭火系统	Fire Detection and Fire Extinguishing System of Auxiliary Power Unit

飞行操纵　Flight Controls

1	方向舵调整片位置指示器	Rudder Trim Tab Position Indicator
2	舵面锁操纵系统	Control Lock Control Mechanism

燃油系统　Fuel

1	软油箱	Bag Tank
2	结构油箱	Structural Tank
3	压力加油接头	Pressure Refueling Adaptor
4	压力加油开关	Pressure Refueling Valve

5	供油泵	Fuel Supply Pump
6	油量表耗油控制器	Fuel Consumption Controller of Fuel Quantity Gauge
7	加油控制器	Refueling Controller
8	结构油箱压力信号器	Pressure Annunciator of Structural Tank
9	加油正常压力信号器	Normal Refueling Pressure Annunciator
10	极限加油压力信号器	Limit Refueling Pressure Annunciator

液压源　Hydraulic Power

1	右系统油泵	Hydraulic Pump of Right System
2	右系统液压蓄压瓶	Hydraulic Pressure Accumulator of Right System
3	左系统油泵	Hydraulic Pump of Left System
4	手摇泵	Hand Pump
5	电动泵	Electric Pump
6	手摇泵-电动泵系统压力表转换装置	Pressure Changeover Device of Hand/Electric Pump System
7	液压油量表	Hydraulic Fluid Quantity Gauge
8	手摇泵-电动泵系统压力表	Pressure Gauge of Hand/Electric Pump System
9	增压系统冷气压力表	Pneumatic Pressure Gauge of Pressurization System
10	右系统蓄压瓶压力表	Pressure Accumulator Pressure Gauge of Right System

除雨和防冰　Ice & Rain Protection

1	风挡玻璃电加温系统	Windshield Glass Electric Heating System
2	风挡雨刷	Windshield Wiper

显示/记录系统　Indicating/Recording Systems

1	航空时钟	Aviation Clock
2	电动加速度表	Electric Accelerometer
3	临界攻角信号系统	Critical AOA Signal System
4	用氧警铃	Oxygen Supply Alarm

起落架　Landing Gear

1	机轮刹车惯性传感器	Wheel Brake Inertial Sensor

| 2 | 刹车压力指示器 | Brake Pressure Indicator |

照明系统　Lights

1	驾驶舱照明	Cockpit Lighting
2	货舱照明	Cargo Cabin Lighting
3	着陆滑行灯	Landing and Taxiing Light
4	防撞灯	Anti-Collision Light
5	航行灯	Navigation Light
6	发动机照明灯	Engine Light
7	编队灯	Formation Light

导航系统　Navigation

1	大气温度表	Air Thermometer
2	升降速度表	Vertical Speed Indicator
3	空速表	Airspeed Indicator
4	高度表	Altimeter
5	M 数表	Mach Indicator
6	大气数据系统	Air Data System
7	航向姿态系统	AHRS
8	地平指示器	Attitude Director Indicator
9	转弯仪	Turn and Slip Indicator
10	气象雷达	Weather Radar
11	无线电罗盘	ADF
12	伏尔/仪表着陆系统	VOR/ILS
13	塔康	TACAN
14	GPS 导航系统	GPS Navigation System
15	DME 测距器	DME
16	激光惯性/卫星组合导航系统	Laser Inertial/Satellite Integrated Navigation System

氧气系统　Oxygen

| 1 | 空勤用氧系统 | Oxygen Supply System for Aircrew |
| 2 | 预备舱、指挥控制舱乘员用氧系统 | Oxygen Supply System for Passengers of Reservation Cabin and Command Control Cabin |

引气系统　Bleed Air

| 1 | 压力调节/关断活门 | Pressure Regulating/Shutoff Valve |
| 2 | 压力传感器 | Pressure Sensor |

机载辅助动力装置　Airborne Aux. Power

| 1 | 涡轮起动发电装置 | Turbine Starter/Generator |
| 2 | 涡轮起动发电装置进气风门电动机构 | Air Inlet Shutter Electrical Mechanism of Turbine Starter/Generator |

舱门　Doors

1	前应急舱门密封系统的压力表	Pressure Gauge for Airtightness System of Bottom Emergency Door of Cockpit
2	货舱大门收放系统	Cargo Cabin Door Up/Down System
3	前应急舱门、货舱大门、登机门信号灯	Signal Lights for Bottom Emergency Door of Cockpit，Cargo Cabin Door and Entry Door

发动机燃油及控制系统　Engine Fuel & Control

| 1 | 燃油耗量表 | Fuel Consumption Indicator |
| 2 | 低压燃油压力表 | Low Fuel Pressure Gauge |

滑油　Oil

1	滑油温度调节盒	Oil Temperature Regulating Box
2	滑油油量表	Oil Quantity Gauge
3	滑油散热风门指位表	Oil Radiator Shutter Position Indicator
4	滑油引射散热开关	Oil Ejecting Radiation Switch

起动　Starting

1	起动电压调节盒	Starting Voltage Regulating Box
2	功率限制器	Power Limiter
3	发动机自动停车器	Engine Automatic Shutdown Device
4	起动自动切油和超温保护器	Starting Automatic Fuel Reducer and Overtemperature Protector

重量和平衡手册　WEIGHT AND BALANCE MANUAL

飞行操纵快速检查程序　QUICK REFERENCE HANDBOOK

维修大纲　MAINTENANCE REVIEW BOARD REPORT

飞机维修计划文件
AIRCRAFT MAINTENANCE PLANNING DOCUMENT

10	附录 B 外部维修舱口盖板图	Appendix B Schematic Diagram of Exterior Maintenance Access Cover Plates
11	附录 C 内部维修舱口盖板图	Appendix C Schematic Diagrams of Interior Maintenance Access Cover Plates
12	附录 D 高压软管清单	Appendix D List of High-Pressure Hoses
13	附录 E 不同气象条件下的维护	Appendix E Aircraft Maintenance in Various Weather Conditions
14	附录 F 试飞	Appendix F Flight Test

飞机维修工作卡(A、C 卡) AIRCRAFT MAINTENANCE WORK CARD (CHECK A AND C CARDS)

机械装配图册 MECHANICAL SYSTEM SCHEMATIC MANUAL

1	滑油系统图	Oil System Drawing
2	螺旋桨顺桨系统图	Propeller Feathering System Drawing
3	发动机操纵系统图	Engine Control System Drawing
4	灭火系统图	Fire Extinguishing System Drawing
5	中性气体系统装配图	Assembly of Neutral Gas System Drawing
6	燃油供油系统图	Fuel Supply System Drawing
7	压力加油系统图	Pressure Refueling System Drawing
8	通气系统图	Air-Vented System Drawing
9	油箱附件安装图	Tank Accessory Installation Drawing
10	液压系统原理图	Hydraulic System Schematic Diagram
11	前起落架操纵半安装图	Installation Drawing (Half) of NLG Control
12	主起落架刹车系统半安装图	Installation Drawing (Half) of MLG Braking System
13	货舱大门操纵半安装图	Installation Drawing (Half) of Cargo Door Control
14	主起落架操纵半安装图	Installation Drawing (Half) of MLG Control
15	自动驾驶仪液压系统半安装图	Installation Drawing (Half) of Autopilot Hydraulic System
16	驾驶舱前应急舱门操纵和起落架应急收放开关操纵半安装图	Installation Drawing (Half) of Bottom Emergency Door and LG Emergency Extension/Retraction Switch Control
17	襟翼液压操纵系统半安装图	Installation Drawing (Half) of Flap Hydraulic Control System
18	风挡雨刷系统半安装图	Installation Drawing (Half) of Windshield Wiper System
19	供压系统半安装图	Installation Drawing (Half) of Pressurization

		System
20	液压顺桨和停车系统半安装图	Installation Drawing（Half）of Hydraulic Feathering and Shutdown System
21	液压油箱增压系统半安装图	Installation Drawing（Half）of Hydraulic Fluid Tank Pressurization System
22	升降舵操纵系统图	Elevator Control System Drawing
23	副翼操纵系统图	Aileron Control System Drawing
24	方向舵操纵系统图	Rudder Control System Drawing
25	襟翼操纵系统图	Flap Control System Drawing
26	升降舵调整片和舵面锁操纵系统图	Control System Drawing for Elevator Trim Tab and Control Surface Lock
27	环控系统图	Environmental Control System Drawing
28	座舱压力调节系统图	Cabin Air Pressure Control System Drawing
29	前应急舱门密封系统图	Airtightness System Drawing of Bottom Emergency Door
30	压差传感器系统图	Differential Pressure Sensor System Drawing
31	机翼防冰系统图	Wing Anti-Icing System
32	氧气系统半安装图	Installation Drawing（Half）of Oxygen System
33	全静压系统原理图	Schematic Diagram of Pitot-Static System
34	螺栓螺母紧度测量图	Bolt and Nut Tightness Measurement Drawing

特设线路图册　WIRING MANUAL

1	直流电源配电图	DC Power Distribution Diagram
2	直流电源系统原理图	Schematic Diagram of DC Power System
3	直流电源系统馈电图	Feed Diagram of DC Power System
4	蓄电池和地面电源原理、馈电图	Schematic and Feed Diagrams of Battery and Ground Power
5	单相交流电源系统原理图	Feed Diagram of Single-Phase AC Power System
6	单相交流电源系统馈电图	Feed Diagram of Single-Phase AC Power System
7	36V 三相静止变流器原理、馈电图	Schematic and Feed Diagrams of 36V Three-Phase Static Inverter
8	单相交流电源系统配电原理图	Schematic Diagram of Single-Phase AC Power System
9	单相电源系统原理图	Schematic Diagram of Single-Phase Power System
10	单相电源系统馈电图	Feed Diagram of Single-Phase Power System
11	机上电源插座原理图	Schematic Diagram of Airborne Power Socket
12	机上电源插座馈电图	Feed Diagram of Airborne Power Socket
13	发动机起动原理图	Engine Starting Schematic Diagram

14	发动机起动馈电图（机身部分）	Engine Starting Feed Diagram (Fuselage)
15	发动机起动馈电图（左机翼部分）	Engine Starting Feed Diagram (Left Wing)
16	发动机起动馈电图（右机翼部分）	Engine Starting Feed Diagram (Right Wing)
17	涡轮起动发电机原理图	Schematic Diagram of Turbine Starter/Generator
18	涡轮起动发电机馈电图	Feed Diagram of Turbine Starter/Generator
19	顺桨系统原理图	Feathering System Schematic Diagram
20	顺桨系统馈电图	Feathering System Feed Diagram
21	滑油温度自动调节原理馈电图	Schematic and Feed Diagram of Oil Temperature Automatic Regulation
22	滑油散热器引射散热原理馈电图	Schematic and Feed Diagram of Oil Radiator Ejector Radiation
23	滑油油量表和液压油油量表原理馈电图	Schematic and Feed Diagram of the Oil and Hydraulic Fluid Quantity Gauges
24	油量表自动耗油压力加油原理图	Schematic Diagram of Automatic Fuel Consumption and Pressure Refueling of Fuel Quantity Gauge
25	油量表自耗油压力加油馈电图	Feed Diagram of Automatic Fuel Consumption and Pressure Refueling of Fuel Quantity Gauge
26	燃油压力表原理馈电图	Schematic and Feed Diagrams of Fuel Pressure Gauge
27	油量表自动耗油压力加油馈电图	Feed Diagram of Automatic Fuel Consumption and Pressure Refueling of the Fuel Quantity Gauge
28	燃油耗量表原理馈电图	Schematic and Feed Diagrams of Fuel Consumption Indicator
29	灭火系统馈电图	Feed Diagram of Fire Extinguishing System
30	灭火系统原理图	Schematic Diagram of Fire Extinguishing System
31	防火开关和连通开关原理馈电图	Schematic and Feed Diagrams of Fuel Shut-off Valve and Communication Valve
32	发动机三用表原理图	Engine Multimeter Schematic Diagram
33	发动机三用表馈电图	Engine Multimeter Feed Diagram
34	发动机排气温度表原理馈电图	Schematic and Feed Diagrams of Engine Exhaust Gas Thermometer
35	辅助燃油系统原理图	Schematic Diagram of Auxiliary Fuel System
36	螺旋桨扭转力矩表原理馈电图	Schematic and Feed Diagrams of Propeller Torque Indicator
37	转速表、油门杆位置指示表原理馈电图	Schematic and Feed Diagrams of Tachometer and Throttle Lever Position
38	发动机振动过载表原理馈电图	Schematic and Feed Diagrams of Engine Vibration

		G-Load Meter
39	中性气体灭火原理图、馈电图	Schematic and Feed Diagrams of Neutral Gas Fire Extinguishing
40	辅助燃油系统馈电图	Feed Diagram of Auxiliary Fuel System
41	调整片操纵原理馈电图	Schematic and Feed Diagrams of Trim Tab Control
42	起落架襟翼操纵和信号原理馈电图	Schematic and Feed Diagrams of LG/Flap Control and Signal
43	机轮刹车和前轮转弯原理馈电图	Schematic and Feed Diagrams of Wheel Brake and Nose Wheel Steering
44	液压泵原理馈电图	Hydraulic Pump Schematic and Feed Diagrams
45	液压压力表原理馈电图	Schematicand Feed Diagrams of Hydraulic Pressure Gauge
46	热空气加温馈电图	Hot-Air Heating Feed Diagram
47	领航员、驾驶员风挡玻璃加温原理图	Schematic Diagram of Navigator and Pilot Windshield Glass Heating
48	电风扇原理馈电图	Schematic and Feed Diagrams of Electric Fan
49	全压和静压受感器加温原理馈电图	Schematic and Feed Diagrams of Pitot-Static Pressure Sensors Heating
50	气密舱温度自动控制原理馈电图	Schematic and Feed Diagrams of Airtight Cabin Temperature Automatic Control
51	气密舱压力调节原理馈电图	Schematic and Feed Diagrams of Airtight Cabin Pressure Regulation
52	结冰信号器原理馈电图	Schematic and Feed Diagrams of Icing Annunciator
53	着陆灯原理馈电图	Schematic and Feed Diagrams of Landing Light
54	领航员、驾驶员风挡玻璃加温馈电图	Feed Diagram of Navigator and Pilot Windshield Glass Heating
55	尾翼防冰系统原理馈电图	Schematic and Feed Diagrams of Tail Anti-Icing System
56	螺旋桨和整流罩加温原理馈电图	Schematic and Feed Diagrams of Propeller and Fairing Heating
57	发动机进气道、导向器加温和结冰信号原理馈电图	Schematic and Feed Diagrams of Engine Air Inlet Guide Vane Heating and Icing Signal
58	燃油压力信号器	Fuel Pressure Annunciator
59	细油滤堵塞信号器	Fine Filter Block Annunciator
60	航行灯和编队灯原理馈电图	Schematic and Feed Diagrams of Navigation and Formation Lights
61	防撞灯原理馈电图	Schematic and Feed Diagrams of Anti-

		Collision Light
62	用氧信号原理馈电图	Schematic and Feed Diagrams of Oxygen Supply Signal
63	舱门信号原理馈电图	Schematic and Feed Diagrams of Cabin Door Signal
64	发动机照明灯原理馈电图	Schematic and Feed Diagrams of Engine Illumination Light
65	货舱、尾舱、服务舱及水平安定面照明设备原理馈电图	Schematic and Feed Diagrams of Cargo Cabin, Tail Cabin, Service Cabin And Horizontal Stabilizer Lighting Equipment
66	前舱及机身前段地板下照明设备原理馈电图	Schematic and Feed Diagrams of Under-Floor Lighting Equipment In the Front Cabin and Fuselage Front Section
67	出口及应急照明原理馈电图	Schematic and Feed Diagrams of Exit and Emergency Lighting
68	大气数据系统原理图	Schematic Diagram of Air Data System
69	大气数据系统馈电图	Feed Diagram of Air Data System
70	大气数据系统继电器盒内部原理图	Schematic Diagram of Interior of Air Data System Relay Box
71	远距加速度表原理馈电图	Schematic and Feed Diagrams of Remote Accelerometer
72	飞行数据记录系统原理图	Schematic Diagram of Flight Data Recording System
73	飞行数据记录系统馈电图	Feed Diagram of Flight Data Recording System
74	磁罗盘原理馈电图	Schematic and Feed Diagrams of Magnetic Compass
75	货舱温度表、大气温度表和襟翼位置指示器原理馈电图	Schematic and Feed Diagrams of Cargo Cabin Thermometer, Air Thermometer and Flap Position Indicator
76	航向姿态系统及应急地平仪原理图	Schematic Diagram of Attitude and Heading Reference System (AHRS) & Emergency Horizon
77	航向姿态系统及应急地平仪馈电图	Feed Diagram of Attitude and Heading Reference System (AHRS) & Emergency Horizon
78	近地告警系统(GPWS)原理图	Schematic Diagram of Ground Proximity Warning System (GPWS)
79	膜盒表原理馈电图	Aneroid Meter Schematic and Feed Diagrams
80	近地告警系统(GPWS)馈电图	Feed Diagram of Ground Proximity Warning System (GPWS)
81	临界状态自动信号装置原理馈电图	Schematic and Feed Diagrams of Automatic Sig-

		nal Device in Critical State
82	航空时钟原理馈电图	Schematic and Feed Diagrams of Aviation Clock
83	激光捷联惯性/卫星组合导航系统原理图	Schematic Diagram of Laser Strapdown Inertial/Satellite Integrated Navigation System
84	激光捷联惯性/卫星组合导航系统馈电图	Feed Diagram of Laser Strapdown Inertial/Satellite Integrated Navigation System
85	信号枪和 M 数表原理馈电图	Schematic and Feed Diagrams of Signal Gun and Mach Indicator
86	自动驾驶仪与转弯仪原理图	Schematic Diagram of Autopilot and Turn Indicator
87	自动驾驶仪与转弯仪馈电图	Feed Diagram of Autopilot and Turn Indicator
88	伞绳回收控制原理馈电图	Schematic and Feed Diagrams of Parachute Cord Retraction Control
89	货舱门操纵及空投电气原理图	Electric Schematic Diagram of Cargo Cabin Door Control and Airdrop
90	货舱门操纵及空投电气馈电图	Electric Feed Diagram of Cargo Cabin Door Control and Airdrop
91	电动绞车原理馈电图	Schematic and Feed Diagrams of Electrical Winch
92	空降信号装置原理馈电图	Schematic and Feed Diagrams of Parachuting Signal Device
93	卫生间供电原理馈电图	Schematic and Feed Diagrams of Washroom Power Supply
94	水箱加温原理馈电图	Schematic and Feed Diagrams of Water Tank Heating
95	电烤箱原理馈电图	Schematic and Feed Diagrams of Electric Oven

无线电、雷达线路图册　RADIO AND RADAR WIRING MANUAL

1	短波电台原理图	Schematic Diagram of HF Radio Set
2	短波电台馈电图	Feed Diagram of HF Radio Set
3	座舱音频记录器系统原理图	Schematic Diagram of Cabin Audio Recording System
4	座舱音频记录器系统馈电图	Feed Diagram of Cabin Audio Recording System
5	超短波电台原理图	Schematic Diagram of VHF/UHF Radio Set
6	超短波电台馈电图	Feed Diagram of VHF/UHF Radio Set
7	机内通话器原理图	Schematic Diagram of Intercom
8	机内通话器馈电图	Feed Diagram of Intercom
9	伏尔/仪表着陆系统原理图	Schematic Diagram of VOR/ILS System
10	伏尔/仪表着陆系统馈电图	Feed Diagram of VOR/ILS System
11	气象雷达原理图	Schematic Diagram of Weather Radar

12	气象雷达馈电图	Feed Diagram of Weather Radar
13	无线电高度表原理图	Schematic Diagram of Radar Altimeter
14	无线电高度表馈电图	Feed Diagram of Radar Altimeter
15	导航仪原理图	Schematic Diagram of GPS Navigator
16	GPS 导航仪馈电图	Feed Diagram of GPS Navigator
17	空中交通告警和防撞系统原理图	Schematic Diagram of Traffic Alert and Collision System
18	空中交通告警和防撞系统馈电图	Feed Diagram of Traffic Alert and Collision Avoidance System
19	无线电罗盘原理图	Schematic Diagram of ADF
20	无线电罗盘馈电图	Feed Diagram of ADF
21	塔康原理图	Schematic Diagram of TACAN
22	塔康馈电图	Feed Diagram of TACAN
23	测距器原理图	Schematic Diagram of DME
24	测距器馈电图	Feed Diagram of DME
25	应急定位仪原理图	Schematic Diagram of Emergency Locator (ELT)
26	应急定位仪馈电图	Feed Diagram of Emergency Locator (ELT)
27	机载应答机馈电图	Feed Diagram of Airborne Transponder

无损检测手册　NONDESTRUCTIVE TESTING MANUAL

1	第一篇　总论	Chapter 1　Introduction
2	X 射线	X-Ray
3	检测方法的选择及准备	Selection of Testing Methods and Testing Preparations
4	超声波	Ultrasonic Wave
5	声振	Acoustic Vibration
6	涡流	Eddy Current
7	磁粉	Magnetic Particle
8	渗透	Penetration
9	第二篇　X 射线检测	Chapter 2　X-Ray Testing
10	辅助动力装置—涡轮发电装置安装支架	Auxiliary Power Unit—Mounting Bracket for Turbine Generating Device
11	第 1 节　斜撑杆	Section 1　Slant Bracing
12	第 2 节　撑杆	Section 2　Bracing
13	机身—中段	Fuselage—Middle Section
14	第 1 节　25 框竖梁上接头及外缘条连接区	Section 1　Connection Zone Between Upper Joints and Outer Edge Strip on Vertical Beam at Frame 25
15	第 2 节　25 框竖梁下接头及外缘条连	Section 2　Connection Zone Between Lower

集装货物装卸操作手册 CONTAINERIZED CARGO HANDLING OPERATION MANUAL

保障设备手册 ILLUSTRATED TOOLS AND EQUIPMENT MANUAL

图解零部件目录 ILLUSTRATED PARTS CATALOG

环控系统 Air Conditioning

3	座舱压力调节系统	Cabin Pressure Regulating System
4	制冷系统	Cooling System
5	温度控制系统	Temperature Control System

自动飞行　Auto Flight

| 1 | 自动驾驶仪 | Autopilot |

通信系统　Communications

1	高频系统	HF System
2	超高频系统	VHF/UHF System
3	音频系统	AF System
4	静电释放	Static Electricity Discharge
5	音响和录相监视系统	Audio and Video Surveillance System

电源系统　Electrical Power

1	交流发电	AC Generation
2	直流发电	DC Generation
3	外部电源	External Power Supply
4	交流负载分配	AC Load Distribution
5	直流负载分配	DC Load Distribution

设备　Equipment/Furnishings

1	驾驶舱设备	Cockpit Equipment
2	生活设备	Living Facilities
3	卫生间	Lavatory
4	货舱设备	Cargo Cabin Equipment
5	应急设备	Emergency Equipment
6	空投空降系统	Airdrop and Parachuting Equipment

防火系统　Fire Protection

1	探测	Detection
2	灭火	Fire Extinguishing
3	防爆	Anti-Explosion

飞行操纵系统　Flight Controls

1	副翼和调整片操纵系统	Aileron and Trim Tab Control System
2	方向舵和调整片操纵系统	Rudder and Trim Tab Control System
3	升降舵和调整片操纵系统	Elevator and Trim Tab Control System
4	襟翼操纵系统	Flap Control System
5	舵面锁操纵	Control Surface Lock Control

燃油系统　Fuel

1	贮存	Storage
2	分配	Distribution
3	排放	Discharge
4	指示	Indication

液压源　Hydraulic Power

1	主液压源	Main Hydraulic Power
2	辅助液压源	Auxiliary Hydraulic Power
3	指示	Indication

除雨和防冰　Ice and Rain Protection

1	翼面	Airfoil
2	进气道	Air Inlet Duct
3	全静压加温系统	Pitot-Static Heating System
4	窗、风挡系统	Windows and Windshield System
5	螺旋桨加温系统	Propeller Heating System
6	探测系统	Detection System

显示/记录系统　Indicating/Recording Systems

1	仪表板和控制板	Instrument Panels and Consoles
2	独立仪表	Independent Instrument
3	记录仪	Recorder
4	中央警告系统	Central Warning System

起落装置　Landing Gear

1	主起落架	Main LG
2	前起落架	Nose LG
3	收放机构	Retraction/Extension Mechanism
4	机轮刹车	Wheel Braking
5	前轮转弯系统	Nose Wheel Steering System
6	位置及警告	Position and Warning

照明系统　Lights

1	驾驶舱照明	Cockpit Lighting
2	中段、尾段及地板下照明	Lighting for Center and Tail Sections and Under Floor
3	外部照明	External Lighting
4	应急照明	Emergency Lighting

导航系统　Navigation

1	飞行环境数据	Flight Environment Data
2	姿态与方位	Attitude and Azimuth
3	着陆和滑行设备	Landing and Taxiing Equipment
4	独立定位	Independent Positioning
5	相关定位	Associated Positioning

氧气系统　Oxygen

1	空勤用氧系统	Oxygen System for Aircrew
2	货舱乘员用氧系统	Oxygen System for Cargo Cabin Passengers
3	便携式氧气系统	Portable Oxygen System

引气系统　Bleed Air

1	分配系统	Distribution System

机载辅助动力装置　Airborne Auxiliary Power

1	动力装置	Power Plant

2	涡轮发电装置	Turbine Starter/Generator
3	起动/点火	Starting/Ignition
4	进气	Air Intake
5	指示	Indication
6	排气	Exhaust
7	滑油	Oil

舱门　Doors

1	舱门	Doors
2	登机门	Entry Door
3	应急门	Emergency Doors
4	货舱门	Cargo Cabin Doors
5	机内舱门	Doors Inside Aircraft
6	舱门警告	Door Warning
7	起落架舱门	LG Doors

机身　Fuselage

1	机身总图	General View of Fuselage
2	机身前段	Front Section of Fuselage
3	机身中段	Center Section of Fuselage
4	机身尾段	Tail Section of Fuselage

发动机短舱　Engine Nacelle

1	发动机短舱	Engine Nacelle
2	发动机短舱前段	Front Section of Engine Nacelle
3	发动机短舱中段	Middle Section of Engine Nacelle
4	承力段	Load-Bearing Segment
5	发动机短舱尾罩	Tail Shroud of Engine Nacelle

尾翼　Tail

1	尾翼	Tail
2	水平安定面	Horizontal Stabilizers
3	升降舵	Elevators
4	垂直安定面	Vertical Stabilizer
5	方向舵	Rudder

窗　Windows

1	窗	Windows
2	飞行员舱观察窗	Cockpit Observation Windows
3	货舱观察窗	Cargo Cabin Observation Windows

机翼　Wings

1	机翼	Wings
2	中央翼	Center Wing
3	中外翼、外翼	Inboard Wing and Outboard Wings
4	外翼翼尖	Outboard Wing Tips
5	机翼前缘	Wing Leading Edges
6	后缘与后缘装置	Trailing Edge and Trailing Edge Devices
7	副翼	Ailerons

螺旋桨　Propellers

1	螺旋桨	Propellers
2	控制装置	Control Device
3	指示	Indication

动力装置　Power Plant

1	发动机的安装	Engine Installation
2	发动机安装支架	Engine Mounting Bracket
3	发动机清洗系统	Engine Cleaning System

发动机燃油及控制　Engine Fuel and Control

1	分配系统	Distribution System
2	燃油调节器静压系统	Static Pressure System of Fuel Control Unit
3	指示	Indication

进气　Air

1	通风冷却系统	Ventilation Cooling System

发动机操纵系统　Engine Control

1	功率操纵系统	Power Control System
2	应急顺桨机构	Emergency Feathering Mechanism

发动机指示　Engine Indicating

1	功率	Power
2	温度	Temperature
3	分析器	Analyzer

排气　Exhaust

1	排气	Exhaust
2	总管/排气管	Manifold/Exhaust Pipe

滑油　Oil

1	滑油系统	Oil System
2	贮存系统	Storage System
3	分配	Distribution
4	指示	Indication

起动　Starting

1	起动系统	Starting System

特种电子设备　Special Electronic Equipment

1	敌我识别系统	Identification Friend or Foe（IFF）

结构修理手册　STRUCTURE REPAIR MANUAL

结构　Structures

1	结构—概述	Structure—General
2	气动光滑度	Aerodynamic Smoothness
3	损伤分类	Damage Classification

4	防腐蚀	Corrosion Prevention
5	腐蚀的消除及表面防护处理	Corrosion Removal and Surface Protection Treatment
6	金属材料	Metal Material
7	金属板材	Metal Sheet
8	铝合金	Aluminum Alloy
9	铝合金挤压型材	Aluminum Alloy Extruded Sections
10	合金结构钢	Alloy Structural Steel
11	紧固件	Fastener
12	螺栓和螺母	Bolt and Nut
13	铆钉	Rivet
14	紧固件孔的准备	Preparation for Fastener Hole
15	飞机修理时的支撑检查程序	Supporting Check Procedures for Aircraft During Repair
16	操纵面平衡	Balance of Control Surface
17	修理程序	Repair Procedures
18	典型修理	Typical Repair
19	钣金零件典型修理	Typical Repair for Sheet Parts
20	整体结构油箱典型修理	Typical Repair for Integral Tank
21	挤压或钣弯制型材典型修理	Typical Repair for Extruding or Plate Bent Sections
22	蒙皮气动光滑度典型修理	Typical Repair for Skin Aerodynamic Smoothness
23	橡胶零件的典型修理	Typical Repair for Rubber Parts
24	漆层的修理	Repair for Paint Coating

舱门　Doors

1	舱门	Doors
2	登机门	Entry Door
3	登机门—结构识别	Entry Door—Structure Identification
4	登机门—允许损伤	Entry Door—Allowable Damage
5	登机门—修理	Entry Door—Repair
6	货舱应急窗—结构识别	Emergency Exit Hatch of Cargo Cabin—Structure Identification
7	货舱应急窗门—允许损伤	Emergency Exit Hatch of Cargo Cabin—Allowable Damage
8	货舱应急窗门—修理	Emergency Exit Hatch of Cargo Cabin—Repair
9	货舱大门—概述	Cargo Door—General
10	货舱后大门—结构识别	Rear Cargo Door—Structure Identification

11	货舱后大门—允许损伤	Rear Cargo Door—Allowable Damage
12	货舱后大门—修理	Rear Cargo Door—Repair
13	货桥大门—结构识别	Loading Ramp Door—Structure Identification
14	货桥大门—允许损伤	Loading Ramp Door—Allowable Damage
15	货桥大门—修理	Loading Ramp Door—Repair
16	59 框出入口盖—结构识别	Exit Hatch at Frame 59—Structure Identification
17	59 框出入口盖—允许损伤	Exit Hatch at Frame 59—Allowable Damage
18	59 框出入口盖—修理	Exit Hatch at Frame 59—Repair
19	主起落架舱门—结构识别	Main Landing Gear Door—Structure Identification
20	主起落架舱门—允许损伤	Main Landing Gear Door—Allowable Damage
21	主起落架舱门—修理	Main Landing Gear Door—Repair

机身　Fuselage

1	机身	Fuselage
2	机身前段—结构识别	Front Fuselage—Structure Identification
3	第 4 框—结构识别	Frame 4—Structure Identification
4	第 4 框—允许的损伤	Frame 4—Allowable Damage
5	第 4 框—修理	Frame 4—Repair
6	第 5 框、6 框、7 框、8 框—结构识别	Frames 5，6，7 and 8—Structure Identification
7	第 5 框、6 框、7 框、8 框—允许的损伤	Frames 5，6，7 and 8—Allowable Damage
8	第 5 框、6 框、7 框、8 框—修理	Frames 5，6，7 and 8—Repair
9	第 9 框—结构识别	Frame 9—Structure Identification
10	第 9 框—允许的损伤	Frame 9—Allowable Damage
11	第 9 框—修理	Frame 9—Repair
12	第 13 框—结构识别	Frame 13—Structure Identification
13	第 13 框—允许的损伤	Frame 13—Allowable Damage
14	第 13 框—修理	Frame 13—Repair
15	登机门门框—结构识别	Entry Door Frame—Structure Identification
16	登机门门框—允许的损伤	Entry Door Frame—Allowable Damage
17	登机门门框—修理	Entry Door Frame—Repair
18	机身中段—结构识别	Mid Fuselage—Structure Identification
19	第 25 框—结构识别	Frame 25—Structure Identification
20	第 25 框—允许的损伤	Frame 25—Allowable Damage
21	第 25 框—修理	Frame 25—Repair
22	第 27 框—结构识别	Frame 27—Structure Identification
23	第 27 框—允许的损伤	Frame 27—Allowable Damage
24	第 27 框—修理	Frame 27—Repair
25	第 30 框—结构识别	Frame 30—Structure Identification

26	第 30 框—允许的损伤	Frame 30—Allowable Damage
27	第 30 框—修理	Frame 30—Repair
28	机身中段普通框—结构识别	Conventional Frame in Mid Fuselage—Structure Identification
29	机身中段普通框—允许的损伤	Conventional Frame in Mid Fuselage—Allowable Damage
30	机身中段普通框—修理	Conventional Frame in Mid Fuselage—Repair
31	机身中段壁板—结构识别	Panel in Mid Fuselage—Structure Identification
32	机身中段壁板—允许的损伤	Panel in Mid Fuselage—Allowable Damage
33	机身中段壁板—修理	Panel in Mid Fuselage—Repair
34	前纵梁(25～33 框之间)—结构识别	Front Longeron（Between Frames 25～33）—Structure Identification
35	前纵梁(25～33 框之间)—允许的损伤	Front Longeron（Between Frames 25～33）—Allowable Damage
36	前纵梁(25～33 框之间)—修理	Front Longeron（Between Frames 25～33）—Repair
37	后纵梁(33～41 框之间)—结构识别	Rear Longeron（Between Frames 33～41）—Structure Identification
38	后纵梁(33～41 框之间)—允许的损伤	Rear Longeron（Between Frames 33～41）—Allowable Damage
39	后纵梁(33～41 框之间)—修理	Rear Longeron（Between Frames 33～41）—Repair
40	第 13 框至第 27 框地板—结构识别	Floor Between Frames 13～27—Structure Identification
41	第 13 框至第 27 框地板—允许的损伤	Floor Between Frames 13～27—Allowable Damage
42	第 13 框至第 27 框地板—修理	Floor Between Frames 13～27—Repair
43	气密地板—结构识别	Airtight Floor—Structure Identification
44	气密地板—允许的损伤	Airtight floor—Allowable Damage
45	气密地板—修理	Airtight Floor—Repair
46	第 33 框至第 41 框地板—结构识别	Floor Between Frames 33～41—Structure Identification
47	第 33 框至第 41 框地板—允许的损伤	Floor Between Frames 33～41—Allowable Damage
48	第 33 框至第 41 框地板—修理	Floor Between Frames 33～41—Repair
49	第 33 框—结构识别	Frame 33—Structure Identification
50	第 33 框—允许的损伤	Frame 33—Allowable Damage
51	第 33 框—修理	Frame 33—Repair
52	主起落架舱盒形梁—结构识别	Box Beam of Main Landing Gear Well—Structure

Identification

53	主起落架舱盒形梁—允许的损伤	Box Beam of Main Landing Gear Well—Allowable Damage
54	主起落架舱盒形梁—修理	Box Beam of Main Landing Gear Well—Repair
55	起落架整流罩—结构识别	Landing Gear Fairing—Structure Identification
56	起落架整流罩—允许的损伤	Landing Gear Fairing—Allowance Damage
57	起落架整流罩—修理	Landing Gear Fairing—Repair
58	中央翼前整流罩—结构识别	Front Fairing of Center Wing—Structure Identification
59	中央翼前整流罩—允许的损伤	Front Fairing of Center Wing—Allowance Damage
60	中央翼前整流罩—修理	Front Fairing of Center Wing—Repair
61	中央翼后整流罩—修理	Rear Fairing of Center Wing—Repair
62	中央翼中整流罩—允许的损伤	Middle Fairing of Center Wing—Allowance Damage
63	中央翼中整流罩—修理	Middle Fairing of Center Wing—Repair
64	中央翼后整流罩—结构识别	Rear Fairing of Center Wing—Structure Identification
65	中央翼后整流罩—允许的损伤	Rear Fairing of Center Wing—Allowance Damage
66	中央翼中整流罩—结构识别	Middle Fairing of Center Wing—Structure Identification
67	机身尾段—结构识别	Rear Fuselage—Structure Identification
68	尾段纵梁—结构识别	Rear Longeron—Structure Identification
69	尾段纵梁—允许的损伤	Rear Longeron—Allowable Damage
70	尾段纵梁—修理	Rear Longeron—Repair
71	第 43 框—结构识别	Frame 43—Structure Identification
72	第 43 框—允许的损伤	Frame 43—Allowable Damage
73	第 43 框—修理	Frame 43—Repair
74	第 49 框—结构识别	Frame 49—Structure Identification
75	第 49 框—允许的损伤	Frame 49—Allowable Damage
76	第 49 框—修理	Frame 49—Repair
77	第 55 框—结构识别	Frame 55—Structure Identification
78	第 55 框—允许的损伤	Frame 55—Allowable Damage
79	第 55 框—修理	Frame 55—Repair
80	第 59 框—结构识别	Frame 59—Structure Identification
81	第 59 框—允许的损伤	Frame 59—Allowable Damage
82	第 59 框—修理	Frame 59—Repair
83	第 62 框—结构识别	Frame 62—Structure Identification
84	第 62 框—允许的损伤	Frame 62—Allowable Damage

85	第 62 框—修理	Frame 62—Repair
86	第 65 框—结构识别	Frame 65—Structure Identification
87	第 65 框—允许的损伤	Frame 65—Allowable Damage
88	第 65 框—修理	Frame 65—Repair
89	机身尾段普通框—结构识别	Conventional frames in Rear Fuselage—Structure Identification
90	机身尾段普通框—允许的损伤	Conventional frames in Rear Fuselage—Allowable Damage
91	机身尾段普通框—修理	Conventional frames in Rear Fuselage—Repair
92	机身尾段壁板—结构识别	Panels in Rear Fuselage—Structure Identification
93	机身尾段壁板—允许的损伤	Panels in Rear Fuselage—Allowable Damage
94	机身尾段壁板—修理	Panels in Rear Fuselage—Repair
95	背鳍—结构识别	Dorsal Fin—Structure Identification
96	背鳍—允许的损伤	Dorsal Fin—Allowable Damage
97	背鳍—修理	Dorsal Fin—Repair
98	尾段地板—结构识别	Floor of Rear Fuselage—Structure Identification
99	尾段地板—允许的损伤	Floor of Rear Fuselage—Allowable Damage
100	尾段地板—修理	Floor of Rear Fuselage—Repair
101	机身尾舱—结构识别	Tail Cabin of Fuselage—Structure Identification
102	机身尾舱普通框—结构识别	Conventional Frame in Tail Cabin of Fuselage—Structure Identification
103	机身尾舱普通框—允许的损伤	Conventional Frame in Tail Cabin of Fuselage—Allowable Damage
104	机身尾舱普通框—修理	Conventional Frame in Tail Cabin of Fuselage—Repair
105	机身尾舱壁板—结构识别	Panel in Tail Cabin of Fuselage—Structure Identification
106	机身尾舱壁板—允许的损伤	Panel in Tail Cabin of Fuselage—Allowable Damage
107	机身尾舱壁板—修理	Panel in Tail Cabin of Fuselage—Repair

发动机短舱　Engine Nacelle

1	发动机短舱	Engine Nacelle
2	发动机短舱前段	Front Section of Engine Nacelle
3	发动机短舱前段—结构识别	Front Section of Engine Nacelle—Structure Identification
4	发动机短舱前段—允许的损伤	Front Section of Engine Nacelle—Allowable Damage

5	发动机短舱前段—修理	Front Section of Engine Nacelle—Repair
6	减速器整流罩—结构识别	Reduction Gear Fairing—Structure Identification
7	减速器整流罩—允许的损伤	Reduction Gear Fairing—Allowable Damage
8	减速器整流罩—修理	Reduction Gear Fairing—Repair
9	发动机短舱中段	Mid Section of Engine Nacelle
10	发动机短舱中段上盖—结构识别	Mid Section Upper Cover of Engine Nacelle—Structure Identification
11	发动机短舱中段上盖—允许的损伤	Mid Section Upper Cover of Engine Nacelle—Allowable Damage
12	发动机短舱中段上盖—修理	Mid Section Upper Cover of Engine Nacelle—Repair
13	发动机短舱中段下盖—结构识别	Mid Section Lower Cover of Engine Nacelle—Structure Identification
14	发动机短舱中段下盖—允许的损伤	Mid Section Lower Cover of Engine Nacelle—Allowable Damage
15	发动机短舱中段下盖—修理	Mid Section Lower Cover of Engine Nacelle—Repair
16	发动机短舱中段侧盖—结构识别	Mid Section Side Cover of Engine Nacelle—Structure Identification
17	发动机短舱中段侧盖—允许的损伤	Mid Section Side Cover of Engine Nacelle—Allowable Damage
18	发动机短舱中段侧盖—修理	Mid Section Side Cover of Engine Nacelle—Repair
19	发动机短舱 4 框—结构识别	Frame 4 of Engine Nacelle—Structure Identification
20	发动机短舱 4 框—允许的损伤	Frame 4 of Engine Nacelle—Allowable Damage
21	发动机短舱 4 框—修理	Frame 4 of Engine Nacelle—Repair
22	发动机短舱尾舱整流罩—结构识别	Tail Cabin Fairing of Engine Nacelle—Structure Identification
23	发动机短舱尾舱整流罩—允许的损伤	Tail Cabin Fairing of Engine Nacelle—Allowable Damage
24	发动机短舱尾舱整流罩—修理	Tail Cabin Fairing of Engine Nacelle—Repair
25	发动机短舱尾舱上包皮—结构识别	Upper Cowling of Tail Cabin of Engine Nacelle—Structure Identification
26	发动机短舱尾舱上包皮—允许的损伤	Upper Cowling of Tail Cabin of Engine Nacelle—Allowable Damage
27	发动机短舱尾舱上包皮—修理	Upper Cowling of Tail Cabin of Engine Nacelle—Repair
28	发动机架后段—结构识别	Rear Section of Engine Mount—Structure Identification

29	发动机架后段—允许的损伤	Rear Section of Engine Mount—Allowable Damage
30	发动机架后段—修理	Rear Section of Engine Mount—Repair
31	发动机短舱尾罩—结构识别	Tail Cowling of Engine Nacelle—Structure Identification
32	发动机短舱尾罩—允许的损伤	Tail Cowling of Engine Nacelle—Allowable Damage
33	发动机短舱尾罩—修理	Tail Cowling of Engine Nacelle—Repair
34	发动机短舱承力段	Load-Bearing Segment of Engine Nacelle

尾翼　Tail

1	安定面	Stabilizers
2	水平安定面—结构识别	Horizontal Stabilizer—Structure Identification
3	水平安定面前梁—结构识别	Front Spar of Horizontal Stabilizer—Structure Identification
4	水平安定面前梁—允许损伤	Front Spar of Horizontal Stabilizer—Allowable Damage
5	水平安定面前梁—修理	Front Spar of Horizontal Stabilizer—Repair
6	水平安定面后梁—结构识别	Rear Spar of Horizontal Stabilizer—Structure Identification
7	水平安定面后梁—允许损伤	Rear spar of horizontal stabilizer—allowable damage
8	水平安定面后梁—修理	Rear Spar of Horizontal Stabilizer—Repair
9	水平安定面壁板—结构识别	Panel of Horizontal Stabilizer—Structure Identification
10	水平安定面壁板—允许损伤	Panel of Horizontal Stabilizer—Allowable Damage
11	水平安定面壁—修理	Panel of Horizontal Stabilizer—Repair
12	水平安定面翼肋—结构识别	Wing Rib of Horizontal Stabilizer—Structure Identification
13	水平安定面翼肋—允许损伤	Wing Rib of Horizontal Stabilizer—Allowable Damage
14	水平安定面翼肋—修理	Wing Rib of Horizontal Stabilizer—Repair
15	水平安定面前缘—结构识别	Leading Edge of Horizontal Stabilizer— Structure Identification
16	水平安定面前缘—允许损伤	Leading Edge of Horizontal Stabilizer—Allowable Damage
17	水平安定面前缘—修理	Leading Edge of Horizontal Stabilizer—Repair

18	水平安定面后缘—结构识别	Trailing Edge of Horizontal Stabilizer—Structure Identification
19	水平安定面后缘—允许损伤	Trailing Edge of Horizontal Stabilizer—Allowable Damage
20	平安定面后缘—修理	Trailing Edge of Horizontal Stabilizer—Repair
21	水平尾翼翼尖整流罩—结构识别	Wing-Tip Fairing of Horizontal Tail—Structure Identification
22	水平尾翼翼尖整流罩—允许损伤	Wing-Tip Fairing of Horizontal Tail—Allowable Damage
23	水平尾翼翼尖整流罩-修理	Wing-Tip Fairing of Horizontal Tail—Repair
24	升降舵—结构识别	Elevator—Structure Identification
25	升降舵—允许损伤	Elevator—Allowable Damage
26	升降舵—修理	Elevator—Repair
27	升降舵调整片—结构识别	Trim Tab of Elevator—Structure Identification
28	升降舵调整片—允许损伤	Elevator Trim Tab—Allowable Damage
29	升降舵调整片—修理	Trim Tab of Elevator—Repair
30	垂直安定面后梁—允许损伤	Rear Spar of Vertical Stabilizer—Allowable Damage
31	垂直安定面前梁—结构识别	Front Spar of Vertical Stabilizer—Structure Identification
32	垂直安定面前梁—允许损伤	Front Spar of Vertical Stabilizer—Allowable Damage
33	垂直安定面前梁—修理	Front Spar of Vertical Stabilizer—Repair
34	垂直安定面后梁—结构识别	Rear Spar of Vertical Stabilizer—Structure Identification
35	垂直安定面	Vertical Stabilizer
36	垂直安定面后梁—修理	Rear Spar of Vertical Stabilizer—Repair
37	垂直安定面壁板—结构识别	Panel of Vertical Stabilizer—Structure Identification
38	垂直安定面壁板—允许损伤	Panel of Vertical Stabilizer—Allowable Damage
39	垂直安定面壁板—修理	Panel of Vertical Stabilizer—Repair
40	垂直安定面前缘—结构识别	Leading Edge of Vertical Stabilizer—Structure Identification
41	垂直安定面前缘—允许损伤	Leading Edge of Vertical Stabilizer—Allowable Damage
42	垂直安定面前缘—修理	Leading Edge of Vertical Stabilizer—Repair
43	垂直安定面翼根整流罩—结构识别	Wing Root Fairing of Vertical Stabilizer—Structure Identification
44	垂直安定面前缘整流罩—允许损伤	Leading Edge Fairing of Vertical Stabilizer—Al-

lowable Damage

45	垂直安定面前缘整流罩—修理	Leading Edge Fairing of Vertical Stabilizer—Repair
46	垂直安定面翼尖整流罩—结构识别	Wing-Tip Fairing of Vertical Stabilizer —Structure Identification
47	垂直安定面翼尖整流罩—允许损伤	Wing-Tip Fairing of Vertical Stabilizer—Allowable Damage
48	垂直安定面翼尖整流罩—修理	Wing-Tip Fairing of Vertical Stabilizer—Repair
49	垂直安定面翼肋—结构识别	Wing Rib of Vertical Stabilizer—Structure Identification
50	垂直安定面翼肋—允许损伤	Wing Rib of Vertical Stabilizer—Allowable Damage
51	垂直安定面翼肋—修理	Wing Rib of Vertical Stabilizer—Repair
52	垂直安定面后缘—结构识别	Trailing Edge of Vertical Stabilizer—Structure Identification
53	垂直安定面后缘—允许损伤	Trailing Edge of Vertical Stabilizer—Allowable Damage
54	直安定面后缘—修理	Trailing Edge of Vertical Stabilizer—Repair
55	机身尾翼整流包皮—结构识别	Fillet of Fuselage and Tail—Structure Identification
56	机身尾翼整流包皮—允许损伤	Fillet of Fuselage and Tail—Allowable Damage
57	机身尾翼整流包皮—修理	Fillet of Fuselage and Tail—Repair
58	方向舵—结构识别	Rudder—Structure Identification
59	方向舵—允许损伤	Rudder—Allowable Damage
60	方向舵—修理	Rudder—Repair
61	方向舵调整片—结构识别	Rudder Trim Tab—Structure Identification
62	方向舵调整片—允许损伤	Rudder Trim Tab—Allowable Damage
63	方向舵调整片—修理	Rudder Trim Tab—Repair
64	方向舵弹簧补偿片—结构识别	Spring Tab of Rudder—Structure Identification
65	方向舵弹簧补偿片—允许损伤	Spring Tab of Rudder—Allowable Damage
66	方向舵弹簧补偿片—修理	Spring Tab of Rudder—Repair

窗　Windows

1	窗	Windows
2	座舱盖观察窗及其固定—结构识别	Cabin Observation Windows and Fixation—Structure Identification
3	座舱盖观察窗及其固定—允许损伤	Cabin Observation Windows and Fixation—Allowable Damage

4	座舱盖观察窗及其固定—修理	Cabin Observation Windows and Fixation—Repair
5	机头罩玻璃及其固定—结构识别	Nose Fairing Glass and Fixation—Structure Identification
6	机头罩玻璃及其固定—允许损伤	Nose Fairing Glass and Fixation—Allowable Damage
7	机头罩玻璃及其固定—修理	Nose Fairing Glass and Fixation—Repair
8	驾驶舱观察窗及其固定—结构识别	Cockpit Observation Windows and Fixation—Structure Identification
9	驾驶舱观察窗及其固定—允许损伤	Cockpit Observation Windows and Fixation—Allowable Damage
10	驾驶舱观察窗及其固定—修理	Cockpit Observation Windows and Fixation—Repair
11	通风窗及其固定—结构识别	Ventilation Windows and Fixation—Structure Identification
12	通风窗及其固定—允许损伤	Ventilation Windows and Fixation—Allowable Damage
13	通风窗及固定—修理	Ventilation Windows and Fixation—Repair
14	货舱观察窗及其固定—结构识别	Observation Windows of Cargo Cabin and Fixation—Structure Identification
15	货舱观察窗及其固定—允许损伤	Observation Windows of Cargo Cabin and Fixation—Allowable Damage
16	货舱观察窗及其固定—修理	Observation Windows of Cargo Cabin and Fixation—Repair

机翼　Wings

1	机翼—总论	Wing—Introduction
2	中央翼—结构识别	Center Wing—Structure Identification
3	中央翼前大梁—修理	Front Spar of Center Wing—Repair
4	中央翼前大梁—允许损伤	Front Spar of Center Wing—Allowable Damage
5	中央翼前大梁—结构识别	Front Spar of Center Wing—Structure Identification
6	中央翼后大梁—结构识别	Rear Spar of Center Wing—Structure Identification
7	中央翼后大梁—允许损伤	Rear Spar of Center Wing—Allowable Damage
8	中央翼后大梁—修理	Rear Spar of Center Wing—Repair
9	中央翼上壁板—结构识别	Upper Panel of Center Wing—Structure Identification

10	中央翼上壁板—允许损伤	Upper Panel of Center Wing—Allowable Damage
11	中央翼上壁板—修理	Upper Panel of Center Wing—Repair
12	中央翼下壁板—结构识别	Lower Panel of Center Wing—Structure Identification
13	中央翼下壁板—允许损伤	Lower Panel of Center Wing—Allowable Damage
14	中央翼下壁板—修理	Lower Panel of Center Wing—Repair
15	中央翼翼肋—结构识别	Center Wing Rib—Structure Identification
16	中央翼翼肋—允许损伤	Center Wing Rib—Allowable Damage
17	中央翼翼肋—修理	Center Wing Rib—Repair
18	中外翼—结构识别	Inboard Wing—Structure Identification
19	中外翼前大梁—结构识别	Front Spar of Inboard Wing—Structure Identification
20	中外翼前大梁—允许损伤	Front Spar of Inboard Wing—Allowable Damage
21	中外翼前大梁—修理	Front Spar of Inboard Wing—Repair
22	中外翼后大梁—结构识别	Rear Spar of Inboard Wing—Structure Identification
23	中外翼后大梁—允许损伤	Rear Spar of Inboard Wing—Allowable Damage
24	中外翼后大梁—修理	Rear Spar of Inboard Wing—Repair
25	中外翼上壁板—结构识别	Upper Panel of Inboard Wing—Structure Identification
26	中外翼上壁板—允许损伤	Upper Panel of Inboard Wing—Allowable Damage
27	中外翼上壁板—修理	Upper Panel of Inboard Wing—Repair
28	中外翼下壁板—结构识别	Lower Panel of Inboard Wing—Structure Identification
29	中外翼下壁板—允许损伤	Lower Panel of Inboard Wing—Allowable Damage
30	中外翼下壁板—修理	Lower Panel of Inboard Wing—Repair
31	中外翼翼肋—结构识别	Inboard Wing Rib—Structure Identification
32	中外翼翼肋—允许损伤	Wing Rib of Inboard Wing—Allowable Damage
33	中外翼翼肋—修理	Wing Rib of Inboard Wing—Repair
34	外翼—结构识别	Outboard Wing—Structure Identification
35	外翼前大梁—结构识别	Front Spar of Outboard Wing—Structure Identification
36	外翼前大梁—允许损伤	Front Spar of Outboard Wing—Allowable Damage
37	外翼前大梁—修理	Front Spar of Outboard Wing—Repair
38	外翼后大梁—结构识别	Rear Spar of Outboard Wing—Structure Identification

39	外翼后大梁—允许损伤	Rear Spar of Outboard Wing—Allowable Damage
40	外翼上壁板—结构识别	Upper Panel of Outboard Wing—Structure Identification
41	外翼后大梁—修理	Rear Spar of Outboard Wing—Repair
42	外翼上壁板—允许损伤	Upper Panel of Outboard Wing—Allowable Damage
43	外翼上壁板—修理	Upper Panel of Outboard Wing—Repair
44	外翼下壁板—结构识别	Lower Panel of Outboard Wing—Structure Identification
45	外翼下壁板—允许损伤	Lower Panel of Outboard Wing—Allowable Damage
46	外翼下壁板—修理	Lower Panel of Outboard Wing—Repair
47	外翼翼肋—结构识别	Wing Rib of Outboard Wing—Structure Identification
48	外翼翼肋—允许损伤	Outboard Wing Rib—Allowable Damage
49	外翼翼肋—修理	Wing Rib of Outboard Wing—Repair
50	外翼翼尖—结构识别	Wing Tip of Outboard Wing—Structure Identification
51	外翼翼尖—允许损伤	Wing Tip of Outboard Wing—Allowable Damage
52	外翼翼尖—修理	Wing Tip of Outboard Wing—Repair
53	中外翼前缘—结构识别	Leading Edge of Inboard Wing—Structure Identification
54	中外翼前缘—允许损伤	Leading Edge of Inboard Wing—Allowable Damage
55	中外翼前缘—修理	Leading Edge of Inboard Wing—Repair
56	外翼前缘—结构识别	Leading Edge of Outboard Wing—Structure Identification
57	外翼前缘—允许损伤	Leading Edge of Outboard Wing—Allowable Damage
58	外翼前缘—修理	Leading Edge of Outboard Wing—Repair
59	襟翼舱—结构识别	Flap—Structure Identification
60	襟翼舱—允许损伤	Flap Bay—Allowable Damage
61	襟翼舱—修理	Flap Bay—Repair
62	襟翼—结构识别	Flap—Structure Identification
63	襟翼—允许损伤	Flap Bay—Allowable Damage
64	襟翼—修理	Flap—Repair
65	副翼舱—结构识别	Aileron Bay—Structure Identification
66	副翼舱—允许损伤	Aileron Bay—Allowable Damage
67	副翼舱—修理	Aileron Bay—Repair

68	副翼及调整片—结构识别	Aileron and Trim Tab—Structure Identification
69	副翼及调整片—允许损伤	Aileron and Trim Tab—Allowable Damage
70	副翼及调整片—修理	Aileron and Trim Tab—Repair

机载设备维修手册
AIRBORNE EQUIPMENT MAINTENANCE MANUAL

随机保障设备配套目录
LIST OF AIRBORNE GROUND SUPPORT EQUIPMENT

1	机械专业地面保障设备	Ground Support Equipment for Airframe Specialty
2	特设专业地面保障设备	Ground Support Equipment for Special Equipment Specialty
3	无线电专业地面保障设备	Ground Support Equipment for Radio Specialty
4	雷达专业地面保障设备	Ground Support Equipment for Radar Specialty

随机工具配套目录　AIRBORNE TOOL LIST

1	机械部分	Airframe Specialty
2	特设部分	Special Equipment Specialty
3	无线部分	Radio Specialty
4	雷达部分	Radar Specialty

随机备件配套目录　AIRBORNE SPARE PART LIST

随机资料目录　AIRBORNE TECHNICAL MATERIAL LIST

| 1 | 飞机类随机技术资料 | Aircraft Airborne Technical Materials |
| 2 | 机载设备类随机技术资料 | Airborne Technical Materials of Airborne Equipment |

随机证明文件目录　AIRBORNE CERTIFICATE LIST

1	电气部分	Electric Part
2	仪表部分	Instrument Part
3	无线电部分	Radio Part
4	军械部分	Armament Part
5	动力、燃油、灭火部分	Power, Fuel and Fire Extinguishing Part
6	环控、氧气部分	ECS and Oxygen Part
7	座椅、生活设备	Seat and High-altitude Living Facility Part
8	液压、起落架、机体部分	Hydraulic, Landing Gear and Airframe Part
9	全机履历本	Aircraft Logbooks

不装机散装件目录　LIST OF BULK PARTS AND COMPONENTS

1	空生部分	High-altitude Living Facility Part
2	照明部分	Lighting Part
3	军械部分	Armament Part

成品备件清单　LIST OF FINISHED PRODUCT SPARE PARTS

1	仪表类	Instrument
2	电气类	Electric
3	无线电类	Radio
4	附件类	Accessories

随机资料用语

飞行手册　AIRCRAFT FLIGHT MANUAL

1	飞机简介	introduction
2	飞机技术数据	aircraft technical data
3	飞机三面图	aircraft three-view arrangement
4	最小飞行机组和飞行使用范围	minimum crew and service range
5	飞行限制数据	flight limitation data
6	速度限制	speed limitations
7	重量、重心数据	weight and C. G. limitations
8	过载限制	G-load limitation
9	动力装置限制	limitations of power plant
10	空调系统使用限制	operational limitation for air conditioning system
11	发动机故障	engine failures
12	大迎角飞行	high angle of attack flight
13	打开舱门飞行和紧急下降	flying with door open and descending in emergency
14	起落架系统发生故障情况下的着陆	landing with the malfunctioned landing gear system
15	轮胎爆破和刹车失效的处置	handling of tire blown-up and brakes failure
16	飞行中航向系统、气压表发生故障的处置	handling of heading system and the barometer failed in flight
17	不放襟翼着陆	landing with flaps up
18	场外迫降	outside forced landing
19	飞行准备	preparations for flight
20	飞行	flight
21	飞机飞行性能计算及换算曲线	aircraft flight performance calculation and conversion curve
22	飞机四发主要性能	main performance of four engines
23	飞机三发主要性能	main performance of three engines
24	非标准条件下的起飞着陆性能	take off and landing performance in non-standard condition
25	不同重量和高度的爬升及下降	climb and descent at different weight and altitude
26	不同重量及高度上的有利平飞速度和油耗	optimum level speed and fuel consumption at diffent weight and altitude

27	动力装置	power plant
28	燃油系统	fuel system
29	滑油系统	oil system
30	液压系统	hydraulic system
31	起落架的正常收放	normal LG extension and retraction
32	灭火和中性气体系统	fire-extinguish and neutral gas system
33	空调系统	air-conditioning system
34	防冰加温系统	anti-icing heating system
35	氧气系统	oxygen system
36	飞控系统	flight control system
37	导航系统	navigation system
38	通信系统	communication system
39	雷达系统	radar system
40	仪表系统	instrument system
41	起动电源与机上电源设备	starting power and onboard power equipment
42	信号、照明设备	signal，illuminating equipment
43	瞄准、空运、空投、空降设备	sighting，airlift，airdrop and parachuting equipment
44	电动信号枪装置	electric signal gun device
45	空速表	(combined) ASI

外形缺损清单　CONFIGURATION DEVIATION LIST

通信系统　Communications

1	放电刷	static discharger

电源系统　Electrical Power

1	地面电源插座盖	ground power supply socket cover

除雨和防冰　Ice and Rain Protection

1	风挡雨刷	windshield wiper

照明系统　Lights

1	发动机照明灯罩	engine light shroud

舱门　Doors

1	下应急舱门检查口盖	access cover of lower emergency door

机身　Fuselage

1	中央翼整流罩上的检查口盖	access cover on the cowling of center wing
2	左、右鼓包检查口盖	access cover of left and right bulges

发动机短舱　Engine nacelles

1	发动机机架接头检查口盖	engine rack joint access cover

尾翼　Stabilizers

1	方向舵调整片拉杆整流罩	rudder trim tab pull rod fairing
2	方向舵补偿片拉杆整流罩	rudder balance tab pull rod fairing
3	升降舵调整片拉杆整流罩	elevator trim tab pull rod fairing
4	垂尾对接处整流片	vertical tail butt joint surface fillet
5	平尾对接处整流片	horizontal tail joint surface fillet

机翼　Wings

1	机身整流罩与后部处的封严板	sealing plate of fuselage fairing and the rear part
2	副翼调整片拉杆整流罩	aileron trim tab pull rod fairing
3	中外翼与外翼后部处的封严板	sealing plate of inboard wing and rear outboard wing

结构修理手册　STRUCTURE REPAIR MANUAL

结构　Structures

1	结构分类	structure classification
2	气动光滑度所要求的区域	areas with requirement of aerodynamic smoothness
3	气动外形值公差及波纹度公差	tolerance of aerodynamic contour value and waviness
4	表面错位控制	surface misplacement control

5	机身各段及窗、窗框、门的阶差和间隙	the step difference and clearance of each section of fuselage，window，window frame and door
6	检查所用的设备、工具	equipment and tools needed in check
7	检查程序	check procedure
8	结论分析	result analysis
9	损伤类型	damage classification
10	损伤的检查	damage check
11	按修理性质对损伤的分类	damage classification based on repair characteristics
12	腐蚀的原因	causes of corrosion
13	腐蚀损伤的种类和外表特征	classification and appearance characteristics of corrosion
14	腐蚀的位置	corrosion location
15	腐蚀损伤的检测及检查工具	corrosion damage check and check tools
16	腐蚀损伤深度的测定	determination of corrosion damage depth
17	腐蚀的去除	corrosion removal
18	表面防护处理	surface protection treatment
19	材料规范或标准	material specification or standard
20	金属材料的化学鉴别	chemical discrimination of metal material
21	成形条件	forming condition
22	金属材板的成形	forming of metal sheet
23	铝合金的热处理工艺	heat treatment technology of aluminum alloy
24	铝合金的典型性能	typical performance of aluminum alloy
25	挤压型材	extruded sections
26	挤压型材下陷	extruded section joggle
27	碳素结构钢	carbon structural steel
28	低合金钢	low alloy steel
29	不锈钢	stainless steel
30	紧固件的种类	classification of fastener
31	紧固件装分解的程序及紧固件的强度值	disassembling procedure and strength value of the fastener
32	紧固件边缘距离	fastener edge distance
33	螺栓和螺母维修准则	repair requirement of bolt and nut
34	铆钉材料	rivet material
35	铆钉长度的计算	calculation of rivet length
36	铆钉的更换和加大要求	requirement of rivet replacing and enlarging
37	铆钉的镦头尺寸及要求	size and requirement of rivet upsetting end
38	铆钉缺陷的种类和排除方法	rivet defect classification and eliminating method
39	螺栓孔	bolt hole

40	锪窝	counterboring
41	铆钉孔的直径及孔的要求	rivet hole diameter and hole requirement
42	防腐	anti-corrosion
43	飞机的顶起设备	jacking equipment of aircraft
44	尾部支撑	tail support
45	机翼固定点支架	wing fixing point bracket
46	发动机固定点支架	engine fixing point bracket
47	操纵面平衡技术要求	technical requirements for control surface balance
48	操纵面平衡方法	methods for control surface balancing
49	修理的分类	repair classification
50	修理的限制	limitations for repair
51	结构支撑	structure supporting
52	校正检查	correction check
53	损伤的清理	clearing of damage
54	确定蒙皮修理线	determination for skin repair line
55	修理步骤	repair procedure
56	注意事项	caution
57	辅助口盖	auxiliary access covers
58	修理程序	repair procedures
59	允许损伤及其小量修理	allowable damage and minor repair
60	蒙皮典型修理程度	typical skin repair degree
61	限制条件	limitations
62	修理材料	repair material
63	工具和设备	tools and equipment
64	整体油箱的故障检查	failure check of integral tank
65	整体油箱的密封修理	sealing repair for integral tank
66	整体油箱的密封试验	sealing test of integral tank
67	排除整体油箱的漏油	eliminate leakage of integral tank
68	漆层去除	removal of paint coating
69	机翼整体油箱可卸壁板周围粘贴橡胶条	bonding rubber tape around the removable panel of wing integral tank
70	漆层的修补	repair for paint coating

舱门　Doors

1	货舱应急窗门的结构	structure of cargo cabin emergency exit hatch
2	货舱应急窗门锁机构	lock mechanism of cargo cabin emergency exit hatch
3	结构说明	structure description

4　　货桥大门结构说明　　　　　　structure description for cargo doors

机身　Fuselage

1　　允许损伤的修理　　　　　　　repair of allowable damage
2　　蒙皮允许损伤的小量修理　　　minor repair of allowable damage of skin

发动机短舱　Engine Nacelle

1　　发动机短舱修理的技术要求　　technical requirements for repair of engine nacelle
2　　蒙皮的允许损伤及其小量修理　allowable damage and simple repair of skin
3　　隔框的允许损伤及其小量修理　allowable damage and simple repair of bulkhead
4　　框腹板的允许损伤及其小量修理　allowable damage and simple repair of frame web
5　　框缘元件允许损伤及其小量修理　allowable damage and simple repair of architrave
6　　型材的允许损伤及其小量修理　allowable damage and simple repair of section
7　　杆件外表面的允许损伤及其小量修理　allowable damage and simple repair on outer surface of rods

尾翼　Tail

1　　水平安定面的拆卸和安装　　　removal and installation of horizontal stabilizer
2　　修理方法及修理程序　　　　　repair methods and procedures
3　　垂直安定面的拆卸和安装　　　removal and installation of vertical stabilizer

窗　Windows

1　　玻璃口框的修理　　　　　　　glass framework repair

机翼　Wings

1　　机翼站位(肋位)分布　　　　　wing station (rib location) distribution
2　　机翼蒙皮分布　　　　　　　　wing skin distribution
3　　机翼长桁分布示意　　　　　　schematic diagram for wing stringer distribution
4　　机翼修理技术要求　　　　　　technical requirements for wing repair
5　　中外翼后大梁下缘条腐蚀排除　removal for corrosion on lower edge strip of rear spar of inboard wing
6　　外翼翼尖的拆卸　　　　　　　removal of wing tip of outboard wing
7　　外翼翼尖的安装　　　　　　　installation of wing tip of outboard wing

重量和平衡手册　WEIGHT & BALANCE MANUAL

1	燃油	fuel
2	工作液体	operating fluid
3	乘员	aircrew
4	装载	loading
5	地面操作	ground operation
6	飞机的载重平衡控制程序	aircraft weight and balance control procedure
7	说明	description
8	飞机报告	aircraft report
9	飞机称重记录	aircraft weighing record

集装货物装卸操作手册　CONTAINERIZED CARGO HANDLING OPERATION MANUAL

1	集装货物装卸设备的组成和功用	composition and functions of containerized cargo handling equipment
2	集装货物装卸设备操作及人员要求	containerized cargo handling operation and personnel requirement
3	飞机部分	aircraft section
4	地面部分	ground section
5	电动装卸程序	electrical handling procedure
6	手动装卸程序	manual handling procedure
7	后挡板锁死程序	backplate locking procedure
8	后挡板恢复程序	backplate restoration procedure
9	电动绞车的功能	electrical winch functions
10	电动绞车的操作	electrical winch operation
11	绞车使用要求及安全注意事项	winch operation requirement and safety precautions

保障设备手册　ILLUSTRATED TOOLS AND EQUIPMENT MANUAL

1	液压千斤顶	hydraulic jack
2	辅助千斤顶	auxiliary jack
3	小千斤顶	small jack
4	充气设备	charging apparatus
5	发动机油封车	engine preservation trolley
6	放油漏斗	fuel drain funnel
7	飞机清水供水车	aircraft clear water feeder
8	排污车	sewage disposal truck
9	机场用直流电缆	airfield DC cable

10	地面用直流过渡电缆	ground DC transition cable
11	机场用交流电缆	airfield AC cable
12	地面通话器	ground communication headset
13	使用吊挂和支承定位设备时的注意事项	caution when using slings and supporting/positioning device
14	发动机吊挂	engine sling
15	螺旋桨吊挂	propeller sling
16	发动机包装箱吊挂	engine packing box sling
17	涡轮发电装置吊挂	turbine generator sling
18	机翼固定点支架	support for fixing point on wing
19	发动机固定点支架	support for fixing point on engine
20	轻便拆胎器	portable tyre remover
21	涡轮发电装置安装车	turbine generator mounting cart
22	螺旋桨放置架	propeller placement stand
23	发动机运输车	engine transporting trolley
24	前起对接车	nose LG butt carriage
25	主起对接车	MLG butt carriage
26	中央翼、中外翼工作梯	center wing/inboard wing service ladder
27	机翼壁板拆卸工作梯	service ladder for dismounting wing panel
28	发动机工作梯	engine service ladder
29	发动机外场工作梯	engine field workbench
30	发动机升降工作梯	engine elevating service ladder
31	飞机维护液压工作梯	hydraulic platform for aircraft maintenance
32	折叠工作梯	foldable service ladder
33	折叠登机门梯	foldable entry door ladder
34	轻便梯	portable ladder
35	伞兵登机梯	paratroopers ladder
36	小工作梯	lower service ladder
37	飞机蒙布和各种堵塞	the aircraft cover and various blockages
38	测量仪器	measuring instrument
39	定力扳手与扭矩值表	torque wrench and torque value table
40	定力扳手的校验夹具	calibration fixture of torque wrench
41	牵引装置	towing device
42	牵引钢索和前轮操纵架	towing cable and nose-wheel control bracket
43	货桥千斤顶	ramp jack
44	撬杠	crowbar
45	地面过渡段	ground transition section

图解零部件目录　ILLUSTRATED PARTS CATALOG

环控系统　**Air conditioning**

1	断流、交叉、分路活门的安装	installation of cut off，cross and shunt valves
2	驾驶舱空调导管的安装	installation of cockpit air conditioner pipe
3	货舱通风系统的安装	installation of cargo cabin ventilation system
4	辅助通风活门的安装	installation of auxiliary vent valve
5	单向活门的安装	installation of check valve
6	流量分配活门的安装	installation of flow distribution valve
7	脚部加温开关的安装	installation of heating switch at foot pedal
8	7～9 框座舱压力调节系统的安装	installation of cabin pressure control system at frames 7～9
9	9～26 框座舱压力调节系统的安装	installation of cabin pressure control system at frames 9～26
10	座舱压力调节器的安装	installation of cabin pressure regulator
11	38～59 框座舱压力调节系统的安装	installation of cabin pressure control system at frames 38～59
12	27～39 框座舱压力调节系统的安装	installation of cabin pressure control system at frames 27～39
13	12 框排气活门的安装	installation of exhaust valve at frame 12
14	38 框排气活门的安装	installation of exhaust valve at frame 38
15	59 框排气活门的安装	installation of exhaust valve at frame 59
16	安全活门的安装	installation of safety valve
17	气滤的安装	installation of air filter
18	飞机制冷系统的安装	installation of aircraft refrigeration system
19	初级散热器的安装	installation of primary radiator
20	次级散热器的安装	installation of secondary radiator
21	冷凝器的安装	installation of condenser
22	水分离器的安装	installation of water separator
23	喷水器的安装	installation of water sprayer
24	涡轮冷却器的安装	installation of turbine cooler
25	地面空调车接头组件的安装	installation of connector assembly of ground air conditioner vehicle
26	座舱温度控制系统的安装	installation of cabin temperature control system
27	温控盒的安装	installation of temperature control box
28	驾驶舱温度传感器的安装	installation of cockpit temperature sensor
29	货舱温度传感器的安装	installation of cargo cabin temperature sensor

30	管路温度传感器的安装	installation of pipeline temperature sensor
31	调温电动活门的安装	installation of temperature-control electric valve
32	货舱温度传感器的安装	installation of cargo cabin temperature sensor

自动飞行　Auto Flight

1	速率陀螺盒的安装	installation of speed gyroscope box
2	中心垂直陀螺的安装	installation of center vertical gyroscope
3	驾驶仪操纵台的安装	installation of autopilot console
4	操纵手柄的安装	installation of control handle
5	飞行控制盒的安装	installation of flight control box
6	飞行控制盒电缆插座板的安装	installation of flight control box cable socket plate
7	高度差传感器的安装	installation of vertical difference sensor
8	航向联系盒的安装	installation of heading liaison box
9	导航系统继电器盒的安装	installation of navigation system relay box
10	导航系统继电器盒	navigation system relay box
11	液压油滤的安装	installation of hydraulic fluid filter

通信系统　Communications

1	短波电台系统	HF radio system
2	收发信机安装图	installation drawing of VHF/UHF radio set transceiver
3	预后选器安装图	installation drawing of pre/post-selector
4	天线调谐器安装图	installation drawing of antenna tuner
5	短波控制盒安装图	installation drawing of HF control box
6	短波继电器盒安装图	installation drawing of HF relay box
7	第一套短波天线安装图	installation drawing of the first HF antenna
8	第二套短波天线安装图	installation drawing of the second HF Antenna
9	电键安装图	installation drawing of key
10	高频穿壁绝缘子安装图	installation drawing of HF wall insulator
11	短波电台天线支柱安装图	installation drawing of HF radio antenna strut
12	超短波电台系统	VHF/UHF radio set system
13	超短波电台收发机安装图	installation drawing of VHF/UHF radio set transceiver
14	超短波电台控制盒安装图	installation drawing of VHF/UHF radio set control box
15	超短波电台收发天线安装图	installation drawing of VHF/UHF radio set transceiver antenna

16	机内通话器系统安装图	installation drawing of interphone system
17	接线盒安装图	installation drawing of terminal box
18	控制盒安装图	installation drawing of control box
19	内通扩展设备安装图	installation drawing of interphone expanding equipment
20	通讯接线板安装图	installation drawing of communication terminal board
21	头戴送受话器组挂放图	hanging drawing of headset monophone
22	左、右驾驶员无线电、通话按钮安装图	installation drawing of pilot/copilot radio/communication button
23	领航员通话脚踏按钮安装图	installation drawing of navigator communication pedal button
24	机械师通话脚踏按钮安装图	installation drawing of flight engineer communication pedal button
25	通讯员脚踏按钮安装图	installation drawing of communicator pedal button
26	领航员无线电按钮安装图	installation drawing of navigator radio button
27	通讯员无线电按钮安装图	installation drawing of communicator radio button
28	应急舱门处耳机话筒插座安装图	installation drawing of headset and microphone sockets at the emergency door
29	左驾驶员耳机、话筒插座安装图	installation drawing of pilot headset and microphone sockets
30	右驾驶员耳机、话筒插座安装图	installation drawing of copilot headset and microphone sockets
31	机械师耳机、话筒插座安装图	installation drawing of flight engineer headset and microphone sockets
32	领航员耳机、话筒插座安装图	installation drawing of navigator headset and microphone sockets
33	通讯员耳机、话筒插座安装图	installation drawing of communicator headset and microphone sockets
34	简易控制盒安装图	installation drawing of simplified control box
35	静电释放	static discharging
36	放电刷安装图	installation drawing of static discharger
37	接地锤安装图	installation drawing of grounding cone
38	座舱音频记录系统	cockpit audio recording system
39	座舱音频记录器安装图	installation drawing of cockpit audio recorder
40	音频监控器安装图	installation drawing of audio monitor
41	拾音器的安装	installation drawing of sound pickup

电源系统　Electrical Power

1	交流发电机的安装	installation of AC generator
2	交流调压器的安装	installation of AC voltage regulator
3	精调盒的安装	installation of fine adjustment box
4	交流过压保护器的安装	installation of AC over-voltage protector
5	交流转换盒的安装	installation of AC changeover box
6	自动定时机构的安装	installation of automatic timing mechanism
7	滤波器的安装	installation of filter
8	电容器和接线柱的安装	installation of capacitor and terminal post
9	单相变流机的安装	installation of single-phase inverter
10	三相变流器的安装	installation of three-phase inverter
11	交流电压表的安装	installation of AC voltmeter
12	交流电流表指示器的安装	installation of AC ammeter indicator
13	频率表的安装	installation of frequency meter
14	单相变流器的安装	installation of single-phase inverter
15	直流调压器在左、右翼根处的安装	installation of DC voltage regulators at left & right wing roots
16	直流调压器在起落架短舱的安装	installation of DC voltage regulator in LG nacelle
17	过压保护器的安装	installation of over-voltage protector
18	差动低限保护器的安装	installation of differential undervoltage relay
19	发电机激磁接触器的安装	installation of generator exciting contactor
20	电流表的安装	installation of ammeter
21	发房搭接线的安装	installation of bonding jumpers in engine nacelle
22	接地装置的安装	installation of grounding device
23	直流发电机电阻盒的安装	installation of DC generator resistance box
24	直流发电机电阻盒	DC generator resistance box
25	发动机接线盒的安装	installation of engine terminal box
26	直流发电机接线盒的安装	installation of DC generator terminal box
27	降压电阻接触器配电盒的安装	installation of PDB for voltage dropping resistor contactor
28	降压电阻接触器配电盒	PDB for voltage dropping resistor contactor
29	差动低限保护器的安装	installation of differential undervoltage relay
30	碱性蓄电池的安装	installation of alkaline battery
31	直流电压表的安装	installation of DC voltmeter
32	直流地面电源插座的安装	installation of DC ground power socket
33	起动电源插座的安装	installation of starting power socket
34	极性继电器的安装	installation of polarized relay

35	起动及电源接触器的安装	installation of starting & electrical power contactor
36	24V～48V 电压转换接触器的安装	installation of 24V～48V voltage changeover contactor
37	交流地面电源插座的安装	installation of AC ground power socket
38	螺旋桨及整流罩防冰加温配电盒的安装	installation of propeller & fairing anti-icing and heating power distribution box
39	螺旋桨及整流罩防冰加温配电盒	propeller & fairing anti-icing and heating power distribution box
40	大汇流条的安装	installation of big bus
41	起动汇流条的安装	installation of starting bus
42	左、右电源连通汇流条的安装	installation of L & R power connecting bus
43	前舱总配电盒的安装	installation of general power distribution box in forward cabin
44	前舱总配电盒	general PDB in forward cabin
45	仪表板和电压表保险丝配电盒的安装	installation of instrument panel and voltmeter fuse distribution box
46	仪表板和电压表保险丝配电盒	instrument panel and voltmeter fuse distribution box
47	通讯员配电盒的安装	installation of communicator distribution box
48	通讯员配电盒	communicator distribution box
49	液压泵配电盒的安装	installation of hydraulic pump distribution box
50	液压泵配电盒	hydraulic pump distribution box
51	起落架、襟翼、调整片配电盒的安装	installation of PDB for landing gears, flaps and trim tabs
52	起落架、襟翼、调整片配电盒	PDB of LG, flap and trim tab
53	23～24 框气密接线柱的安装	installation of airtight terminal post between frames 23～24
54	24～25 框气密接线柱的安装（典型）	installation of airtight terminal post between frames 24～25 (typical)
55	密封套的典型图	typical drawing of sealing sleeve
56	防撞灯配电盒的安装	installation of anti-collision light distribution box
57	防撞灯配电盒	anti-collision light power distribution box
58	42 框左电源接线盒的安装	installation of left power terminal box at frame 42
59	42 框电源接线盒	power terminal box at frame 42
60	42 框右电源接线盒的安装	installation of right power terminal box at frame 42
61	垂直尾翼防冰电源接线盒的安装	installation of vertical fin anti-icing power terminal box

62	水平尾翼防冰电源接线盒的安装	installation of horizontal tail anti-icing power terminal box
63	尾翼防冰电源接线盒	tail anti-icing power terminal box
64	57 框左复合供电配电盒的安装	installation of left compound PDB at frame 57
65	左复合供电配电盒	left compound PDB
66	57 框右复合供电配电盒的安装	installation of right compound PDB at frame 57
67	右复合供电配电盒	right compound PDB
68	空投配电盒的安装	installation of airdrop power distribution box
69	空投配电盒	airdrop power distribution box
70	应急空投配电盒的安装	installation of emergency airdrop power distribution box
71	应急空投配电盒	emergency airdrop power distribution box
72	中性气体瓶加温配电盒的安装	installation of neutral gas bottle heating power distribution box
73	中性气体瓶加温配电盒	neutral air bottle heating power distribution box
74	主起落架舱接线盒的安装	installation of MLG well terminal box
75	接线盒的安装	installation of terminal box
76	起动发电机配电盘	starter/generator power distribution box
77	上、下接触器板的安装	installation of upper & lower contactor board
78	上接触器板	upper contactor board
79	下接触器板	lower contactor board
80	右发动机电源配电盘	right engine power distribution board
81	左发动机电源配电盘	left engine power distribution board

设备　Equipment

1	驾驶员座椅安装	installation of pilot seat
2	领航员座椅安装	installation of navigator seat
3	机械员座椅安装	installation of flight engineer seat
4	通讯员座椅安装	installation of communicator seat
5	滤光镜、把手、文件袋、水杯座、保温水瓶座安装	installation of light filter, handle, document bag, cup holder and thermos bottle holder
6	水箱、烤箱加温配电板的安装	installation of water tank, oven heating power distribution board
7	水箱、烤箱加温配电板	water tank, oven heating power distribution board
8	卫生间安装	installation of lavatory
9	尾门厕所安装	installation of lavatory at rear cargo door
10	货运系统的安装	installation of cargo transportation system

11	电动绞车的安装	installation of electric winch
12	货运导轨安装	installation of cargo transportation guide rail
13	前限动装置	forward limit unit
14	双向限动锁	bidirectional limit lock
15	后向限动锁	backward limit lock
16	导向滑轮	guide pulley
17	装卸滑轮	loading/unloading pulley
18	分力接头	load distributing adapter
19	带编系留网	webbing tie-down net
20	拉紧带	fastening strap
21	带快卸锁系留锦丝带	tie-down polyamide webbing with quick-release lock
22	系留锦丝带	tie-down polyamide webbing
23	系留钢索	tie-down steel cable
24	伞兵座椅、指挥员座椅安装	installation of paratrooper seat and commander seat
25	医务工作桌安装	installation of medical work table
26	医务工作台安装	installation of medical work bench
27	电动信号枪安装	installation of electric signal gun
28	应急定位仪安装	installation of emergency locator
29	引伞挂钩、抛放装置安装	installation of extractor hook，and airdrop unit
30	空降挡风板安装	installation of parachuting wind-shield plate
31	空投滚棒系统安装	installation of airdrop roller system
32	侧导轨安装	installation of side guide rail
33	货舱滚道安装	installation of roller rail in cargo cabin
34	货桥滚道安装	installation of loading ramp roller rail
35	空投员电气操纵台的安装	installation of airdrop operator electric console
36	空投员电气操纵台	airdrop operator electric console
37	空投继电器盒的安装	installation of airdrop relay box
38	空投继电器盒	airdrop relay box
39	战术投放延时继电器盒的安装	installation of tactics airdrop time delay relay box
40	战术投放延时继电器盒	tactics airdrop time delay relay box
41	应急投放延时继电器盒的安装	installation of emergence airdrop time delay relay box
42	应急投放延时继电器盒	emergency airdrop time delay relay box
43	继电器盒的安装	installation of relay box

防火系统　Fire Protection

1	火警传感器在舱内的安装	installation of fire warning sensor in compartment
2	机身副油箱舱火警传感器的安装	installation of fire warning sensor in fuselage auxiliary fuel tank compartment
3	火警控制盒的安装	installation of fire warning control box
4	火警传感器在机翼和发动机短舱内的安装	installation of fire warning sensor in wings and engine nacelles
5	发动机内部火警传感器的安装	installation of fire warning sensor in engine
6	机身地板下灭火系统的安装	installation of fire extinguishing system under the floor of fuselage
7	机身中段地板上灭火系统的安装	installation of fire extinguishing system on middle fuselage floor
8	尾段灭火系统的安装	installation of fire extinguishing system of tail section
9	涡轮发电装置短舱灭火导管的安装	installation of fire extinguishing pipe in turbine generator nacelle
10	中外翼、外翼灭火系统的安装	installation of fire extinguishing system of inboard wing and outboard wing
11	1、2 发外部灭火系统的安装	installation of external fire extinguishing system of engines 1 and 2
12	3、4 发外部灭火系统的安装	installation of external fire extinguishing system of engines 3 and 4
13	发动机内部灭火系统的安装	installation of internal fire extinguishing system of engine
14	发动机及机翼火警控制盒的安装	installation of fire warning control boxes for engines and wings
15	WDZ 和副油箱火警控制盒的安装	installation of fire warning control boxes of WDZ and auxiliary fuel tank
16	迫降灭火开关的安装	installation of forced landing fire switch
17	灭火继电器盒的安装	installation of fire extinguishing relay box
18	单向阀的安装	installation of check valve
19	手提灭火瓶在驾驶舱内的安装	installation of portable fire bottle in cockpit
20	手提灭火瓶在 9 框上的安装	installation of portable fire bottle at frame 9
21	手提灭火瓶在 25 框上的安装	installation of portable fire bottle at frame 25
22	手提灭火瓶在 30 框上的安装	installation of portable fire bottle at frame 30
23	球形灭火瓶的安装	installation of ball fire bottle
24	8L 灭火瓶的安装	installation of fire bottle（8L）

25	电磁阀的安装	installation of solenoid valve
26	中性气体系统安装图	installation of neutral gas system
27	机身、中央翼中性气体系统的安装	installation of neutral gas system of fuselage and center wing
28	带开关的沉淀池	sedimentation tank with switch
29	中央翼后电磁开关组的安装	installation of rear solenoid switch group in center wing
30	中外翼、外翼中性气体系统的安装	installation of neutral gas system of inboard wing and outboard wing
31	中性气体瓶安装托架	installation bracket of neutral gas bottle
32	尾段中性气体系统的安装	installation of neutral gas system of tail section

飞行操纵系统　Flight Controls

1	副翼操纵系统的安装	installation of aileron control system
2	驾驶舱地板上摇臂及支架的安装	installation of rocker and bracket on the floor in cockpit
3	驾驶舱地板下摇臂及支架的安装	installation of rocker and bracket under the floor in cockpit
4	驾驶舱地板下摇臂及支架的安装	installation of rocker and bracket under the floor in cockpit
5	9框下摇臂及支架的安装	installation of lower rocker and bracket at frame 9
6	脚操纵台	foot pedal control stand
7	9框上摇臂及支架的安装	installation of upper rocker and bracket at frame 9
8	副翼中介摇臂及支架的安装	installation of aileron idler rocker and bracket
9	中央翼前梁中介摇臂及支架的安装	installation of idler rocker and bracket on center wing front spar
10	16肋副翼操纵摇臂及支架的安装	installation of aileron control rocker and bracket at rib 16
11	18肋处摇臂及支架的安装	installation of rocker and bracket at rib 18
12	20肋处摇臂及支架的安装	installation of rocker and bracket at rib 20
13	22肋处摇臂及支架的安装	installation of rocker and bracket at rib 22
14	副翼操纵系统拉杆(典型)	aileron control system pull rod (typical)
15	内扰流板操纵装置的安装	installation of internal spoiler control device
16	外扰流板操纵装置的安装	installation of external spoiler control device
17	II大梁上副翼立轴的安装	installation of upper aileron vertical shaft on spar II
18	副翼补偿片与调整片操纵机构	aileron balance tab and trim tab control mechanism

19	方向舵操纵系统安装图	installation drawing of rudder control system
20	脚蹬机构	pedal mechanism
21	右扇形件	right sector part
22	带摇臂的脚蹬	pedal with rocker
23	脚蹬	pedal
24	带摇臂的轴	shaft with rocker
25	带止动件的卡子	clamp with stop part
26	左扇形件	left sector part
27	脚蹬制动器	pedal brake
28	摇臂	rocker
29	带摇臂的轴	shaft with rocker
30	弹簧拉杆	spring pull rod
31	9 框地板下支架安装	installation of bracket under floor at frame 9
32	62 框舵机支架及升降舵制动机构的安装	installation of actuator bracket and elevator braking mechanism at frame 62
33	62 框后摇臂支架的安装	installation of rear rocker bracket at frame 62
34	下弹簧拉杆	lower spring pull rod
35	方向舵弹簧拉杆	rudder spring pull rod
36	方向舵偏角限制器	rudder deflection angle limiter
37	方向舵调整片操纵机构的安装	installation of rudder trim tab control mechanism
38	升降舵操纵系统的安装	installation of elevator control system
39	驾驶杆的安装	installation of control column
40	65 框摇管的安装	installation of rocker tube at frame 65
41	59 框气密接头	airtight joint at frame 59
42	59 框钢索分配器	steel cable distributor at frame 59
43	滑轮支架的安装	installation of pulley brackets
44	升降舵调整片操纵机构的安装	installation of elevator trim tab control mechanism
45	升降舵调整片钢索、舵面锁钢索的安装	installation of elevator trim tab steel cable and control lock steel cable
46	升降舵调整片偏转机构的安装	installation of elevator trim tab deflection mechanism
47	襟翼操纵系统的安装	installation of flap control system
48	襟翼液压操纵系统	flap hydraulic control system
49	液压传动机构的安装	installation of hydraulic transmission mechanism
50	传动杆	transmission rod
51	支架的安装	installation of bracket
52	中央翼 II 大梁后 2～3 肋间支架的安装	installation of bracket between ribs 2～3 behind center wing spar II

53	中央翼 II 大梁后 3～4 肋间支架的安装	installation of bracket between ribs 3～4 behind center wing spar II
54	中央翼 II 大梁后 6～7 肋间支架的安装	installation of bracket between ribs 6～7 behind center wing spar II
55	中央翼 II 大梁后 7A～8 肋间支架的安装	installation of bracket between ribs 7A～8 behind center wing spar II
56	中央翼 II 大梁后 9～10 肋间支架的安装	installation of bracket between ribs 9～10 behind center wing spar II
57	襟翼收放机构(内、外)	flap extension/retraction mechanism (internal and external)
58	襟翼位置机构的安装	installation of flap position mechanism
59	电磁阀的安装	installation of solenoid valve
60	液压马达的安装	installation of hydraulic motor
61	液压锁的安装	installation of hydraulic lock
62	转换活门的安装	installation of changeover valve
63	舵面锁操纵机构的安装	installation of control lock control mechanism
64	方向舵锁机构	rudder lock mechanism
65	副翼制动钢索支架安装	installation of aileron brake steel cable bracket

燃油系统　Fuel

1	机翼软油箱的安装	installation of wing bag tank
2	机身软油箱的安装	installation of fuselage bag tank
3	整体油箱的安装(油箱附件安装)	installation of integral fuel tanks (installation of tank accessories)
4	左 1 号软油箱的安装	installation of bag tank 1 at the left
5	右 1 号软油箱的安装	installation of bag tank 1 at the right
6	2 号软油箱的安装	installation of bag tank 2
7	3 号软油箱的安装	installation of bag tank 3
8	4 号软油箱的安装	installation of bag tank 4
9	5 号软油箱的安装	installation of bag tank 5
10	6 号软油箱的安装	installation of bag tank 6
11	7 号软油箱的安装	installation of bag tank 7
12	8 号软油箱的安装	installation of bag tank 8
13	9 号软油箱的安装	installation of bag tank 9
14	10 号软油箱的安装	installation of bag tank 10
15	11 号软油箱的安装	installation of bag tank 11
16	12 号软油箱的安装	installation of bag tank 12
17	13 号软油箱的安装	installation of bag tank 13

18	16 号软油箱的安装	installation of bag tank 16
19	机翼通气系统的安装	installation of wing vent system
20	机身通气系统的安装	installation of fuselage vent system
21	油塞	fuel plug
22	余油箱的安装	installation of residual fuel tank
23	压力加油系统的安装	installation of pressure refueling system
24	加油接头的安装	installation of filler connection
25	单向活门的安装	installation of check valve
26	起动供油系统的安装	installation of starting fuel supply system
27	机翼供油系统的安装	installation of wing fuel supply system
28	机身副油箱供油系统的安装	installation of fuel supply system of fuselage auxiliary fuel tank
29	输油泵的安装	installation of fuel transfer pump
30	连通开关的安装	installation of crossfeed valve
31	防火开关的安装	installation of fire shut off valve
32	单向活门的安装	installation of check valve
33	油滤的安装	installation of fuel filter
34	燃油电动开关的安装	installation of fuel electrical valve
35	放油开关的安装	installation of fuel drain valve
36	油箱放油漏斗支架的安装	installation of fuel tank defueling funnel bracket
37	油量表遥控开关的安装	installation of remote switch for fuel quantity gauge
38	耗油控制器的安装	installation of fuel consumption controller
39	加油控制器的安装	installation of refueling controller
40	逆变电源盒的安装	installation of inverted power box
41	压力信号器的安装	installation of pressure annunciator

液压源　Hydraulic Power

1	右液压系统的安装	installation of right hydraulic system
2	液压油箱的安装	installation of hydraulic reservoir
3	液压泵的安装	installation of hydraulic pump
4	安全阀的安装	installation of safety valve
5	液压蓄压瓶的安装	installation of hydraulic accumulator
6	缓冲蓄压瓶的安装	installation of buffer accumulator
7	单向阀的安装	installation of check valve
8	散热器的安装	installation of radiator
9	最低流量节流阀的安装	installation of minimum flow throttle valve
10	分离接头的安装	installation of separation joint

11	左液压系统的安装	installation of left hydraulic system
12	液压泵的安装	installation of hydraulic pump
13	卸荷阀的安装	installation of unload valve
14	刹车蓄压瓶的安装	installation of brake accumulator
15	压力转换活门的安装	installation of pressure changeover valve
16	连通开关的安装	installation of crossfeed valve
17	单向阀的安装	installation of check valve
18	液压油滤的安装	installation of hydraulic filter
19	分离接头的安装	installation of separation joint
20	油箱空气增压系统的安装	installation of hydraulic reservoir air pressurization system
21	减压阀的安装	installation of unload valve
22	节流阀的安装	installation of restrictor valve
23	干燥过滤器的安装	installation of desiccation filter
24	气滤的安装	installation of air filter
25	安全阀的安装	installation of safety valve
26	放气活门的安装	installation of air bleed valve
27	单向阀的安装	installation of check valve
28	空气单向阀在发动机短舱内的安装	installation of air check valve in engine nacelle
29	空气单向阀在附件板上的安装	installation of air check valve on accessory plate
30	地面液压接头的安装	installation of ground hydraulic joint
31	手摇泵-电动泵系统的安装	installation of hand/electric pump system
32	手摇泵的安装	installation of hand pump
33	电动泵的安装	installation of electric pump
34	三位分配开关的安装	installation of three-position distribution cock
35	七位分配开关的安装	installation of seven-position distribution cock
36	安全阀的安装	installation of safety valve
37	单向阀的安装	installation of check valve
38	油量传感器的安装	installation of hydraulic fluid quantity sensor
39	右系统压力传感器的安装	installation of pressure sensor of right system
40	驾驶舱 7 框左侧附件板压力表的安装	installation of pressure gauge on left accessory plate at frame 7 in cockpit
41	42 框地面操纵台压力表的安装	installation of pressure gauge on ground console at frame 42

除雨和防冰　Ice and Rain Protection

1	机翼前缘防冰系统的安装	installation of anti-icing system on wing leading edge

2	机翼前缘防冰活门的安装	installation of anti-icing valve on wing leading edge
3	补偿器的安装	installation of compensator
4	尾翼防冰定时器的安装	installation of tail anti-icing timing mechanism
5	尾翼加温信号控制盒的安装	installation of tail heating signal control box
6	进气道加温系统的安装	installation of air intake heating system
7	进气道防冰活门的安装	installation of air intake anti-icing valve
8	全压加温控制盒的安装	installation of pitot heating control box
9	玻璃加温控制盒的安装	installation of windshield heating control box
10	玻璃加温定时机构的安装	installation of windshield heating timing mechanism
11	玻璃加温自耦变压器的安装	installation of windshield heating auto-transformer
12	玻璃加温配电盒的安装	installation of windshield heating PDB
13	玻璃加温配电盒	windshield heating PDB
14	风挡雨刷限流开关的安装	installation of windshield wiper restrictor valve
15	风挡雨刷转动装置的安装	installation of windshield wiper rotating device
16	螺旋桨及整流罩加温定时机构的安装	installation of propeller and fairing heating timing mechanism
17	螺旋桨加温信号控制盒的安装	installation of propeller heating signal control box
18	12～13 框顶棚结冰信号器随动器的安装	installation of icing annunciator follower on ceiling between frames 12～13
19	结冰信号器受感器的安装	installation of icing annunciator sensor
20	结冰信号器控制盒的安装	installation of icing signal control box

显示/记录系统　Indicating/Recording Systems

1	右仪表板	copilot instrument panel
2	中央仪表板	center instrument panel
3	顶部操纵台	overhead console
4	顶部操纵台上面板及发动机振动仪表板	upper panel and engine vibration instrument panel on the overhead console
5	115V 交流配电盒	115V AC power distribution box
6	36V 交流配电盒	36V AC power distribution box
7	电气员仪表板	electric mechanic instrument panel
8	通讯员自动开关板	communicator circuit breaker board
9	通讯员直流仪表板	communicator DC instrument panel
10	通讯员交流仪表板	communicator AC instrument panel
11	领航员仪表板	navigator instrument panel

12	通讯员操纵板	communicator control panel
13	加油配电盒的安装	installation of refueling PDB
14	加油配电盒	refueling distribution box
15	右操纵台电气设备的安装	installation of electrical equipment on copilot console
16	左操纵台电气设备的安装	installation of electrical equipment on pilot console
17	领航员侧面盖板	navigator side cover plate
18	领航员右操纵台	navigator right console
19	无线电罗盘控制板	ADF control panel
20	加速度传感器的安装	installation of acceleration sensor
21	飞行参数采集器的安装	installation of flight data collector
22	飞行参数记录器的安装	installation of flight data recorder
23	三轴加速度计的安装	installation of three-axis accelerometer
24	副翼角位移传感器电缆插座的安装	installation of aileron angular displacement sensor cable socket
25	升降舵角位移传感器电缆插座的安装	installation of elevator angular displacement sensor cable socket
26	方向舵角位移传感器电缆插座的安装	installation of rudder angular displacement sensor cable socket
27	飞参外场检测处理机插座的安装	installation of flight data field test processor socket
28	飞行参数记录器接线盒的安装	installation of flight data recorder terminal box
29	前舱用氧高度信号器的安装	installation of oxygen supply altitude annunciator in forward cabin
30	货舱用氧高度信号器的安装	installation of oxygen supply altitude annunciator in cargo cabin

起落装置　Landing Gears

1	主起落架	main landing gear
2	主起落架缓冲支柱	damper strut of MLG
3	前起落架	nose landing gear
4	前起落架可折斜支柱	foldable inclined strut of NLG
5	前轮转弯随动系统	nose wheel steering follow-up system
6	扭力臂组件	torque arm assembly
7	四连杆机构组件	assembly of four link mechanism
8	驾驶舱应急放起落架液压操纵系统的安装	installation of hydraulic control system for emergency LG extension in cockpit

9	前起落架及舱门收放系统的安装	installation of NLG and door retraction/extension system
10	前起落架舱门收放机构	retraction/extension mechanism of NLG door
11	主起落架收放系统的安装	installation of MLG retraction/extension system
12	主起落架舱门收放机构	retraction/extension mechanism of MLG doors
13	导管在主起落架上的安装	installation of pipe on MLG
14	下位锁作动筒导管组件	pipe assembly of downlock actuator
15	电磁阀的安装	installation of solenoid valve
16	前起落架舱门收放作动筒的安装	installation of retraction/extension actuator for NLG door
17	主起落架上位锁安装	installation of MLG uplock
18	前起落架上位锁组件	assembly of NLG uplock
19	电磁阀的安装	installation of solenoid valve
20	单向阀安装图	installation drawing of check valve
21	单向阀的安装	installation of check valve
22	转换活门的安装	installation of changeover valve
23	电磁阀在主起舱的安装	installation of solenoid valve in MLG well
24	电磁阀在前起落架舱的安装	installation of solenoid valve in NLG Well
25	协调活门的安装	installation of coordination valve
26	协调活门的安装	installation of coordination valve
27	液压锁的安装	installation of hydraulic lock
28	主起舱门作动筒的安装	installation of MLG door actuators
29	液压手动阀的安装	installation of hydraulic handle valve
30	液压开关的安装	installation of hydraulic valve
31	回油转换活门的安装	installation of fluid return changeover valve
32	前起落架应急放下上锁系统	locking system for NLG emergency extension
33	前起落架应急放下开锁系统	unlocking system for NLG emergency extension
34	支架及应急放下开锁机构	bracket and emergency extension unlocking mechanism
35	主起落架放下应急上锁系统安装	installation of locking system for MLG emergency extension
36	主起落架上位锁应急开锁系统的安装	installation of emergency unlocking system of MLG uplock
37	刹车系统的安装	installation of braking system
38	刹车系统导管和软管在主起落架上的安装	installation of braking system pipe and hose on MLG
39	减压阀的安装	installation of pressure reducing valve
40	转换活门的安装	installation of changeover valve
41	定量器的安装	installation of proportioner

42	防滞刹车电磁活门的安装	installation of solenoid valve for braking with anti-jamming
43	刹车活门的安装	installation of brake valve
44	惯性传感器安装图的安装	installation of inertial sensor
45	应急刹车活门的安装	installation of emergency brake valve
46	电磁阀的安装	installation of solenoid valve
47	主起机轮组件的安装	installation of MLG wheel assembly
48	前起机轮组件的安装	installation of NLG wheel assembly
49	转弯操纵系统的安装	installation of steering control system
50	转弯操纵系统在前起落架舱的安装	installation of steering control system in NLG well
51	前轮转弯钢索系统的安装	installation of nose wheel steering cable system
52	电磁阀的安装	installation of solenoid valve
53	舵板操纵机构的安装	installation of rudder pedal control mechanism
54	分配活门的安装	installation of distribution valve
55	安全阀的安装	installation of safety valve
56	转弯作动筒的安装	installation of steering actuator
57	前轮操纵装置的安装	installation of nose wheel steering device
58	转弯操纵系统在前起落架上的安装	installation of steering control system on nose landing gear
59	手柄在左操纵台上的安装	installation of handle on pilot console
60	滑轮组件在5框前的安装	installation of pulley assembly in front of frame 5
61	滑轮组件在5框的安装	installation of pulley assembly at frame 5
62	滑轮组件在8框的安装	installation of pulley assembly at frame 8
63	滑轮组件在9框的安装	installation of pulley assembly at frame 9
64	终点开关在前起上防扭臂上的安装	installation of limit switch on NLG upper torque arm
65	前起落架终点开关的安装	installation of NLG limit switch
66	终点开关在前起可折斜支柱上的安装	installation of limit switch on NLG foldable inclined strut
67	终点开关在主起落架防扭臂上的安装	installation of limit switch on main landing gear torque arm
68	主起落架终点开关的安装	installation of main landing gear limit switch
69	起落架整流器信号盒的安装	installation of LG rectifier signal box
70	断开舱门自动关闭操纵开关的安装	installation of control switch for disconnecting automatic door closing
71	压力传感器的安装	installation of pressure sensor

照明系统　Lights

1	领航员仪表板和通讯员右侧荧光灯的安装	installation of fluorescent lights on navigator instrument panel and at the right of communicator
2	领航员右操纵台和 7 框左侧荧光灯的安装	installation of fluorescent lights on navigator right console and at the left of frame 7
3	领航员左操纵台荧光灯的安装	installation of fluorescent light on navigator left console
4	驾驶杆荧光灯的安装	installation of fluorescent lights on control stick
5	通讯员左侧荧光灯的安装	installation of fluorescent light at the left of communicator
6	7 框右侧荧光灯的安装	installation of fluorescent light at the right of frame 7
7	上配电盘荧光灯的安装	installation of fluorescent lights on upper power distribution board
8	2 框左侧、右侧、8 框右侧座舱灯的安装	installation of cabin light at the left & right sides of frame 2 and right side of frame 8
9	货舱灯开关盒	switch box of cargo cabin lights
10	2a 框左侧座舱灯的安装	installation of cabin light at the left of frame 2a
11	3 框领航员仪表板座舱灯的安装	installation of cabin light on navigator instrument panel at frame 3
12	3 框上部座舱灯的安装	installation of cabin light on the upper section of frame 3
13	侧风挡滑轨上座舱灯的安装	installation of cabin light on side windshield sliding rail
14	8 框右上部座舱灯和荧光灯的安装	installation of cabin light and fluorescent light at the upper right of frame 8
15	2a 框左侧顶灯的安装	installation of dome light at the left of frame 2a
16	右驾驶员座椅下工作灯的安装	installation of operating light under copilot seat
17	11 框两侧顶灯的安装	installation of dome lights at the left & right of frame 11
18	11 框中间顶灯的安装	installation of dome light at the middle of frame 11
19	货舱白顶灯的安装	installation of white dome lights in cargo cabin
20	货舱蓝顶灯的安装	installation of blue dome lights in cargo cabin
21	货舱灯开关盒的安装	installation of switch box for cargo cabin lights
22	货舱配电盒照明灯的安装	installation of light for cargo cabin PDB
23	43～44 框左侧工作灯的安装	installation of operating lights at the left of frames

		43～44
24	53～64 框顶灯的安装	installation of dome lights between frames 53～64
25	43 框右侧、44 框左侧装货照明灯的安装	installation of cargo loading light at the right of frame 43 and left of frame 44
26	49 框空降信号灯和尾舱顶灯开关的安装	installation of paradrop signal lights and tail cabin dome light switch at frame 49
27	25 框空投信号灯的安装	installation of airdrop signal lights at frame 25
28	48～49 框两侧空投信号灯的安装	installation of airdrop signal lights at the left & right of frames 48～49
29	护送舱地板下照明灯的安装	installation of lights under floor in escort cabin
30	25 框地板下工作灯的安装	installation of operating light under floor of frame 25
31	机身中段地板下照明灯的安装	installation of lights under the floor in the middle fuselage
32	左、右起落架短舱照明灯的安装	installation of lights for left & right LG well
33	右起落架短舱加油配电盒照明灯的安装	installation of refuelling PDB light in right LG well
34	主起落架舱照明灯的安装	installation of lights in MLG well
35	驾驶舱应急门着陆滑行灯的安装	installation of cockpit emergency door landing & taxiing light
36	左、右着陆滑行灯的安装	installation of left & right landing & taxiing lights
37	下防撞灯的安装	installation of lower anti-collision light
38	上防撞灯的安装	installation of upper anti-collision light
39	后防撞灯的安装	installation of rear anti-collision light
40	下防撞灯电源控制盒的安装	installation of power control box for lower anti-collision light
41	上防撞灯电源控制盒的安装	installation of power control box for upper anti-collision light
42	后防撞灯电源控制盒的安装	installation of power control box for rear anti-collision light
43	翼尖航行灯的安装	installation of wing-tip navigation lights
44	尾部航行灯的安装	installation of tail navigation light
45	航行灯、防撞灯控制盒的安装	installation of control box for navigation light & anti-collision light
46	航行灯、防撞灯控制盒	control box for navigation light & anti-collision light
47	1、4 发照明灯的安装	installation of lights for engines 1 and 4
48	2、3 发照明灯的安装	installation of lights for engines 2 and 3

49	平尾照明灯的安装	installation of light for horizontal tail
50	机翼编队灯及插座的安装	installation of wing formation lights and sockets
51	机身上壁板编队灯的安装	installation of formation lights on upper fuselage panels
52	机身下壁板编队灯的安装	installation of formation lights on lower fuselage panels
53	9 框应急灯的安装	installation of emergency light at frame 9
54	39~40 框应急灯的安装	installation of emergency light at frames 39~40
55	出口标志灯的安装	installation of exit lights

导航系统　Navigation

1	全静压导管安装	installation of pitot-static tube
2	全静压转换开关的安装	installation of pitot-static changeover switch
3	沉淀器	precipitator
4	静压 I 孔的安装	installation of static port I
5	静压 II 孔的安装	installation of static port II
6	静压系统防护装置的安装	installation of protective device in static system
7	全压受感器的安装	installation of pitot sensor
8	大气温度传感器的安装	installation of air temperature sensor
9	大气数据计算机的安装	installation of air data computer
10	温度传感器的安装	installation of temperature sensor
11	航姿放大器的安装	installation of attitude heading amplifier
12	全姿态组合陀螺的安装	installation of all-attitude combined gyroscope
13	大气数据系统继电器盒的安装	installation of relay box in air data system
14	磁航向传感器的安装	installation of magnetic heading sensor
15	电源变压器的安装	installation of power transformer
16	配电盒的安装	installation of power distribution box
17	继电器盒的安装	installation of relay box
18	航姿接线盒的安装	installation of attitude heading terminal box
19	驾驶舱磁罗盘的安装	installation of magnetic compass in cockpit
20	领航员磁罗盘的安装	installation of navigator magnetic compass
21	伏尔/仪表着陆系统	VOR/instrument landing system (ILS)
22	接收机的安装	installation of receiver
23	控制盒的安装	installation of control box
24	下滑天线的安装	installation of gliding antenna
25	天线分配器的安装	installation of antenna distributor
26	信标天线的安装	installation of beacon antenna
27	伏尔/航向天线的安装	installation of VOR/LOC antenna

28	耦合器的安装	installation of coupler
29	气象雷达系统	weather radar system
30	天线的安装	installation of antenna
31	收发机的安装	installation of transceiver
32	显示器的安装	installation of display
33	波导的安装	installation drawing of waveguide
34	惯性导航部件的安装	installation of inertial navigation component
35	激光惯性导航系统天线的安装	installation of antenna for laser inertial navigation system
36	近地告警计算机的安装	installation of ground proximity warning computer
37	近地告警天线的安装	installation of GPWS antenna
38	无线电高度表系统	radio altimeter system
39	高度表收发机的安装	installation of altimeter transceiver
40	天线的安装	installation of antenna
41	2101 I/O GPS 导航仪	2101 I/O GPS navigation instrument
42	接收/显示单元的安装	installation of receiver/display unit
43	GPS 天线的安装	installation of GPS antenna
44	空中交通告警和防撞系统	Traffic alert and collision avoidance system
45	TCAS 收发机的安装	installation of TCAS transceiver
46	TCAS 指示器的安装	installation of TCAS indicator
47	TCAS 控制盒的安装	installation of TCAS control box
48	空管应答机收发机的安装	installation of transceiver of ATC transponder
49	天线转换器的安装	installation of antenna adapter
50	空管应答机控制盒的安装	installation of control box of ATC transponder
51	TCAS 顶部天线的安装	installation of TCAS top antenna
52	TCAS 底部天线的安装	installation of TCAS bottom antenna
53	空管应答机天线的安装	installation of ATC transponder antenna
54	扬声器的安装	installation of loudspeaker
55	无线电罗盘系统	ADF system
56	罗盘接收机的安装	installation of compass receiver
57	控制盒的安装	installation of control box
58	组合天线的安装	installation of combined antenna
59	信号阻尼器的安装	installation of signal damper
60	塔康系统	TACAN system
61	收发机的安装	installation of transceiver
62	控制盒的安装	installation of control box
63	天线的安装	installation of antenna
64	综合显示器的安装	installation of comprehensive display

65	测距器系统	distance measuring equipment（DME）system
66	收发机的安装	installation of transceiver
67	转换器的安装	installation of converter
68	指示器的安装	installation of indicator
69	天线的安装	installation of antenna

氧气系统　Oxygen

1	驾驶舱氧气设备的安装	installation of oxygen equipment in cockpit
2	软管	hose
3	尾段氧气设备的安装	installation of oxygen equipment in tail section
4	左、右驾驶员氧气调节器的安装	installation of pilot and copilot oxygen regulators
5	领航员、通讯员氧气调节器的安装	installation of navigator and communicator oxygen regulators
6	机械师氧气调节器的安装	installation of flight engineer oxygen regulator
7	氧气开关的安装	installation of oxygen valve
8	玻璃钢氧气瓶的安装	installation of oxygen bottle made of glass fiber reinforced plastic
9	货舱氧气设备的安装	installation of oxygen equipment in cargo cabin
10	充氧仪表板	oxygen charging instrument panel
11	货舱氧气仪表板的安装	installation of oxygen instrument panel in cargo cabin
12	机上充氧软管	onboard oxygen charging hose
13	氧气减压设备的安装	installation of oxygen pressure reducing equipment
14	氧气调节器的安装	installation of oxygen regulator
15	氧气减压器的安装	installation of oxygen pressure reducer
16	氧气开关的安装	installation of oxygen valve
17	氧源开关的安装	installation of oxygen source valve
18	氧气压力表的安装	installation of oxygen pressure gauge
19	氧源压力表的安装	installation of oxygen source pressure gauge
20	氧气压力表的安装	installation of oxygen pressure gauge
21	货舱活动氧气瓶的安装	installation of portable oxygen bottle in cargo cabin

引气系统　Bleed Air

1	引气系统的安装	installation of air bleed system
2	压力调节/关断活门的安装	installation of pressure regulating/shut off valve

3	引气超压开关的安装	installation of air bleed overpressure valve
4	单向活门的安装	installation of check valve
5	压力传感器的安装	installation of pressure sensor

机载辅助动力装置　　Airborne Auxiliary Power

1	撑杆、斜撑杆及减震接头的安装	installation of bracing, inclined bracing and shock joint
2	电动机构的安装	installation for electrical mechanism
3	涡轮发电装置的安装	installation of turbine generating device
4	起动箱的安装	installation of starting box
5	电机通风罩的安装	installation of generator ventilation hood
6	转速传感器、排气温度传感器的安装	installation of rotating speed sensor and exhaust gas temperature sensor
7	排气管的安装	installation of exhaust pipe
8	通气油箱的安装	installation of air-vented tank
9	起动机加温管路的安装	installation of starter heating pipeline

舱门　　Doors

1	舱门	doors
2	登机门	entry door
3	登机门安装图	installation drawing of entry door
4	登机门锁机构	entry door lock mechanism
5	下应急舱门	bottom emergency door
6	下应急舱门安装图	installation drawing of bottom emergency door
7	下应急舱门锁机构	lock mechanism of bottom emergency door
8	上应急口盖	top emergency exit hatch
9	上应急口盖锁机构	lock mechanism of top emergency exit hatch
10	前应急门	front emergency door
11	后应急门	rear emergency door
12	前应急门锁机构	lock mechanism of front emergency door
13	后应急门锁机构	lock mechanism of rear emergency door
14	下应急舱门操纵系统的安装	installation of control system of bottom emergency door
15	手动阀的安装	installation of manual valve
16	分配活门的安装	installation of distribution valve
17	延迟活门的安装	installation of delay valve
18	下应急舱门收放作动筒的安装	installation of lower emergency door actuators

19	地板门锁作动筒的安装	installation of floor door lock actuator
20	下应急舱门密封系统的安装	installation of sealing system of bottom emergency door
21	冷气操纵开关的安装	installation of pneumatic control valve
22	空气减压器的安装	installation of air pressure reducer
23	冷气瓶的安装	installation of pneumatic bottle
24	压力表的安装	installation of pressure gauge
25	单向活门的安装	installation of check valve
26	货舱后大门	rear cargo door
27	货舱后大门上位锁安装图	installation drawing of rear cargo door uplock
28	货舱后大门下位锁系统	rear cargo door downlock system
29	货舱大门收放系统的安装	installation of cargo door extension/retraction system
30	电磁阀的安装	installation of solenoid valve
31	分配活门的安装	installation of distribution valve
32	延迟活门的安装	installation of delay valve
33	液压锁的安装	installation of hydraulic lock
34	减压活门的安装	installation of pressure reducing valve
35	流量调节阀的安装	installation of flow control valve
36	安全阀的安装	installation of safety valve
37	节流活门的安装	installation of restrictor valve
38	货桥锁作动筒的安装	installation of cargo ramp lock actuator
39	货桥作动筒的安装	installation of cargo ramp actuator
40	后大门作动筒的安装	installation of rear door actuator
41	转换活门的安装	installation of changeover valve
42	手动阀的安装	installation of manual valve
43	七位分配开关的安装	installation of seven-position distributing cock
44	单向阀的安装	installation of check valve
45	节流阀的安装	installation of throttle valve
46	货桥大门	loading ramp door
47	货桥大门锁机构	lock mechanism of loading ramp door
48	机械指示杆	mechanical indicating bar
49	9框门	frame 9 door
50	9框门安装图	installation drawing of frame 9 door
51	9框门锁安装图	installation drawing of frame 9 door lock
52	59框出入口盖安装图	installation of access cover of frame 59
53	登机门微动开关的安装	installation of entry door microswitch
54	压差传感器及其管路的安装	installation of differential pressure sensor and its pipeline

55	驾驶员下应急舱门前微动开关的安装	installation of cockpit bottom emergency door front microswitch
56	主起落架舱门	main LG door
57	驾驶员下应急舱门后微动开关的安装	installation of cockpit bottom emergency door rear microswitch
58	主起落架舱门安装图	installation drawing of main LG door
59	前起落架舱门	nose LG door
60	前起落架舱门安装图	installation of nose LG door

机身　Fuselage

1	机身总图(工艺分离面图)	general drawing of fuselage（technology separation surface）
2	机身口盖分布总图	general view of fuselage access cover distribution
3	机身前段总图	general drawing of front fuselage
4	机身前段整流包皮	front fuselage fairing
5	前雷达罩	front radome
6	下应急门框	bottom emergency door frame
7	前起支座	NLG support abutment
8	前起斜支座	NLG inclined support abutment
9	9 框	frame 9
10	13 框对接	butt joint at frame 13
11	驾驶台	control stand
12	驾驶舱地板口盖	cockpit floor access cover
13	9~13 框货舱地板口盖	floor access covers in cargo cabin at frames 9~13
14	9~13 框顶部装饰板	top decorative panel at frames 9~13
15	9~13 框左侧装饰板	decorative panel at the left of frames 9~13
16	9~13 框右侧装饰板	decorative panel at the right of frames 9~13
17	机身中段总图	general drawing of middle fuselage
18	25 框	frame 25
19	27 框主起接头	MLG joint at frame 27
20	30 框	frame 30
21	13~30 框顶部装饰板	top decorative panel at frames 13~30
22	30~41 框顶部装饰板	top decorative panel at frames 30~41
23	13~41 框左侧装饰板	left decorative panel at frames 13~41
24	13~41 框右侧装饰板	right decorative panel at frames 13~41
25	前应急窗门门框	front emergency exit hatch frame
26	后应急窗门门框	rear emergency exit hatch frame
27	16~17 框和 34~35 框货舱地板口盖	cargo cabin floor access covers at frames 16~17

		and frames 34～35
28	主起应急观察口盖	main LG emergency observation access cover
29	36～37 框货舱地板口盖	cargo cabin floor access cover at frames 36～37
30	抗压油箱舱口盖	anti-stress fuel tank cabin access cover
31	左主起落架短舱口盖示意图	schematic diagram of left MLG well access cover
32	口盖	access cover
33	散热器口盖	radiator access cover
34	充氧口盖	oxygen charging access cover
35	搭地线口盖	grounding wire access cover
36	支点加油口盖	fulcrum filling access cover
37	护罩	shield
38	主起落架护板	MLG door protective plate
39	口盖(左二)	access cover (No. 2 at the left)
40	口盖(左三)	access cover (No. 3 at the left)
41	起动机检查口盖	starter inspection access cover
42	起动机口盖固定锁	starter access cover fixing lock
43	起动机尾喷管口框	starter tailpipe access frame
44	口盖(左四)	access cover (No. 4 at the left)
45	口盖(左五)	access cover (No. 5 at the left)
46	蓄电瓶口盖	battery access cover
47	右主起落架短舱口盖示意图	schematic diagram of right MLG well access cover
48	大口盖	large access cover
49	中央翼整流罩前段	front fairing of center wing
50	中央翼整流罩中段	middle fairing of center wing
51	中央翼整流罩后段	rear fairing of center wing
52	机身尾段总图	general view of tail fuselage
53	尾段侧大梁	side spar in the tail section
54	41 框对接	butt joint at frame 41
55	尾段背鳍	dorsal fin in tail section
56	机尾尾段内蒙皮	tail fuselage inner skin
57	尾段地板口盖	tail floor access cover
58	尾舱	tail cabin
59	尾舱地板口盖	tail cabin floor access cover
60	65 框对接	butt joint at frame 65

发动机短舱　Engine Nacelle

| 1 | 发动机短舱 | engine nacelle |
| 2 | 发动机短舱前段 | front section of engine nacelle |

3	减速器整流罩	reduction gear fairing
4	发动机短舱中段上盖	upper cover at middle section of engine nacelle
5	发动机短舱中段上盖在发动机上的固定支架	fixed support of upper cover at mid section on engine nacelle
6	发动机短舱中段下盖	lower cover at middle section of engine nacelle
7	发动机短舱中段下盖在发动机上的固定支架(左侧)	fixed support of lower cover at mid section on engine nacelle (left side)
8	发动机短舱中段下盖在发动机上的固定支架(右侧)	fixed support of lower cover at mid section on engine nacelle (right side)
9	散热器排气风门	radiator exhaust shutter
10	承力段	load-bearing segment
11	发动机短舱中段侧盖连杆锁	side cover link lock at middle section of engine nacelle
12	发动机短舱中段侧盖	side cover at middle section of engine nacelle
13	尾舱整流罩	tail cabin fairing
14	尾舱整流罩上包皮	upper cowling of tail cabin fairing
15	发动机架后段安装图(在I大梁上的安装)	installation drawing of engine rear mount on spar I
16	发动机架后段安装图(在发动机短舱4框上的安装)	installationdrawing of engine rear mount on engine nacelle at frame 4
17	发动机短舱内发尾罩	tail cowling of inboard engine nacelle
18	发动机短舱外发尾罩	tail cowling of outboard engine nacelle

尾翼　Tail

1	尾翼	tail
2	水平尾翼与机身对接	butt joint between horizontal tail and fuselage
3	垂直尾翼与机身对接	butt joint between vertical tail and fuselage
4	水平尾翼	horizontal tail
5	水平安定面前缘	leading edge of horizontal stabilizer
6	平尾翼尖	horizontal tail tip
7	水平安定面后部口盖	rear access covers of horizontal stabilizer
8	升降舵	elevator
9	升降舵悬挂支架	elevator suspension arm
10	升降舵调整片	elevator trim tab
11	垂直安定面	vertical stabilizer
12	垂直安定面前缘	vertical stabilizer leading edge
13	垂尾整流包皮	vertical fin fairing
14	垂直安定面翼尖整流罩	vertical stabilizer wing tip fairing

15	垂直安定面后部口盖	rear access cover of vertical stabilizer
16	方向舵	rudder
17	方向舵悬挂支臂	rudder suspension arm
18	方向舵弹簧补偿片	rudder spring balance tab
19	方向舵调整片	rudder trim tab

窗　Windows

1	窗分布图	window distribution drawing
2	机头罩玻璃的安装	installation of nose fairing glass
3	领航舱玻璃的安装	installation of navigation cabin glass
4	驾驶舱天窗骨架玻璃的安装	installation of cockpit skylight framework glass
5	通风窗的安装	installation of ventilation window
6	通风窗	ventilation window
7	9～13框侧壁观察窗的安装（典型）	installation of observation window on side panel of frames 9～13（typical）
8	9～13框侧壁圆观察窗玻璃组件	round observation window organic glass assembly on side panel of frames 9～13
9	13～41框侧壁圆观察窗的安装（典型）	installation of round observation window on side panel of frames 13～41（typical）
10	13～41框侧壁圆观察窗玻璃组件	round observation window organic glass assembly on side panel of frames 13～41

机翼　Wings

1	机翼长桁分布	wing stringer distribution
2	中央翼	center wing
3	中央翼放油口盖	center wing fuel drain access cover
4	中外翼	inboard wing
5	外翼	outboard wing
6	发动机架与吊挂接头	engine rack and hanger fitting
7	中央翼与中外翼对接	butt joint between center wing and inboard wing
8	中外翼可卸壁板	inboard wing detachable panel
9	中外翼与外翼对接	butt joint between inboard wing and outboard wing
10	中外翼可卸壁板对接	butt joint of inboard wing detachable panel
11	中外翼放油口盖	inboard wing fuel drain access cover
12	外翼可卸壁板	outboard wing detachable panel
13	机翼翼尖	wing tip

14	中外翼前缘	inboard wing leading edge
15	外翼前缘	outboard wing leading edge
16	襟翼	flap
17	襟翼滑轨安装	installation of flap sliding rail
18	襟翼前缘口盖	flap leading edge access cover
19	内副翼	inboard aileron
20	外副翼	outboard aileron
21	外副翼摇臂	outboard aileron rocker
22	副翼悬挂接头	aileron suspension joint
23	副翼调整片	aileron trim tab

螺旋桨　　Propellers

1	桨叶在桨毂上的安装	installation of blade on the hub
2	螺旋桨在发动机上的安装	installation of propeller on the engine
3	端面刷架组件和径向刷架组件的安装	installation of endface brush holder assembly and radial brush holder assembly
4	桨毂整流罩的安装	installation of hub fairing
5	桨根整流罩	propeller root fairing
6	螺旋桨顺桨系统的安装	installation of propeller feathering system
7	顺桨泵的安装	installation of feathering pump
8	螺旋桨制动继电器盒	propeller brake relay box
9	螺旋桨制动继电器盒的安装	installation of propeller brake relay box
10	顺桨继电器盒	feathering relay box
11	顺桨继电器盒的安装	installation of feathering relay box
12	顺桨按钮的安装	installation of feathering button
13	定时机构的安装	installation of timing mechanism
14	扭矩顺桨检查开关的安装	installation of torque feathering check button
15	负拉力顺桨检查开关的安装	installation of negative thrust feathering check switch
16	部分顺桨按钮的安装	installation of partial feathering button
17	螺旋桨解除制动操纵盒	propeller brake-release control box
18	螺旋桨解除制动操纵盒的安装	installation of propeller brake-release control box
19	定距油路压力信号器的安装	installation of constant pitch line pressure annunciator
20	负拉力自动顺桨压力信号器的安装	installation of negative thrust auto-feathering pressure annunciator
21	小距油路压力信号器的安装	installation of fine pitch line pressure annunciator

动力装置　Power Plant

1	发动机的安装	engine installation
2	发动机支架的安装	installation of engine mount
3	主撑杆	main bracing
4	下撑杆	lower bracing
5	上中撑杆	upper middle bracing
6	减震支柱	damper strut
7	发动机清洗系统的安装	installation of engine cleaning system

发动机燃油及控制　Engine Fuel and Control

1	发动机燃油分配系统的安装	installation of engine fuel distribution system
2	粗油滤的安装	installation of coarse filter
3	细油滤的安装	installation of fine filter
4	燃油调节器静压系统	fuel control unit static system
5	右机翼静压系统的安装	installation of right wing static system
6	左机翼静压系统的安装	installation of left wing static system
7	机身静压系统的安装	installation of fuselage static system
8	燃油耗量表传感器的安装	installation of fuel consumption indicator sensor
9	低压燃油压力表传感器的安装	installation of low fuel pressure gauge sensor
10	细油滤堵塞信号器的安装	installation of fine filter blocking annunciator
11	燃油压力传感器的安装	installation of fuel pressure sensor

进气　Air

1	通风冷却系统的安装	installation of ventilation cooling system
2	涡轮吹风罩冷却进气口的安装	installation of turbine blowing cover cooling inlet

发动机操纵系统　Engine Controls

1	发动机操纵系统安装图	installationdrawing of engine control system
2	左油门手柄图解图	illustration of left throttle handle
3	左油门手柄面盖图解图	illustration of left throttle handle cover
4	右油门手柄图解图	illustration of right throttle handle
5	右面盖图解图	illustration of right cover
6	右操纵台	copilot console
7	左操纵台	pilot console

8	左联锁支架图解图	illustration of left interlocking bracket
9	拉紧支架图解图	illustration of tension bracket
10	下联锁支架图解图	illustration of lower interlocking bracket
11	右联锁支架图解图	illustration of right interlocking bracket
12	滑轮支架安装图	installation of pulley bracket
13	微动电门组合件图解图	illustration of microswitch assembly
14	舵面锁滑轮支架安装图	installation of control surface lock pulley bracket
15	拉杆图解图	illustration of pull rod
16	油门杆位置传感器安装图	installationdrawing of throttle lever position sensor
17	应急顺桨操纵系统安装图	installation drawing of emergency feathering control system
18	减压活门安装图	installation drawing of decompression valve
19	顺桨开关安装图	installation drawing of feathering switch
20	断油活门安装图	installationdrawing of fuel cut off valve
21	漏油杯安装图	installationdrawing of drain cup

发动机指示　Engine Indicating

1	螺旋桨扭矩传感器的安装	installation of propeller torque sensor
2	发动机转速传感器的安装	installation of engine rotating speed sensor
3	转速信号转换器的安装	installation of rotating signal converter
4	发动机排气温度传感器的安装	installation of engine exhaust temperature sensor
5	振动加速度放大器的安装	installation of vibration acceleration amplifier
6	振动速度传感器的安装	installation of vibration speed sensor

排气　Exhaust

1	发动机排气系统的安装	installation of engine exhaust system
2	排气管	exhaust pipe

滑油系统　Oil

1	滑油系统	oil system
2	滑油箱的安装	installation of oil reservoir
3	滑油箱	oil reservoir
4	通气油箱的安装	installation of air-vented tank
5	通气油箱	air-vented tank
6	滑油散热器的安装	installation of oil radiator

7	散热风门电动机构的安装	installation of electric mechanism of radiator shutter
8	温调盒的安装	installation of temperature regulating box
9	引射散热活门的安装	installation of ejector radiation valve
10	引射器的安装	installation of ejector
11	滑油油量传感器的安装	installation of oil quantity sensor
12	滑油压力传感器的安装	installation of oil pressure sensor
13	滑油温度传感器的安装	installation of oil temperature sensor

起动　Starting

1	定时机构的安装	installation of timing mechanism
2	起动箱的安装	installation of starting box
3	分流器与起动检查继电器的安装	installation of current divider and starting inspection relay
4	起动断开接通发电机输出继电器的安装	installation of starting OFF/ON generator output relay
5	起动电压调节盒的安装	installation of starting voltage regulating box
6	功率限制器的安装	installation of power limiter
7	发动机自动停车器的安装	installation of engine automatic shutdown device
8	自动切油和超温保护器的安装	installation of automatic fuel reducer and over-temperature protector
9	起动继电器盒的安装	installation of starting relay box
10	导航系统继电器盒	relay box of navigation system

特种电子设备　Special Electronic Equipment

1	应答机系统	transponder system
2	应答机(2分机)的安装	installation of transponder (extension 2)
3	控制盒(3分机)的安装	installation of control box (extension 3)
4	信号交联盒(5分机)的安装	installation of signal crosslinking box (extension 5)
5	电源滤波盒(25D分机)的安装	installation of power filtering box (extension 25D)
6	天线(I分机)的安装	installation of antenna (extension I)

无损检测手册　NONDESTRUCTIVE TESTING MANUAL

| 1 | 检测人员的资格鉴定 | determination of testing personnel qualification |
| 2 | 检测方法的选择 | selection of testing methods |

3	飞机结构、部件的位置及其可接近程度	location and accessibility degree of aircraft structure and components
4	检测前的准备	preparations before testing
5	标准试块	standard testing blocks
6	射线的透照作业	radiograph operation
7	脉冲反射法的说明	description of pulse reflection method
8	作业	operation
9	标准试块	standard testing block
10	应用和检测方法	application and testing methods
11	标准试件	standard testing piece
12	检测前的清理	cleaning before inspection
13	检测程序	test procedure
14	退磁	demagnetization
15	设备及辅助材料	equipment and auxiliary materials
16	设备调整	instrument debugging
17	工作程序	operation procedure
18	后处理	post-treatment
19	仪器及试块	instrument and testing block
20	几点说明	description
21	准备及清洗	preparation and cleaning
22	仪器调试	instrument debugging
23	检测用仪器和配件	testing instrument and parts

飞行操纵快速检查程序　QUICK REFERENCE HANDBOOK

1	外部检查	exterior check
2	飞机内部检查	interior check
3	开车前	before startup
4	滑行前	before taxiing
5	滑行	taxiing
6	起飞前	before takeoff
7	起飞后	after takeoff
8	下降校场压后	after descending & airfield barometric altitude calibration
9	进入下滑道前	before entering the glide slope
10	着陆后关车前	after landing & before engine shutdown
11	关车后	after shutdown
12	发动机故障	engine failure
13	失火	fire
14	燃油系统故障	fuel system failure

15	增压系统故障	pressurization system failure
16	滑油系统故障	oil system failure
17	液压系统故障	hydraulic system failure
18	应急放起落架	emergency LG extension
19	场外迫降	forced landing

维修手册章节

时限/维护　TIME LIMITS/MAINTENANCE

| 22 | 剧力牵引后的检查—检验/检查 | Check After Severe Towing—Inspection/Check |
| 23 | 过量刹车后的检查—检验/检查 | Check After Excessive Braking—Inspection/Check |

尺寸和区域   DIMENSIONS AND AREAS

1	尺寸和区域—总论	Dimensions and Areas—Introduction
2	基本尺寸—说明	Principal Dimension—Description
3	区域划分—说明	Area Division—Description
4	飞机框位/肋位—说明	Locations of Aircraft Frames/Ribs—Description
5	接近装置—说明/工作	Access Facility—Description/Operation
6	飞机外部接近装置—说明/工作	Aircraft External Access Facility —Description/Operation
7	飞机内部接近装置—说明/工作	Aircraft Internal Access Facility —Description/Operation

顶起和支撑   JACKING & SHORING

1	顶起和支撑—总论	Jacking and Shoring—Introduction
2	顶起—说明	Jacking—Description
3	为进行维修工作而顶起飞机—说明/工作	Jacking the Aircraft for Maintenance—Description/Operation
4	机翼千斤顶—说明/工作	Wing Jack—Description/Operation
5	机翼千斤顶—维修实施	Wing Jack—Maintenance Practice
6	尾部千斤顶—说明/工作	Tail Jack—Description/Operation
7	尾部千斤顶—维修实施	Tail Jack—Maintenance Practice
8	机头千斤顶—说明/工作	Nose Jack—Description/Operation
9	机头千斤顶—维修实施	Nose Jack—Maintenance Practice
10	机轮千斤顶—说明/工作	Wheel Jack—Description/Operation
11	为更换机轮而顶起飞机—说明/工作	Jacking Aircraft for Replacing Landing Gear Wheels—Description/Operation
12	机轮千斤顶—维修实施	Wheel Jack—Maintenance Practice
13	支撑—说明	Shoring—Description
14	尾部支撑—说明/工作	Tail Support—Description/Operation
15	尾部支撑—维修实施	Tail Support—Maintenance Practice

调平和称重   LEVELING AND WEIGHING

1	调平和称重—总论	Leveling and Weighing— Introduction
2	飞机调平和水平测量—说明/工作	Aircraft Leveling and Leveling Measurement—Description/Operation
3	称重和平衡—说明/工作	Weighing and Balancing—Description/Opera-

tion

牵引　TOWING AND TAXIING

停放和系留　PARKING & MOORING

标牌和标志　PLACARDS AND MARKINGS

保养　SERVICING

8	供水/排污物设备加注—说明/工作	Filling of Water Supply/Waste Disposal Equipment—Description/Operation
9	辅助动力装置加注—说明/工作	Filling of Auxiliary Power Unit（APU）—Description/Operation
10	滑油系统加注—说明/工作	Filling of Oil System—Description/Operation
11	定期保养—说明/工作	Scheduled Servicing—Description/Operation
12	环控系统定期保养—说明/工作	Scheduled Servicing of Environmental Control System—Description/Operation
13	通讯系统定期保养—说明/工作	Scheduled Servicing of Communication System—Description/Operation
14	飞机除冰雪—说明/工作	De-icing and Rain Removal—Description/Operation
15	飞行操纵系统定期保养—说明/工作	Scheduled Servicing of Flight Control System—Description/Operation
16	防冰、防雨系统定期保养—说明/工作	Scheduled Servicing for Ice and Rain Protection System—Description/Operation
17	导航系统定期保养—说明/工作	Scheduled Servicing of Navigation System—Description/Operation
18	辅助动力装置定期保养—说明/工作	Scheduled Servicing of Auxiliary Power Unit（APU）—Description/Operation
19	舱门定期保养—说明/工作	Scheduled Servicing of Doors—Description/Operation
20	机身定期保养—说明/工作	Scheduled Servicing of Fuselage—Description/Operation
21	尾翼定期保养—说明/工作	Scheduled Servicing of Stabilizers—Description/Operation
22	窗定期保养—说明/工作	Scheduled Servicing of Windows—Description/Operation
23	机翼定期保养—说明/工作	Scheduled Servicing of Wings—Description/Operation
24	动力装置定期保养—说明/工作	Scheduled Servicing of Power Plant—Description/Operation
25	发动机操纵定期保养—说明/工作	Scheduled Servicing of Engine Controls—Description/Operation
26	滑油系统定期保养—说明/工作	Scheduled Servicing of Oil System—Description/Operation
27	不定期保养—说明/工作	Unscheduled Servicing—Description/Operation
28	飞机的外部清洁—说明/工作	Cleaning of Aircraft Exterior—Description/Operation

31	涡轮冷却器—说明/工作	Turbine Cooler—Description/Operation
32	涡轮冷却器—拆卸/安装	Turbine Cooler—Removal/Installation
33	座舱温度控制系统—说明/工作	Cabin Temperature Control System—Description/Operation
34	温度控制盒—说明/工作	Temperature Control Box—Description/Operation
35	温度控制盒—拆卸/安装	Temperature Control Box—Removal/Installation
36	温度传感器—说明/工作	Temperature Sensor—Description/Operation
37	温度传感器—拆卸/安装	Temperature Sensor—Removal/Installation
38	温调电动活门—说明/工作	Temperature Regulating Electric Valve—Description/Operation
39	温调电动活门—拆卸/安装	Temperature Regulating Electric Valve—Removal/Installation
40	双金属温度表—说明/工作	Bimetal Thermometer—Description/Operation
41	双金属温度表—拆卸/安装	Bimetal Thermometer—Removal/Installation
42	温度表—说明/工作	Thermometer—Description/Operation
43	温度表—拆卸/安装	Thermometer—Removal/Installation
44	货舱温度表—说明/工作	Cargo Cabin Thermometer—Description/Operation
45	货舱温度表—拆卸/安装	Cargo Cabin Thermometer—Removal/Installation

自动飞行　AUTO FLIGHT

1	自动飞行—总论	Auto Flight—Introduction
2	自动驾驶仪—说明/工作	Autopilot—Description/Operation
3	自动驾驶仪—故障分析	Autopilot—Failure Analysis
4	自动驾驶仪—调整/试验	Autopilot—Adjustment/Test
5	中心垂直陀螺—说明/工作	Central Vertical Gyro—Description/Operation
6	中心垂直陀螺—拆卸/安装	Central Vertical Gyro—Removal/Installation
7	速率陀螺盒—说明/工作	Rate Gyro Box—Description/Operation
8	速率陀螺盒—拆卸/安装	Rate Gyro Box—Removal/Installation
9	操纵台—说明/工作	Console—Description/Operation
10	操纵台—拆卸/安装	Console—Removal/Installation
11	操纵手柄—说明/工作	Control Handle—Description/Operation
12	操纵手柄—拆卸/安装	Control Handle—Removal/Installation
13	飞行控制盒—说明/工作	Flight Control Box—Description/Operation
14	飞行控制盒—拆卸/安装	Flight Control Box—Removal/Installation
15	高度差传感器—说明/工作	Altitude Difference Sensor—Description/Operation

通信系统　COMMUNICATIONS

14	机内通话器—调整/试验	Intercom—Adjustment/Test
15	头戴送受话器组—说明/工作	Headset—Description/Operation
16	头戴送受话器组—调整/试验	Headset—Adjustment/Test
17	头戴送受话器组—维修实施	Headset—Maintenance Practice
18	头戴送受话器组—故障分析	Headset—Failure Analysis
19	静电释放—说明/工作	Static Discharging—Description/Operation
20	静电释放—拆卸/安装	Static Discharging—Removal/Installation
21	静电释放—检验/检查	Static Discharging—Inspection/Check
22	座舱音频记录系统—说明/工作	Cabin Audio Recording System—Description/Operation
23	座舱音频记录系统—故障分析	Cabin Audio Recording System—Failure Analysis
24	座舱音频记录系统—拆卸/安装	Cabin Audio Recording System—Removal/Installation
25	座舱音频记录系统—调整/试验	Cabin Audio Recording System—Adjustment/Test

电源系统　ELECTRICAL POWER

1	飞机电源—总论	Aircraft Power Supply—Introduction
2	飞机电源—维修实施	Aircraft Power Supply—Maintenance Practice
3	交流电源系统—说明/工作	AC Power Supply System—Description/Operation
4	交流电源系统—检验/检查	AC Power Supply System—Inspection/Check
5	交流发电机—说明/工作	AC Generator—Description/Operation
6	交流发电机—故障分析	AC Generator—Failure Analysis
7	交流发电机—维修实施	AC Generator—Maintenance Practice
8	交流调压器—说明/工作	AC Voltage Regulator—Description/Operation
9	交流调压器—故障分析	AC Voltage Regulator—Failure Analysis
10	交流调压器—维修实施	AC Voltage Regulator—Maintenance Practice
11	精调盒—说明/工作	Fine Adjustment Box—Description/Operation
12	精调盒—故障分析	Fine Adjustment Box—Failure Analysis
13	精调盒—维修实施	Fine Adjustment Box—Maintenance Practice
14	过压保护器—故障分析	Overvoltage Protector—Failure Analysis
15	交流过压保护器—说明/工作	AC Overvoltage Protector—Description/Operation
16	过压保护器—维修实施	Overvoltage Protector—Maintenance Practice
17	交流转换盒—说明/工作	AC Changeover Box—Description/Operation
18	交流转换盒—维修实施	AC Changeover Box—Maintenance Practice
19	定时机构—说明/工作	Timing Device—Description/Operation
20	定时机构—故障分析	Timing Device—Failure Analysis

48	碳片调压器—维修实施	Carbon Pile Voltage Regulator—Maintenance Practice
49	直流过压保护器—说明/工作	DC Overvoltage Protector—Description/Operation
50	直流过压保护器—故障分析	DC Overvoltage Protector—Failure Analysis
51	直流过压保护器—维修实施	DC Overvoltage Protector—Maintenance Practice
52	差动低限保护器—说明/工作	Differential Undervoltage Relay—Description/Operation
53	差动低限保护器—故障分析	Differential Undervoltage Relay—Failure Analysis
54	差动低限保护器—维修实施	Differential Undervoltage Relay—Maintenance Practice
55	发电机激磁接触器—说明/工作	Generator Exciting Contactor—Description/Operation
56	发电机激磁接触器—维修实施	Generator Exciting Contactor—Maintenance Practice
57	碱性蓄电池—说明/工作	Alkaline Battery—Description/Operation
58	碱性蓄电池—故障分析	Alkaline Battery—Failure Analysis
59	碱性蓄电池—维修实施	Alkaline Battery—Maintenance Practice
60	碱性蓄电池—保养	Alkaline Battery—Servicing
61	应急供电接触器—说明/工作	Emergency Power Supply Contactor—Description/Operation
62	应急供电接触器—维修实施	Emergency Power Supply Contactor—Maintenance Practice
63	直流电压表—说明/工作	DC Voltmeter—Description/Operation
64	直流电压表—维修实施	DC Voltmeter—Maintenance Practice
65	直流电流表—说明/工作	DC Amperemeter—Description/Operation
66	直流电流表—维修实施	DC Amperemeter—Maintenance Practice
67	外部电源—说明/工作	External Power Supply—Description/Operation
68	直流地面电源插座—说明/工作	DC Ground Power Supply Receptacle—Description/Operation
69	直流地面电源插座—维修实施	DC Ground Power Supply Receptacle—Maintenance Practice
70	70V 起动电源插座—说明/工作	70V Starting DC Ground Power Supply Receptacle—Description/Operation
71	70V 起动电源插座—维修实施	70V Starting DC Ground Power Supply Receptacle—Maintenance Practice
72	极性继电器—说明/工作	Polarity Relay—Description/Operation
73	极性继电器—维修实施	Polarity Relay—Maintenance Practice
74	起动及电源接触器—说明/工作	Starting and Power Supply Contactor—Descrip-

16	救生绳—说明/工作	Escape Rope—Description/Operation
17	救生绳—维修实施	Escape Rope—Maintenance Practice
18	电动信号枪—说明/工作	Electric Signal Gun—Description/Operation
19	电动信号枪—维修实施	Electric Signal Gun— Maintenance Practice
20	多人救生船—说明/工作	Liferaft—Description/Operation
21	多人救生船—维修实施	Liferaft— Maintenance Practice
22	救生伞—说明/工作	Emergency Parachute—Description/Operation
23	救生伞—维修实施	Emergency Parachute—Maintenance Practice
24	救护设备—说明/工作	Life-saving Device—Description/Operation
25	救护设备—维修实施	Life-saving Device—Maintenance Practice
26	应急斧头、铁棍—说明/工作	Emergency Axe and Iron Stick—Description/Operation
27	应急斧头、铁棍—维修实施	Emergency Axe and Iron Stick—Maintenance Practice
28	应急定位仪—说明/工作	Emergency Locator—Description/Operation
29	应急定位仪—故障分析	Emergency Locator—Failure Analysis
30	应急定位仪—拆卸/安装	Emergency Locator—Removal/Installation
31	应急定位仪—检验/检查	Emergency Locator—Inspection/Check
32	应急定位仪—清理/涂覆	Emergency Locator—Cleaning/Coating
33	空投空降设备—说明/工作	Airdrop & Parachuting Equipment—Description/Operation
34	跳伞钢索—说明/工作	Bailout Cable—Description/Operation
35	跳伞钢索—检验/检查	Bailout Cable—Inspection/Check
36	空降信号装置—说明/工作	Parachuting Signal Device—Description/Operation
37	空降信号装置—维修实施	Parachuting Signal Device—Maintenance Practice
38	电动伞绳回收装置—说明/工作	Electric Parachute Cord Recovery Device—Description/Operation
39	电动伞绳回收装置—检验/检查	Electric Parachute Cord Recovery Device—Inspection/Check

防火　FIRE PROTECTION

1	防火系统—总论	Fire Protection System—Introduction
2	探测—说明/工作	Detection—Description/Operation
3	火警传感器—说明/工作	Fire Warning Sensor—Description/Operation
4	火警传感器—故障分析	Fire Warning Sensor— Failure Analysis
5	火警传感器—维修实施	Fire Warning Sensor—Maintenance Practice
6	灭火—说明/工作	Extinguishing—Description/Operation
7	灭火—故障分析	Extinguishing—Failure Analysis

飞行操纵　FLIGHT CONTROLS

4	副翼操纵系统—故障分析	Aileron Control System—Failure Analysis
5	副翼操纵系统—维修实施	Aileron Control System—Maintenance Practice
6	驾驶盘—维修实施	Control Wheel—Maintenance Practice
7	无声链条、带耳环接头的钢索及钢索滑轮—维修实施	Cable with Noiseless Chain and Lug Joint and Cable Pulley—Maintenance Practice
8	扇形摇臂—维修实施	Sector Rocker—Maintenance Practice
9	副翼拉杆—维修实施	Aileron Pull Rod—Maintenance Practice
10	导向滑轮—维修实施	Guide Pulley—Maintenance Practice
11	扰流板操纵装置—维修实施	Spoiler Control Device—Maintenance Practice
12	副翼操纵摇臂组—维修实施	Aileron Control Rocker Group—Maintenance Practice
13	副翼摇臂及拉杆摇臂支架—说明/工作	Brackets of Aileron Rocker and Pull Rod Rocker—Description/Operation
14	副翼摇臂及拉杆摇臂支架—维修实施	Brackets of Aileron Rocker and Pull Rod Rocker—Maintenance Practice
15	副翼立轴—维修实施	Vertical Shaft of Aileron—Maintenance Practice
16	副翼调整片(随动补偿片)偏转机构—说明/工作	Aileron Trim Tab (Balancing Tab) Deflection Mechanism—Description/Operation
17	副翼调整片(随动补偿片)偏转机构—故障分析	Aileron Trim Tab (Balancing Tab) Deflection Mechanism—Failure Analysis
18	副翼调整片(随动补偿片)偏转机构—维修实施	Aileron Trim Tab (Balancing Tab) Deflection Mechanism—Maintenance Practice
19	副翼调整片操纵开关—说明/工作	Control Switch of Aileron Trim Tab—Description/Operation
20	副翼调整片操纵开关—拆卸/安装	Control Switch of Aileron Trim Tab—Removal/Installation
21	副翼调整片电动机构—通电检查	Electric Mechanism of Aileron Trim Tab—Power-on Check
22	方向舵和方向舵调整补偿片操纵系统—说明/工作	Control System of Rudder and Rudder Trim Balance Tab—Description/Operation
23	方向舵操纵系统—说明/工作	Rudder Control System—Description/Operation
24	方向舵操纵系统—维修实施	Rudder Control System—Maintenance Practice
25	脚操纵机构—说明/工作	Foot Control Mechanism—Description/Operation
26	脚操纵机构—维修实施	Foot Control Mechanism—Maintenance Practice
27	方向舵拉杆—维修实施	Rudder Pull Rod—Maintenance Practice
28	弹簧拉杆—说明/工作	Spring Pull Rod—Description/Operation
29	弹簧拉杆—维修实施	Spring Pull Rod—Maintenance Practice
30	下弹簧拉杆—说明/工作	Lower Spring Pull Rod—Description/Operation
31	下弹簧拉杆—维修实施	Lower Spring Pull Rod—Maintenance Practice

55	升降舵调整片偏转机构—维修实施	Elevator Trim Tab Deflection Mechanism—Maintenance Practice
56	钢索分配器—维修实施	Cable Distributor—Maintenance Practice
57	襟翼操纵系统—说明/工作	Flap Control System—Description/Operation
58	襟翼操纵系统—故障分析	Flap Control System—Failure Analysis
59	襟翼操纵系统—维修实施	Flap Control System—Maintenance Practice
60	襟翼机械传动系统—说明/工作	Flap Mechanical Driving System—Description/Operation
61	襟翼机械传动组件—维修实施	Flap Mechanical Driving Assembly—Maintenance Practice
62	襟翼组件支座—维修实施	Flap Assembly Support—Maintenance Practice
63	襟翼收放机构—说明/工作	Flap Retraction/Extension Mechanism— Description/Operation
64	襟翼收放机构—拆卸/安装	Flap Retraction/Extension Mechanism— Removal/Installation
65	襟翼电气操纵系统—说明/工作	Flap Electrical Control System—Description/Operation
66	襟翼位置电气讯号机构—说明/工作	Flap Position Electrical Signal Mechanism—Description/Operation
67	襟翼位置电气讯号机构—拆卸/安装	Flap Position Electrical Signal Mechanism—Removal/Installation
68	襟翼位置指示器—说明/工作	Flap Position Indicator—Description/Operation
69	襟翼位置指示器—拆卸/安装	Flap Position Indicator—Removal/Installation
70	传感器—拆卸/安装	Sensor—Removal/Installation
71	襟翼收放开关—说明/工作	Flap Retraction/Extension Switch—Description/Operation
72	襟翼收放开关—拆卸/安装	Flap Retraction/Extension Switch—Removal/Installation
73	襟翼液压传动系统—说明/工作	Flap Hydraulic Driving System—Description/Operation
74	襟翼液压传动系统—故障分析	Flap Hydraulic Driving System—Failure Analysis
75	襟翼液压传动系统—调整/试验	Flap Hydraulic Driving System—Adjustment/Test
76	襟翼液压传动系统—检验/检查	Flap Hydraulic Driving System—Inspection/Check
77	电磁阀—拆卸/安装	Solenoid Valve—Removal/Installation
78	液压传动机构—拆卸/安装	Hydraulic Driving Mechanism—Removal/Installation
79	转换活门—拆卸/安装	Changeover Valve—Removal/Installation

燃油系统　FUEL

19	单向活门—故障分析	Check Valve—Failure Analysis
20	单向活门—维修实施	Check Valve—Maintenance Practice
21	供油系统—说明/工作	Fuel Supply System—Description/Operation
22	直流电动离心式燃油泵—说明/工作	DC Electric Centrifugal Fuel Pump—Description/Operation
23	直流电动离心式燃油泵—维修实施	DC Electric Centrifugal Fuel Pump—Maintenance Practice
24	连通开关(防火开关)—说明/工作	Crossfeed Valve (Fire Shut off Valve)—Description/Operation
25	连通开关(防火开关)—维修实施	Crossfeed Valve (Fire Shut off Valve)—Maintenance Practice
26	防火电磁开关—说明/工作	Fire-Proof Solenoid Valve—Description/Operation
27	防火电磁开关—维修实施	Fire-Proof SolenoidValve—Maintenance Practice
28	燃油电动开关—说明/工作	Electric Fuel Valve—Description/Operation
29	燃油电动开关—故障分析	Electric Fuel Valve—Failure Analysis
30	燃油电动开关—维修实施	Electric Fuel Valve—Maintenance Practice
31	单向活门—说明/工作	Check Valve—Description/Operation
32	单向活门—故障分析	Check Valve—Failure Analysis
33	单向活门—维修实施	Check Valve—Maintenance Practice
34	双单向活门—说明/工作	Two-Way Check Valve—Description/Operation
35	双单向活门—故障分析	Two-Way Check Valve—Failure Analysis
36	双单向活门—维修实施	Two-Way Check Valve—Maintenance Practice
37	燃油滤—说明/工作	Fuel Filter—Description/Operation
38	燃油滤—故障分析	Fuel Filter—Failure Analysis
39	燃油滤—维修实施	Fuel Filter—Maintenance Practice
40	放油系统—说明/工作	Fuel Drain System—Description/Operation
41	放油开关—说明/工作	Fuel Drain Cock—Description/Operation
42	放油开关—故障分析	Fuel Drain Cock—Failure Analysis
43	放油开关—拆卸/安装	Fuel Drain Cock—Removal/Installation
44	燃油油量表—说明/工作	Fuel Quantity Gauge—Description/Operation
45	燃油油量表—故障分析	Fuel Quantity Gauge—Failure Analysis
46	燃油油量表—拆卸/安装	Fuel Quantity Gauge—Removal/Installation
47	燃油油量表—调整/试验	Fuel Quantity Gauge—Adjustment/Test
48	燃油油量表—检验/检查	Fuel Quantity Gauge—Inspection/Check
49	压力信号器—说明/工作	Pressure Annunciator—Description/Operation
50	压力信号器—拆卸/安装	Pressure Annunciator—Removal/Installation

液压系统　HYDRAULIC SYSTEM

34	油箱增压系统—调整/试验	Reservoir Pressurization System—Adjustment/Test
35	油箱增压系统—检验/检查	Reservoir Pressurization System—Inspection/Check
36	减压阀—说明/工作	Pressure Reducing Valve—Description/Operation
37	减压阀—拆卸/安装	Pressure Reducing Valve—Removal/Installation
38	干燥过滤器—维修实施	Dryer Filter—Maintenance Practice
39	气滤—维修实施	Air Filter—Maintenance Practice
40	放气活门—拆卸/安装	Air Bleed Valve—Removal/Installation
41	手摇泵—电动泵系统—说明/工作	Hand-Electric Pump System—Description/Operation
42	手摇泵—电动泵系统—故障分析	Hand-Electric Pump System—Failure Analysis
43	手摇泵—电动泵系统—调整/试验	Hand-Electric Pump System—Adjustment/Test
44	手摇泵—电动泵系统—检验/检查	Hand-Electric Pump System—Inspection/Check
45	手摇泵—说明/工作	Hand Pump—Description/Operation
46	手摇泵—维修实施	Hand Pump—Maintenance Practice
47	电动泵—说明/工作	Electric Pump—Description/Operation
48	电动泵—维修实施	Electric Pump—Maintenance Practice
49	三位分配开关—说明/工作	Three-Position Distribution Valve—Description/Operation
50	七位分配开关—说明/工作	Seven-Position Distribution Valve—Description/Operation
51	分配开关—维修实施	Distribution Valve—Maintenance Practice
52	安全阀—说明/工作	Safety Valve—Description/Operation
53	安全阀—拆卸/安装	Safety Valve—Removal/Installation
54	压力转换装置—说明/工作	Pressure Changeover Device—Description/Operation
55	压力转换装置—拆卸/安装	Pressure Changeover Device—Removal/Installation
56	液压指示系统—说明/工作	Hydraulic Indication System—Description/Operation
57	液压系统油量表—说明/工作	Fluid Quantity Gauge of Hydraulic System—Description/Operation
58	液压系统油量表—故障分析	Fluid Quantity Gauge of Hydraulic System—Failure Analysis
59	液压系统油量表—维修实施	Fluid Quantity Gauge of Hydraulic System—Maintenance Practice
60	液压系统压力表—说明/工作	Hydraulic System Pressure Gauge—Description/Operation

除雨和防冰　ICE AND RAIN PROTECTION

		Operation
23	尾翼加温信号控制盒—拆卸/安装	Tail Heating Signal Control Box—Removal/Installation
24	发动机进气道防冰系统—说明/工作	Engine Air Intake Anti-icing System—Description/Operation
25	发动机进气道防冰系统—故障分析	Engine Air Intake Anti-icing System—Failure Analysis
26	发动机进气道防冰系统—调整/试验	Engine Air Intake Anti-icing System—Adjustment/Test
27	发动机进气道防冰开关—说明/工作	Engine Air Intake Anti-icing Valve—Description/Operation
28	发动机进气道防冰开关—拆卸/安装	Engine Air Intake Anti-icing Valve—Removal/Installation
29	全压静压加温系统—说明/工作	Pitot and Static Pressure Heating System—Description/Operation
30	全压静压加温系统—故障分析	Pitot and Static Pressure Heating System—Failure Analysis
31	全压静压加温系统—检验/检查	Pitot and Static Pressure Heating System—Inspection/Check
32	加温报警控制盒—拆卸/安装	Heating Warning Control Box—Removal/Installation
33	窗及风挡系统—说明/工作	Windows and Windshield System—Description/Operation
34	风挡玻璃加温系统—说明/工作	Windshield Glass Heating System—Description/Operation
35	风挡玻璃加温系统—故障分析	Windshield Glass Heating System—Failure Analysis
36	风挡玻璃加温系统—检验/检查	Windshield Glass Heating System—Inspection/Check
37	玻璃加温温度控制盒—说明/工作	Glass Heating Temperature Control Box—Description/Operation
38	自耦变压器—说明/工作	Autotransformer—Description/Operation
39	自耦变压器—拆卸/安装	Autotransformer—Removal/Installation
40	风挡刷系统—说明/工作	Windshield Wiper System—Description/Operation
41	风挡刷系统—故障分析	Windshield Wiper System—Failure Analysis
42	风挡刷系统—调整/试验	Windshield Wiper System—Adjustment/Test
43	风挡刷开关—说明/工作	Windshield Wiper Cock—Description/Operation
44	风挡刷开关—拆卸/安装	Windshield Wiper Cock—Removal/Installation

显示/记录系统　INDICATING/RECORDING SYSTEMS

16	飞行数据记录系统—调整/试验	Flight Data Recording System—Adjustment/Test
17	用氧信号警告系统—说明/工作	Oxygen Supply Signal Warning System—Description/Operation
18	用氧信号警告系统—故障分析	Oxygen Supply Signal Warning System—Failure Analysis
19	用氧信号警告系统—调整/试验	Oxygen Supply Warning System—Adjustment/Test
20	临界攻角信号系统—说明/工作	Critical AOA Signal System—Description/Operation
21	临界攻角信号系统—故障分析	Critical AOA Signal System—Failure Analysis
22	临界攻角信号系统—拆卸/安装	Critical AOA Signal System—Removal/Installation
23	临界攻角信号系统—调整/试验	Critical AOA Signal System—Adjustment/Test

起落架　LANDING GEAR

1	起落架—总论	Landing Gear—Introduction
2	主起落架和舱门—说明/工作	Main Landing Gear and Doors—Description/Operation
3	主起落架—拆卸/安装	Main Landing Gear—Removal/Installation
4	主起落架—调整/试验	MLG—Adjustment/Test
5	主起落架—检验/检查	MLG—Inspection/Check
6	主起落架缓冲支柱—说明/工作	MLG Shock Strut—Description/Operation
7	飞起落架带锁可折撑杆—说明/工作	MLG Foldable Brace With Lock—Description/Operation
8	主起落架带锁可折撑杆—拆卸/安装	MLG Foldable Bracing with Lock—Removal/Installation
9	机轮组—说明/工作	Wheel Group—Description/Operation
10	稳定缓冲器—说明/工作	Stable Buffer—Description/Operation
11	主起落架舱门操纵机构—说明/工作	Control Mechanism of MLG Door—Description/Operation
12	前起落架—说明/工作	Nose Landing Gear—Description/Operation
13	前起落架—拆卸/安装	NLG — Removal/Installation
14	前起落架—调整/试验	NLG—Adjustment/Test
15	前起落架—检验/检查	NLG—Inspection/Check

		nance Practice
40	液压锁—说明/工作	Hydraulic Lock—Description/Operation
41	液压锁—拆卸/安装	Hydraulic Lock—Removal/Installation
42	主起舱门作动筒—说明/工作	MLG Doors Actuator—Description/Operation
43	主起舱门作动筒（带转换活门）—维修实施	MLG Doors Actuator（with changeover valve）—Maintenance Practice
44	液压手动阀—说明/工作	Hydraulic Handle Valve—Description/Operation
45	液压手动阀—拆卸/安装	Hydraulic Handle Valve—Removal/Installation
46	主起落架下位锁作动筒—说明/工作	MLG Downlock Actuator—Description/Operation
47	主起落架下位锁作动筒—维修实施	MLG Downlock Actuator—Maintenance Practice
48	液压开关—说明/工作	Hydraulic Switch—Description/Operation
49	液压开关—拆卸/安装	Hydraulic Switch—Removal/Installation
50	回油转换活门—说明/工作	Fluid Return Changeover Valve—Description/Operation
51	回油转换活门—拆卸/安装	Fluid Return Changeover Valve—Removal/Installation
52	起落架应急放下系统—说明/工作	Emergency LG Extension System—Description/Operation
53	起落架应急放下系统—故障分析	Emergency LG Extension System—Failure Analysis
54	起落架应急放下系统—调整/试验	Emergency LG Extension System—Adjustment/Test
55	正常刹车系统—说明/工作	Normal Braking System—Description/Operation
56	前起落架应急放下系统—说明/工作	Emergency Extension System of NLG—Description/Operation
57	前起落架应急放下系统—拆卸/安装	Emergency Extension System of NLG—Removal/Installation
58	主起落架应急放下系统—说明/工作	Emergency Extension System of MLG—Description/Operation
59	主起落架应急放下系统—拆卸/安装	Emergency Extension System of MLG—Removal/Installation
60	正常刹车系统—故障分析	Normal Braking System—Failure Analysis
61	正常刹车系统—调整/试验	Normal Braking System—Adjustment/Test
62	正常刹车系统—检验/检查	Normal Braking System—Inspection/Check
63	刹车活门—拆卸/安装	Brake Valve—Removal/Installation
64	定量器—说明/工作	Proportioner—Description/Operation
65	定量器—拆卸/安装	Proportioner—Removal/Installation
66	减压阀—拆卸/安装	Pressure Reducing Valve—Removal/Installation

98	终点开关—拆卸/安装	Terminal Switch—Removal/Installation
99	起落架收放转换开关—说明/工作	LG Retraction/Extension Changeover Switch—Description/Operation
100	起落架收放转换开关—拆卸/安装	LG Retraction/Extension Changeover Switch—Removal/Installation
101	终点开关—说明/工作	Terminal Switch—Description/Operation
102	终点开关—拆卸/安装	Terminal Switch—Removal/Installation
103	着陆信号灯盒—说明/工作	Landing Signal Light Box—Description/Operation
104	着陆信号灯盒—拆卸/安装	Landing Signal Box—Removal/Installation
105	压力传感器—说明/工作	Pressure Sensor—Description/Operation
106	压力传感器—拆卸/安装	Pressure Sensor—Removal/Installation

照明系统　LIGHTS

1	照明—总论	Lights—Introduction
2	驾驶舱照明—总论	Cockpit Lighting—Introduction
3	荧光照明—说明/工作	Fluorescent Lights—Description/Operation
4	荧光照明—故障分析	Fluorescent Lights—Failure Analysis
5	荧光照明—维修实施	Fluorescent Lights—Maintenance Practice
6	座舱灯照明—说明/工作	Cabin Lights—Description/Operation
7	座舱灯照明—故障分析	Cabin Lights—Failure Analysis
8	座舱灯照明—维修实施	Cabin Lights—Maintenance Practice
9	顶灯照明—说明/工作	Ceiling Lights—Description/Operation
10	顶灯照明—故障分析	Ceiling Lights—Failure Analysis
11	顶灯照明—维修实施	Ceiling Lights—Maintenance Practice
12	仪表灯—说明/工作	Instrument Lights—Description/Operation
13	仪表灯—故障分析	Instrument Lights — Failure Analysis
14	仪表灯—维修实施	Instrument Lights—Maintenance Practice
15	伸缩灯—说明/工作	Retractable Lights—Description/Operation
16	伸缩灯—故障分析	Retractable Lights—Failure Analysis
17	伸缩灯—维修实施	Retractable Lights—Maintenance Practice
18	工作灯—说明/工作	Operating Lights—Description/Operation
19	工作灯—故障分析	Operating Lights—Failure Analysis
20	工作灯—维修实施	Operating Lights—Maintenance Practice
21	货舱、服务舱照明—总论	Cargo and Service Cabin Lighting—Introduction
22	货舱照明—说明/工作	Cargo Cabin Lights—Description/Operation
23	货舱照明—故障分析	Cargo Cabin Lights—Failure Analysis
24	货舱照明—维修实施	Cargo Cabin Lights—Maintenance Practice
25	服务舱照明—说明/工作	Service Cabin Lights—Description/Operation

导航系统　NAVIGATION

4	全静压系统—调整/试验	Pitot-Static System—Adjustment/Test
5	全静压系统—检验/检查	Pitot-Static System—Inspection/Check
6	全压受感器—说明/工作	Pitot Sensor—Description/Operation
7	全压受感器—拆卸/安装	Pitot Sensor—Removal/Installation
8	大气温度表—说明/工作	Air Thermometer—Description/Operation
9	大气温度表—故障分析	Air Thermometer—Failure Analysis
10	大气温度表—拆卸/安装	Air Thermometer—Removal/Installation
11	升降速度表—说明/工作	Vertical Speed Indicator—Description/Operation
12	升降速度—故障分析	Vertical Speed Indicator—Failure Analysis
13	升降速度表—维修实施	Vertical Speed Indicator—Maintenance Practice
14	空速表—说明/工作	Airspeed Indicator—Description/Operation
15	空速表—故障分析	Airspeed Indicator—Failure Analysis
16	空速表—拆卸/安装	Airspeed Indicator—Removal/Installation
17	空速表—调整/试验	Airspeed Indicator—Adjustment/Test
18	高度表—说明/工作	Altimeter—Description/Operation
19	高度表—故障分析	Altimeter—Failure Analysis
20	高度表—维修实施	Altimeter—Maintenance Practice
21	超速告警系统—维修实施	Overspeed Warning System—Maintenance Practice
22	超速告警系统—说明/工作	Overspeed Warning System—Description/Operation
23	大气数据系统—说明/工作	Air Data System—Description/Operation
24	大气数据系统—故障分析	Air Data System—Failure Analysis
25	大气数据系统—调整/试验	Air Data System—Adjustment/Test
26	大气数据计算机—说明/工作	Air Data Computer—Description/Operation
27	大气数据计算机—维修实施	Air Data Computer—Maintenance Practice
28	温度传感器—说明/工作	Temperature Sensor—Description/Operation
29	高度指示器—说明/工作	Altitude Indicator—Description/Operation
30	航向姿态系统—说明/工作	Attitude Heading Reference System—Description/Operation
31	航向姿态系统—故障分析	Attitude Heading Reference System—Failure Analysis
32	航向姿态系统—维修实施	Attitude Heading Reference System—Maintenance Practice
33	应急地平仪—说明/工作	Emergency Horizon—Description/Operation
34	应急地平仪—故障分析	Emergency Horizon—Failure Analysis
35	应急地平仪—维修实施	Emergency Horizon—Maintenance Practice
36	磁罗盘—说明/工作	Magnetic Compass—Description/Operation
37	磁罗盘—故障分析	Magnetic Compass—Failure Analysis

nance Practice

67	无线电高度表—说明/工作	Radio Altimeter—Description/Operation
68	无线电高度表—故障分析	Radio Altimeter—Failure Analysis
69	无线电高度表—拆卸/安装	Radio Altimeter—Removal/Installation
70	无线电高度表—调整/试验	Radio Altimeter—Adjustment/Test
71	Free Flight 2101 I/O GPS 导航仪—说明/工作	Free Flight 2101 I/O GPS—Description/Operation
72	Free Flight 2101 I/O GPS 导航仪—故障分析	Free Flight 2101 I/O GPS—Failure Analysis
73	Free Flight 2101 I/O GPS 导航仪—拆卸/安装	Free Flight 2101 I/O GPS—Removal/Installation
74	Free Flight 2101 I/O GPS 导航仪—调整/试验	Free Flight 2101 I/O GPS—Adjustment/Test
75	空中交通告警和防撞系统—说明/工作	Traffic Alert and Collision Avoidance System—Description/Operation
76	空中交通告警和防撞系统—故障分析	Traffic Alert and Collision Avoidance System—Failure Analysis
77	空中交通告警和防撞系统—拆卸/安装	Traffic Alert and Collision Avoidance System—Removal/Installation
78	空中交通告警和防撞系统—调整/试验	Air Traffic Alert and Collision Avoidance System—Adjustment/Test
79	无线电罗盘—说明/工作	Radio Compass—Description/Operation
80	无线电罗盘—故障分析	Radio Compass—Failure Analysis
81	无线电罗盘—调整/试验	Radio Compass—Adjustment/Test
82	无线电罗盘—检验/检查	Radio Compass—Inspection/Check
83	罗盘接收机—说明/工作	Compass Receiver—Description/Operation
84	罗盘接收机—拆卸/安装	Compass Receiver—Removal/Installation
85	控制盒—说明/工作	Control Box— Description/Operation
86	控制盒—拆卸/安装	Control Box— Removal/Installation
87	组合天线—说明/工作	Antenna Integrated—Description/Operation
88	组合天线—拆卸/安装	Integrated Antenna—Removal/Installation
89	信号阻尼器—说明/工作	Signal Damper—Description/Operation
90	信号阻尼器—拆卸/安装	Signal Damper—Removal/Installation
91	塔康—说明/工作	TACAN—Description/Operation
92	塔康—故障分析	TACAN—Failure Analysis
93	塔康—调整/试验	TACAN—Adjustment/Test
94	塔康收发机—说明/工作	TACAN Transceiver—Description/Operation
95	塔康收发机—拆卸/安装	TACAN Transceiver—Removal/Installation
96	塔康控制盒—说明/工作	TACAN Control Box—Description/Operation

氧气系统　OXYGEN

23	氧源开关—拆卸/安装	Oxygen Supply Valve—Removal/Installation
24	氧源压力表—拆卸/安装	Oxygen Supply Pressure Gauge—Removal/Installation
25	氧气接嘴—说明/工作	Oxygen Filler—Description/Installation
26	便携式氧气系统—说明/工作	Portable Oxygen System—Description/Operation
27	便携式氧气系统—故障分析	Portable Oxygen System—Failure Analysis
28	便携式氧气系统—维修实施	Portable Oxygen System—Maintenance Practice
29	活动氧气瓶—说明/工作	Movable Oxygen Bottle—Description/Operation

引气系统　BLEED AIR

1	引气系统—总论	Bleed Air System—Introduction
2	压力调节/关断活门—说明/工作	Pressure Regulating/Shut off Valve—Description/Operation
3	压力调节/关断活门—拆卸/安装	Pressure Regulating/Shut off Valve—Removal/Installation
4	引气超压开关—说明/工作	Bleed Air Overpressure Valve—Description/Operation
5	引气超压开关—拆卸/安装	Bleed Air Overpressure Valve—Removal/Installation
6	单向活门—说明/工作	Check Valve—Description/Operation
7	单向活门—拆卸/安装	Check Valve—Removal/Installation

供水/排污物设备　WATER/WASTE

1	供水/排污物设备—总论	Water/Waste—Introduction
2	饮用水设备—说明/工作	Drinking Water Equipment—Description/Operation
3	饮用水设备—故障分析	Drinking Water Equipment—Failure Analysis
4	饮用水设备—维修实施	Drinking Water Equipment—Maintenance Practice
5	洗涤用水—总论	Washing Water—Introduction
6	洗涤用水—说明/工作	Washing Water—Description/Operation
7	洗涤用水—故障分析	Washing Water—Failure Analysis
8	洗涤用水—维修实施	Washing Water—Maintenance Practice
9	洗涤用水—调整/试验	Washing Water—Adjustment/Test
10	污废物处理—说明/工作	Waste Disposal—Description/Operation
11	污废物处理—故障分析	Waste Disposal—Failure Analysis
12	污废物处理—维修实施	Waste Disposal—Maintenance Practice
13	污废物处理—调整/试验	Waste Disposal—Adjustment/Test

机载辅助动力装置　AIRBORNE AUXILIARY POWER UNIT

舱门　DOORS

2	登机门—说明/工作	Entry Door—Description/Operation
3	登机门—维修实施	Entry Door—Maintenance Practice
4	前应急舱门—说明/工作	Front Emergency Door of Cockpit—Description/Operation
5	前应急舱门—维修实施	Front Emergency Door of Cockpit—Maintenance Practice
6	驾驶舱上应急口盖—说明/工作	Top Emergency Exit Hatch of Cockpit—Description/Operation
7	驾驶舱上应急口盖—维修实施	Top Emergency Exit Hatch of Cockpit—Maintenance Practice
8	货舱应急窗门—说明/工作	Emergency Exit Hatch of Cargo Cabin—Description/Operation
9	货舱应急窗门—维修实施	Emergency Exit Hatch of Cargo Cabin—Maintenance Practice
10	前应急舱门操纵系统—说明/工作	Control System for Front Emergency Door of Cockpit—Description/Operation
11	前应急舱门操纵系统—故障分析	Control System for Front Emergency Door of Cockpit—Failure Analysis
12	前应急舱门操纵系统—调整/试验	Control System for Front Emergency Door of Cockpit—Adjustment/Test
13	前应急舱门操纵系统—检验/检查	Control System for Front Emergency Door of Cockpit—Inspection/Check
14	分配转换活门—拆卸/安装	Distribution Changeover Valve—Removal/Installation
15	前应急舱门作动筒—拆卸/安装	Actuator for Front Emergency Door of Cockpit—Removal/Installation
16	前舱地板门锁作动筒—拆卸/安装	Lock Actuator for Floor Doors of Cockpit—Removal/Installation
17	前应急舱门密封系统—说明/工作	Sealing System for Front Emergency Door—Description/Operation
18	前应急舱门密封系统—故障分析	Sealing System for Front Emergency Door—Failure Analysis
19	冷气操纵开关—说明/工作	Pneumatic Control Valve—Description/Operation
20	空气减压器—说明/工作	Air Pressure Reducer—Description/Operation
21	货舱大门收放系统—说明/工作	Retraction/Extension System for Cargo Cabin Doors—Description/Operation
22	货舱大门液压操纵系统—说明/工作	Hydraulic Control System for Cargo Cabin Doors—Description/Operation
23	货舱大门液压操纵系统—故障分析	Hydraulic Control System for Cargo Cabin

53	货舱大门—检验/检查	Cargo Cabin Door—Inspection/Check
54	9 框门—说明/工作	Door at Frame 9—Description/Operation
55	9 框门—维修实施	Door at Frame 9—Maintenance Practice
56	59 框出入口盖—说明/工作	Exit Hatch at Frame 59—Description/Operation
57	59 框出入口盖—维修实施	Exit Hatch at Frame 59—Maintenance Practice
58	主起落架舱门—说明/工作	Main Landing Gear Door—Description/Operation
59	主起落架舱门—维修实施	Main Landing Gear Door—Maintenance Practice
60	前起落架舱门—说明/工作	Nose Landing Gear Door—Description/Operation
61	前起落架舱门—维修实施	Nose Landing Gear Door—Maintenance Practice

机身　FUSELAGE

1	机身—总论	Fuselage—Introduction
2	机身—维修实施	Fuselage—Maintenance Practice
3	机身前段—说明/工作	Front Fuselage—Description/Operation
4	驾驶舱盖—说明/工作	Cockpit Canopy—Description/Operation
5	驾驶舱盖—检验/检查	Cockpit Canopy—Inspection/Check
6	雷达罩—说明/工作	Radome—Description/Operation
7	雷达罩—维修实施	Radome—Maintenance Practice
8	机身前段上壁板—说明/工作	Upper Panel of Front Fuselage—Description/Operation
9	机身前段上壁板—检验/检查	Upper Panel of Front Fuselage—Inspection/Check
10	机身前段侧壁板—说明/工作	Side Panel of Front Fuselage—Description/Operation
11	机身前段侧壁板—检验/检查	Side Panel of Front Fuselage—Inspection/Check
12	机身前段下壁板—说明/工作	Lower Panel of Front Fuselage—Description/Operation
13	机身前段下壁板—检验/检查	Lower Panel of Front Fuselage—Inspection/Check
14	机身中段—说明/工作	Middle Fuselage—Description/Operation
15	机身中段前上壁板—说明/工作	Upper Front Panel of Mid Fuselage—Description/Operation
16	机身中段前上壁板—检验/检查	Upper Front Panel of Mid Fuselage—Inspection/Check
17	机身中段前侧壁板—说明/工作	Front Side Panel of Mid Fuselage—Description/Operation
18	机身中段前侧壁板—检验/检查	Front Side Panel of Mid Fuselage—Inspection/Check
19	机身中段前下壁板—说明/工作	Lower Front Panel of Mid Fuselage—Descrip-

		Operation
43	机身尾段前侧壁板—检验/检查	Front Side Panel of Rear Fuselage—Inspection/Check
44	机身尾段后侧壁板—说明/工作	Rear Side Panel of Rear Fuselage—Description/Operation
45	机身尾段后侧壁板—检验/检查	Rear Side Panel of Rear Fuselage—Inspection/Check
46	59～65 框间后下壁板—说明/工作	Lower Rear Panel Between Frames 59～65—Description/Operation
47	58～65 框间后下壁板—检验/检查	Lower Rear Panel Between Frames 59～65—Inspection/Check
48	尾段背鳍—说明/工作	Dorsal Fin of Tail Section—Description/Operation
49	尾段背鳍—检验/检查	Dorsal Fin of Tail Section—Inspection/Check
50	吊车梁安装支架—说明/工作	Crane Framework Mounting Bracket—Description/Operation
51	吊车梁安装支架—检验/检查	Crane Framework Mounting Bracket—Inspection/Check
52	机身尾段地板—说明/工作	Floor of Tail Section—Description/Operation
53	机身尾段地板—检验/检查	Floor of Tail Section—Inspection/Check
54	机身尾舱—说明/工作	Tail Cabin—Description/Operation
55	机身尾舱—检验/检查	Tail Cabin—Inspection/Check

发动机短舱　ENGINE NACELLE

1	短舱—总论	Nacelle—Introduction
2	短舱前段—说明/工作	Nacelle Front Section—Description/Operation
3	短舱前段—检验/检查	Nacelle Front Section—Inspection/Check
4	短舱中段—说明/工作	Nacelle Mid Section—Description/Operation
5	短舱中段—检验/检查	Nacelle Mid Section—Inspection/Check
6	短舱承力段—说明/工作	Nacelle Load-bearing Section—Description/Operation
7	短舱承力段—检验/检查	Nacelle Load-bearing Section—Inspection/Check
8	短舱尾罩—说明/工作	Nacelle Tail Cowling—Description/Operation
9	短舱尾罩—维修实施	Nacelle Tail Cowling—Maintenance Practice

尾翼　STABILIZERS

1	尾翼—总论	Tail—Introduction
2	水平安定面—说明/工作	Horizontal Stabilizer—Description/Operation
3	水平安定面前缘—说明/工作	Leading Edge of Horizontal Stabilizer—Descrip-

窗　WINDOWS

| 7 | 货舱观察窗—维修实施 | Cargo Cabin Observation windows—Maintenance Practice |

机翼　WINGS

1	机翼—总论	Wings—Introduction
2	中央翼—说明/工作	Center Wing—Description/Operation
3	中央翼前大梁—说明/工作	Center Wing Front Spar—Description/Operation
4	中央翼前大梁—检验/检查	Center Wing Front Spar—Inspection/Check
5	中央翼后大梁—说明/工作	Center Wing Rear Spar—Description/Operation
6	中央翼后大梁—检验/检查	Center Wing Rear Spar—Inspection/Check
7	中央翼可卸壁板—说明/工作	Center Wing Removable Panel—Description/Operation
8	中央翼可卸壁板—拆卸/安装	Center Wing Removable Panel—Removal/Installation
9	中央翼可卸壁板—检验/检查	Center Wing Removable Panel—Inspection/Check
10	中央翼上翼面第 I、II 工艺壁板—说明/工作	Process Panels I and II of Center Wing Upper Airfoil—Description/Operation
11	中央翼上翼面第 I、II 工艺壁板—检验/检查	Process Panels I and II of Center Wing Upper Airfoil—Inspection/Check
12	中央翼上翼面第 II 工艺壁板上油量表舱口盖密封垫板—说明/工作	Hatch Cover Sealing Pad of Fuel Quantity Gauge on Process panel II of Center Wing Upper Airfoil—Description/Operation
13	中央翼上翼面第 II 工艺壁板上油量表舱口盖密封垫板—拆卸/安装	Hatch Cover Sealing Pad of Fuel Quantity Gauge on ProcessPanel II of Center Wing Upper Airfoil—Removal/Installation
14	中央翼下壁板—说明/工作	Center Wing Lower Panel—Description/Operation
15	中央翼上翼面第 II 工艺壁板上油量表舱口盖密封垫板—检验/检查	Hatch Cover Sealing Pad of Fuel Quantity Gauge on Process Panel II of Center Wing Upper Airfoil—Inspection/Check
16	中央翼下壁板—检验/检查	Center Wing Lower Panel—Inspection/Check
17	中央翼 0 号翼肋筒道—说明/工作	Rib 0 Tube Channel of Center Wing—Description/Operation
18	中央翼 0 号翼肋筒道—检验/检查	Rib 0 Tube Channel of Center Wing—Inspection/Check
19	中外翼—说明/工作	Inboard Wing—Description/Operation
20	发动机架吊挂接头—说明/工作	Engine Mount Sling Joint—Description/Operation

21	发动机架吊挂接头—检验/检查	Engine Mount Sling Joint—Inspection/Check
22	中央翼与中外翼对接螺栓—说明/工作	Butt Bolts of Center Wing and Inboard Wing—Description/Operation
23	中央翼和中外翼对接螺栓—维修实施	Butt Bolts of Center Wing and Inboard Wing—Maintenance Practice
24	中外翼与外翼对接螺栓—说明/工作	Butt Bolts of Inboard Wing and Outboard Wing—Description/Operation
25	中外翼与外翼对接螺栓—维修实施	Butt Bolts of Inboard Wing and Outboard Wing—Maintenance Practice
26	中外翼上壁板—说明/工作	Inboard Wing Upper Panel—Description/Operation
27	中外翼上壁板—检验/检查	Inboard Wing Upper Panel—Inspection/Check
28	中外翼下部可卸壁板—说明/工作	Inboard Wing Lower Removable Panel—Description/Operation
29	中外翼下部可卸壁板—拆卸/安装	Inboard Wing Lower Removable Panel—Removal/Installation
30	中外翼下部可卸壁板—检验/检查	Inboard Wing Lower Removable Panel—Inspection/Check
31	中外翼下部可卸壁板对接螺栓—说明/工作	Removable Panel Butt Bolts of Lower Inboard Wing—Description/Operation
32	中外翼下部可卸壁板对接螺栓—维修实施	Removable Panel Butt Bolts of Lower Inboard Wing—Maintenance Practice
33	中外翼下部第I、II工艺壁板—说明/工作	Process Panels I and II of Lower Inboard Wing—Description/Operation
34	中外翼下部第I、II艺壁板—检验/检查	Process Panel I and II of Lower Inboard Wing—Inspection/Check
35	外翼—说明/工作	Outboard Wing—Description/Operation
36	外翼上翼面可卸壁板—说明/工作	Outboard Wing Upper Removable Panel—Description/Operation
37	外翼上翼面可卸壁板—检验/检查	Outboard Wing Upper Removable Panel—Inspection/Check
38	外翼上翼面工艺壁板—说明/工作	Outboard Wing Upper Process Panel—Description/Operation
39	外翼上翼面工艺壁板—检验/检查	Outboard Wing Upper Process Panel—Inspection/Check
40	外翼下翼面工艺壁板—说明/工作	Outboard Wing Lower Process Panel—Description/Operation
41	外翼下翼面工艺壁板—检验/检查	Outboard Wing Lower Process Panel—Inspection/Check

螺旋桨　PROPELLERS

动力装置　POWER PLANT

1	动力装置—总论	Power Plant—Introduction
2	动力装置—拆卸/安装	Power Plant—Removal/Installation
3	动力装置—调整/试验	Power Plant—Adjustment/Test
4	发动机安装支架—说明/工作	Engine Mount—Description/Operation
5	发动机安装支架—维修实施	Engine Mount—Maintenance Practice
6	发动机余油系统—说明/工作	Engine Draining Fuel System—Description/Operation
7	发动机余油系统—维修实施	Engine Draining Fuel System—Maintenance Practice

发动机燃油及控制　ENGINE FUEL AND CONTROL

1	发动机燃油及控制—总论	Engine Fuel and Control—Introduction
2	分配系统—说明/工作	Fuel Distribution System—Description/Operation
3	粗油滤—说明/工作	Coarse Filter—Description/Operation
4	粗油滤—维修实施	Coarse Filter—Maintenance Practice
5	细油滤—说明/工作	Fine Filter—Description/Operation
6	细油滤—维修实施	Fine Filter—Maintenance Practice
7	指示系统—说明/工作	Indicating System—Description/Operation
8	燃油耗量指示系统—说明/工作	Fuel Consumption Indicating System—Description/Operation
9	燃油耗量指示系统—故障分析	Fuel Consumption Indicating System—Failure Analysis
10	燃油耗量传感器—说明/工作	Fuel Consumption Sensor—Description/Operation
11	燃油耗量传感器—维修实施	Fuel Consumption Sensor—Maintenance Practice
12	低压燃油压力指示系统—说明/工作	Low Fuel Pressure Indicating System—Description/Operation
13	低压燃油压力传感器—说明/工作	Low Fuel Pressure Gauge Sensor—Description/Operation
14	低压燃油压力传感器—维修实施	Low Fuel Pressure Gauge Sensor—Maintenance Practice
15	细油滤堵塞信号器—说明/工作	Fine Filter Blocking Annunciator—Description/Operation
16	细油滤堵塞信号器—维修实施	Fine Filter Blocking Annunciator—Maintenance Practice
17	燃油压力指示系统—说明/工作	Fuel Pressure Indicating System—Description/Operation
18	燃油压力传感器—说明/工作	Fuel Pressure Sensor—Description/Operation
19	燃油压力传感器—维修实施	Fuel Pressure Sensor—Maintenance Practice

点火　IGNITION

进气　AIR

16	空气封严—说明/工作	Air Sealing—Description/Operation
17	中、后轴承肋形封严腔通气—调整/试验	Ventilation of Mid-Rear Bearing Rib Sealing Cavity—Adjustment/Test
18	中、后轴承肋形封严腔通气—说明/工作	Ventilation of Mid-Rear Bearing Rib Sealing Cavity—Description/Operation
19	空气封严—调整/试验	Air Sealing—Adjustment/Test

发动机操纵系统　ENGINE CONTROL

发动机指示　ENGINE INDICATING

		Description/Operation
8	发动机排气温度表—故障分析	Engine Exhaust Gas Thermometer—Failure Analysis
9	发动机排气温度表—维修实施	Engine Exhaust Gas Thermometer—Maintenance Practice
10	发动机振动过载表—说明/工作	Engine Vibration G-load Indicator—Description/Operation
11	发动机振动过载表—故障分析	Engine Vibration G-load Indicator—Failure Analysis
12	发动机振动过载表—维修实施	Engine Vibration G-load Indicator—Maintenance Practice

排气系统　EXHAUST

1	排气—总论	Air Exhaust—Introduction
2	排气管—说明/工作	Exhaust Pipe—Description/Operation
3	排气管—维修实施	Exhaust Pipe—Maintenance Practice
4	排气管—检验/检查	Exhaust Pipe—Inspection/Check
5	延伸管外罩—说明/工作	Extension Pipe Shroud—Description/Operation
6	延伸管外罩—拆卸/安装	Extension Pipe Shroud—Removal/Installation
7	尾喷管—说明/工作	Tailpipe—Description/Operation
8	尾喷管—拆卸/安装	Tailpipe—Removal/Installation

滑油系统　OIL

1	滑油系统—总论	Oil System—Introduction
2	滑油系统—维修实施	Oil System—Maintenance Practice
3	贮存系统—说明/工作	Storage System—Description/Operation
4	滑油箱—说明/工作	Oil Tank—Description/Operation
5	滑油箱—维修实施	Oil Tank—Maintenance Practice
6	通气油箱—说明/工作	Ventilation Tank—Description/Operation
7	通气油箱—维修实施	Ventilation Tank—Maintenance Practice
8	分配—说明/工作	Distribution—Description/Operation
9	滑油温度调节系统—说明/工作	Oil Temperature Auto Control System—Description/Operation
10	滑油温度自动调节系统—故障分析	Oil Temperature Auto Control System—Failure Analysis
11	滑油温度自动调节系统—检验/检查	Oil Temperature Auto Control System—Inspection/Check
12	滑油散热器—说明/工作	Oil Radiator—Description/Operation
13	滑油散热器—维修实施	Oil Radiator—Maintenance Practice

起动　STARTING

2	起动系统—说明/工作	Starting System—Description/Operation
3	起动系统—故障分	Starting System—Failure Analysis
4	起动系统—维修实施	Starting System—Maintenance Practice
5	定时机构—说明/工作	Timing Mechanism—Description/Operation
6	定时机构—故障分析	Timing Mechanism—Failure Analysis
7	定时机构—维修实施	Timing Mechanism—Maintenance Practice
8	起动箱—说明/工作	Starting Box—Description/Operation
9	起动箱—故障分析	Starting Box—Failure Analysis
10	起动箱—维修实施	Starting Box—Maintenance Practice
11	起动检查继电器—说明/工作	Starting Check Relay—Description/Operation
12	起动检查继电器—故障分析	Starting Check Relay—Failure Analysis
13	起动检查继电器—维修实施	Starting Check Relay—Maintenance Practice
14	起动断开接通发电机输出继电器—说明/工作	Generator Output Cutting off Relay in Starting—Description/Operation
15	起动断开接通发电机输出继电器—故障分析	Generator Output Cutting off Relay in Starting—Failure Analysis
16	起动断开接通发电机输出继电器—维修实施	Generator Output Cutting off Relay in Starting—Maintenance Practice
17	起动接触器—说明/工作	Starting Contactor—Description/Operation
18	起动接触器—故障分析	Starting Contactor—Failure Analysis
19	起动接触器—维修实施	Starting Contactor—Maintenance Practice
20	起动电压调节盒—说明/工作	Starting Voltage Adjusting Box—Description/Operation
21	起动电压调节盒—故障分析	Starting Voltage Adjusting Box—Failure Analysis
22	起动电压调节盒—维修实施	Starting Voltage Regulating Box—Maintenance Practice
23	功率限制器—说明/工作	Power Limiter—Description/Operation
24	功率限制器—故障分析	Power Limiter—Failure Analysis
25	功率限制器—维修实施	Power Limiter—Maintenance Practice
26	发动机自动停车器—说明/工作	Automatic Shutdown Device of Engine—Description/Operation
27	发动机自动停车器—故障分析	Automatic Shutdown Device of Engine—Failure Analysis
28	发动机自动停车器—维修实施	Automatic Shutdown Device of Engine—Maintenance Practice
29	起动自动切油和超温保护器—说明/工作	Starting Auto Fuel Reducer and Over-temperature Protector—Description/Operation
30	起动自动切油和超温保护器—故障分析	Starting Auto Fuel Reducer and Over-temperature Protector—Failure Analysis

特种电子设备 SPECIAL ELECTRONIC EQUIPMENT

维修手册用语

时限/维护 TIME LIMITS/MAINTENANCE

1	概述	general
2	检验/检查	inspection/check
3	第一步检查内容	contents for first step check
4	复查	recheck
5	外部检查	exterior check
6	内部检查	interior check
7	动力装置检查	powerplant check
8	雷击损伤报告	report for lightning strike damage
9	检查程序和内容	check procedures and contents
10	一个轮胎故障	one tyre failure
11	检查步骤	check procedures

尺寸和区域 DIMENSIONS AND AREAS

1	尺寸	dimension
2	面积	area
3	主要区域	major area
4	次级区域	secondary area
5	区域	area
6	标记	identification

顶起和支撑 JACKING AND SHORING

1	用千斤顶顶起飞机	jacking the aircraft with jack
2	降下飞机	lowering the aircraft
3	构造	configuration
4	液压系统工作原理	operating principle of hydraulic system
5	日常维护	routine maintenance
6	工作一个月后必须进行的工作	required maintenance after operating for one month
7	冬季防护和长途运输	protection in winter and long-distance transportation
8	维护	maintenance

9	存放	storage
10	用机轮千斤顶更换机轮	replacing landing gear wheels with wheel jack
11	使用说明	operating description

调平和称重　LEVELING AND WEIGHING

1	所需的设备和材料	required equipment and materials
2	飞机水平测量要求	leveling measurement requirements of aircraft
3	飞机调平要求	requirements for aircraft leveling
4	飞机调平	aircraft leveling
5	测量数据	measurement data
6	注意事项	precautions
7	准备工作	preparations
8	称重步骤	weighing procedure
9	称重后空机重量、重心的计算	calculation of empty weight and C. G. after weighing
10	飞机称重报告	weighing report for aircraft
11	飞机履历本的填写	fill-in of aircraft logbook
12	飞机载重平衡控制程序	loading balance control procedures of aircraft
13	燃油指数表	fuel index sheet

牵引　TOWING

1	牵引飞机的安全规则	safety rules for towing aircraft
2	牵引飞机前的准备	preparations before towing aircraft
3	机头朝前牵引	towing the aircraft with nose forward
4	机尾朝前牵引	towing aircraft with tail forward
5	卸下发动机后的飞机牵引	towing aircraft after engines dismounted
6	机轮损坏情况下的飞机牵引	towing aircraft with wheel damaged
7	主要技术数据	primary technical data
8	维修实施	maintenance practice
9	维护和保养	maintenance and servicing

停放和系留　PARKING & MOORING

1	短停后的飞机再启用程序	aircraft return to service procedures after parking for a short time
2	未油封存放的飞机再启用程序	return to service procedures after storage without preservation
3	油封存放的飞机再启用程序	return to service procedures for preserved aircraft

保养　SERVICING

1	灭火瓶充填	charging of fire bottle
2	灭火瓶的清洗	cleaning of fire bottle
3	中性气体瓶的充填	charging of neutral gas bottle
4	燃油系统各组油箱的容积和加油量	volume and refueling quantity of each tank group of fuel system
5	使用的燃油牌号	type No. for fuel
6	燃油加注	filling of fuel
7	液压油箱加注	filling of hydraulic reservoir
8	液压蓄压瓶充氮	nitrogen charging of hydraulic accumulator
9	起落架缓冲支柱基本技术数据	principle technical data of damper strut
10	前起落架缓冲支柱充填	filling for NLG damper strut
11	主起落架缓冲支柱充填	filling for MLG damper strut
12	轮胎充压	inflating of tires
13	氧气瓶的主要技术数据	primary technical data of oxygen bottle
14	充氧注意事项	precautions for oxygen charging
15	高压玻璃钢氧气瓶、活动氧气瓶的充灌	charging of fiber glass reinforced plastic oxygen bottle and movable oxygen bottle
16	保温水瓶注水	filling of thermal water bottles
17	电加温水箱的注水	filling of electric heating water tank
18	保温水箱注水	filling of thermal water tank
19	滑油箱技术数据	technical data of oil tank
20	滑油加注	filling of oil
21	滑油的放出	draining of oil
22	滑油箱的技术数据	technical data of oil tank
23	滑油种类	type of oil
24	滑油系统放油	draining of oil system
25	润滑剂的说明	description of lubricant
26	润滑时所用的工具和材料	tools and materials for lubricating
27	使用方法	operating method
28	润滑注意事项	precautions of lubricating
29	润滑周期所使用的符号	symbols for lubricating period
30	润滑图例说明	description of lubricating drawing
31	起落架定期保养	scheduled servicing for landing gear
32	清洁飞机外部	cleaning of aircraft exterior
33	清洗飞机窗户和风挡	cleaning of aircraft windows and windshield
34	清洗飞机螺旋桨	cleaning of aircraft propellers
35	飞机停留防结冰	anti-icing during parking

| 36 | 起飞前飞机除冰雪 | ice and snow removal before take off |

环控系统　ENVIRONMENTAL CONTROL SYSTEM

1	环控系统常见故障的分析及排除	analysis and troubleshooting of common failures of environmental control system
2	输气管路气密性检查	airtightness check for air-delivery pipeline
3	增压管路的气密性检查	airtightness check for pressurizing pipeline
4	环控系统的功能检查	check for function of environmental control system
5	环控系统外部检查及维护	external check and maintenance of environmental control system
6	元件位置	component installation position
7	拆卸	removal
8	安装	installation
9	元件说明	component description
10	收尾工作	winding up
11	工作	operation
12	消耗器材	consumption equipment and material
13	主要技术指标	principle technical data
14	工作原理	operating principle
15	过滤器的清洗	cleaning of filter
16	构造及工作	configuration and operation
17	消耗材料	consumable material
18	滤蕊的清洗	cleaning of the filter element
19	指示数据	indicated data
20	组成及安装位置	composition and installation position
21	元件组成及安装位置	component and installation position

自动飞行　AUTO FLIGHT

1	空中使用	operation in flight
2	调整/试验前的检查	check before adjustment/test
3	通电前的准备	power-on preparation
4	通电检查	power-on check
5	垂直陀螺和速率陀螺的装机调整检查	adjusting check for installation of the vertical gyro and rate gyro on aircraft
6	地面定中心检查	perform the ground centering check
7	调整飞行的典型任务书	typical mission record for adjusting flight
8	副翼舵机的拆卸/安装	removal/installation of aileron actuator
9	升降舵机和方向舵机的拆卸/安装	removal/installation of elevator and rudder actua-

tors

10	位置反馈传感器的拆卸/安装	removal/installation of position feedback sensor
11	位置反馈传感器的拆卸/安装	removal/installation of position feedback sensor
12	接通按钮、改平按钮的拆卸/安装	removal/installation of engagement button and level- off button
13	接通指示灯的拆卸/安装	removal/installation of engagement indicating light
14	交联部件及安装位置	crosslink components and installation position
15	交联部件说明	description of crosslink components
16	自动导航信号选择的拆卸/安装	removal/installation of automatic navigation signal selector
17	导航选择指示的拆卸/安装	removal/installation of navigation select indication
18	导航系统继电器盒的拆卸/安装	removal/installation of navigation system relay box
19	航向联系盒的拆卸/安装	removal/installation of heading linkage box
20	液压油滤的拆卸/安装	removal/installation of hydraulic filter
21	清洗液压油滤滤芯	hydraulic filter core cleaning

通信系统　COMMUNICATIONS

1	短波通信电台	HF communication radio set
2	超短波通信电台	VHF/UHF communication radio set
3	音频系统	audio system
4	静电释放	static discharging
5	座舱音频记录系统	cabin audio recording system
6	系统检查	system check
7	各单元部件的更换检查	replacement and check of all unit components
8	简易故障寻找及应急维修	simple failure search and emergency maintenance
9	常见故障处理方法	common failure troubleshooting methods
10	外场拆装	field Removal/installation
11	铜索天线的拆卸/安装	removal/installation of copper-cable antenna
12	支柱内钢索的拆卸/安装	removal/installation of cable inside the strut
13	超短波电台天线	VHF/UHF radio set antenna
14	超短波电台故障排除步骤	troubleshooting procedures for VHF/UHF radio set
15	常见故障模式及排除方法	common failure mode and troubleshooting method
16	超短波电台通电前准备	preparations before VHF/UHF radio set power-on
17	电台加电	radio set power-on
18	超短波电台断电操作	operation at VHF/UHF radio set power-off

19	超短波电台自检操作	self-test operation of VHF/UHF radio set
20	超短波电台功能检查	functional check of VHF/UHF radio set
21	单音检查	tone check
22	外场故障诊断和排除	field failure diagnosis and troubleshooting methods
23	维修过程中的注意事项	maintenance precautions
24	外场故障诊断和排除	field failure diagnosis and troubleshooting
25	内场故障诊断和排除	workshop failure diagnosis and troubleshooting
26	运输和贮存	transportation and storage
27	放电刷的拆卸/安装	removal/installation of discharge brush
28	接地锤的拆卸/安装	removal/installation of grounding cone
29	接地刷的拆卸/安装	removal/installation of grounding brush
30	与机上设备信号交联关系	signal crosslinking relation with onboard equipment
31	设备拆卸、安装前的准备和安装后的检查	preparation before equipment removal/installation and check after installation
32	回放检查	replay check

电源系统 ELECTRICAL POWER

1	电气设备的维护	electrical equipment maintenance
2	电气网路的维护	electrical network maintenance
3	主要技术数据及使用条件	main technical data and operation conditions
4	直流电源系统供电原理说明	power supply principle description for DC power supply system
5	地面电源供电功能检查	functional check of ground power supply
6	蓄电池供电的操纵与检查	control and check of battery power supply
7	直流起动发电机供电的操纵与检查	power supply control and check of DC starter/generator
8	碳刷的修理	repair of carbon brush

设备 EQUIPMENT/FURNISHINGS

1	装弹与卸弹	loading and unloading of signal flare
2	信号枪发射线路的检查	checking of signal gun shooting track
3	维护要求及方法	maintenance requirement and method
4	清洗与修理	cleaning and repair
5	工作原理及使用方法	operating principle and method
6	维护要求及方法	maintenance requirement and method
7	系统配套组成及安装位置	system composition and installation position
8	故障分析	failure analysis

9	搭接测试	bonding test
10	通电检查	power-on check
11	清理/涂覆	cleaning/coating

防火　FIRE PROTECTION

1	手提式灭火瓶的充填	filling of the portable fire bottle
2	灭火瓶和球形灭火瓶的充填	charging of fire bottle and spherical fire bottle
3	中性气体瓶的充填	filling of the neutral gas bottle

飞行操纵　FLIGHT CONTROLS

1	拉杆的更换	replacement of pull rod
2	调整/试验	adjustment/test
3	拆卸/安装（典型）	removal/installation（typical)
4	电气控制线路说明	sescription of electrical control circuit
5	方向舵操纵系统的调整	adjustment of rudder control system
6	润滑保养	lubrication maintenance
7	电气控制线路说明/工作	description/operation of electrical control circuit
8	清理收尾	cleaning andwinding up
9	襟翼液压收放系统的主要技术数据	principle technical data of flap retraction/extension system
10	停机制动系统的调整	adjustment of control surface lock system

燃油系统　FUEL SYSTEM

1	软油箱的保管	storage of bag tank
2	软油箱折叠	folding of bag tank
3	软油箱的拆卸/安装	removal/installation of bag tank
4	整体油箱的清洗	cleaning of integral fuel tank
5	整体油箱的密封试验	sealing test of integral fuel tank
6	通气系统的检验/检查	inspection/check of vent system
7	系统组成及安装位置	system composition and installation position
8	指示器的拆卸/安装	removal/installation of indicator
9	逆变电源盒的拆卸/安装	removal/installation of inverter power box
10	转换开关的拆卸/安装	removal/installation of changeover switch
11	油量表遥控开关的拆卸/安装	removal/installation of remote control switch of fuel quantity gauge
12	加油控制器的拆卸/安装	removal/installation of refueling controller
13	耗油控制器的拆卸/安装	removal/installation of fuel consumption controller
14	传感器的拆卸/安装	removal/installation of sensor

15	油量表零位调整(在空油箱中进行)	adjusting the zero position of the fuel quantity indicator (carry out adjustment under the condition of the empty fuel tank)
16	油量表满值误差调整(在油箱加满油时进行)	adjusting the full value error of the fuel quantity indicator (when the fuel tank is filled up)
17	测量系统的检查	check of measuring system
18	人工耗油检查(发动机开车时进行)	manual fuel consumption check (perform when engine running)
19	自动耗油检查(发动机开车时进行)	automatic fuel consumption check (perform when engine running)
20	油量平衡情况的检查(发动机开车时进行)	fuel quantity balance check (perform when engine running)

液压源 HYDRAULIC POWER

1	增压系统管路的检查	check on reservoir pressurization system pipeline
2	液压系统的使用数据	hydraulic system operation data
3	干燥过滤器的检查	check on dryer filter
4	安全阀和放气活门的检查	check on safety valveand air bleed valve
5	减压器压力的检查	pressure check on pressure reducer
6	电动泵电路故障分析	circuit failure analysis of electric pump circuit
7	指示器的拆卸/安装	removal/installation of the indicator
8	右系统压力传感器的拆卸/安装	removal/installation of pressure gauge sensor of right system
9	左系统和左、右系统蓄压瓶压力传感器的拆卸/安装	removal/installation of pressure sensor of left system，accumulator pressure sensors of left/right systems

除雨和防冰 ICE AND RAIN PROTECTION

1	飞机的防冰系统	aircraft anti-icing system
2	飞机的防雨系统	aircraft rain protection system
3	机翼前缘热空气防冰系统电气控制线路说明/工作	description/operation of electrical controlling circuit for hot air anti-icing system of wing leading edge
4	工作原因	operation cause
5	电气原理及工作说明	electrical principle and operation description
6	电气控制线路说明/工作	description/operation of electrical control circuit
7	加温报警控制盒的主要技术数据	main technical data for heating warning control box
8	左侧全静压受感器加温检查	heating inspection for left pitot-static

		pressure sensor
9	右侧全静压受感器加温检查	heating inspection for right pitot-static pressure sensor
10	检查结束	inspection finished
11	螺旋桨加温系统的通电检查	power-on check for propeller heating system
12	螺旋桨加温信号控制盒的拆卸/安装	removal/installation of propeller heating signal control box
13	调换灯泡	bulb replacement

显示/记录系统　INDICATING/RECORDING SYSTEMS

1	卡箍式仪表指示器的拆卸/安装	removal/installation of the clamp-type instrument indicator
2	前固定式和后固定式仪表指示器的拆卸/安装	removal/installation of the front-mounting-type and rear-mounting-type instrument indicators
3	更换电池	battery replacement
4	亮度调节检查	brightness adjustment check
5	现时时间的显示及调整	display and adjustment of real time
6	航时的起动、停止及回零	starting, stop and return-to-zero of flight time
7	测时的起动、停止及回零	starting, stop and return-to-zero of measurement time
8	组成及简要原理	composition and simplified principle
9	系统工作原理及与其它设备交联简介	system operating principle and introduction to crosslinking with other devices
10	各部件功能说明	component function description
11	飞参外场检测处理机时钟正确性说明	clock accuracy description of the flight parameter field check processor
12	水下定位信标	underwater locator beacon
13	采集器电池	battery of flight parameter collector
14	快取记录器(含数据卡)	quick access recorder (including data card)
15	飞行参数采集器	flight parameter collector
16	飞行参数记录器	flight data recorder
17	三轴加速度计	three-axis accelerometer
18	角位移传感器	angular displacement sensor
19	角位移传感器的调整	adjustment of angular displacement sensor
20	系统调整检查	system adjustment check
21	维护说明	maintenance description

起落架　LANDING GEAR

1	调整	adjustment

2	主起落架的试验	MLG test
3	试验	test
4	前起落架上位锁应急开锁系统	emergency unlocking system of NLG uplock
5	前起落架下位锁应急上锁系统	emergency locking system of NLG downlock
6	主起落架上位锁应急开锁系统	emergency unlocking system of MLG uplock
7	主起落架下位锁应急上锁系统	emergency locking system of MLG downlock
8	排除系统内空气	air bleed in system
9	应急刹车压力的调整	adjustment of emergency brake pressure
10	主机轮轴承轴向间隙的调整	adjustment of axial clearance of main wheel bearing
11	机轮维护	wheel maintenance
12	前轮转弯操纵系统主要技术性能	main technical performance for nose wheel steering control system
13	前轮转弯系统的组成及安装	composition and installation position of nose wheel steering system
14	元件说明	component description
15	钢索目视检查	visual check of steel cable
16	随动系统检查	follow-up system check
17	电路组成及安装位置	circuit composition and installation position

照明系统　LIGHTS

1	照明灯具的通用技术要求	general technical requirements for lights
2	飞机上调整/试验	onboard adjustment/test

导航系统　NAVIGATION

1	全静压系统提供气压的设备	equipment with pressure supplied by pitot-static system
2	全、静系统发生故障的处置	troubleshooting for pitot-static system
3	全压、静压系统故障的原因和现象	cause and symptom of trouble in pitot-static system
4	判断全、静系统发生故障的方法	method for judging trouble in pitot-static system
5	试验结束	winding up test
6	空速管的检查	check for pitot tube
7	全、静压管路检查	check for pitot-static pipeline
8	沉淀器的检查	check for precipitator
9	安装后的检查	check after installation
10	M 数表	Mach indicator
11	速压信号器	dynamic pressure annunciator
12	系统配套组成及安装位置	composition and installation position of system kit

13	外场故障诊断和排除	field failure diagnosis and troubleshooting
14	内场故障诊断和排除	workshop fault diagnosis and troubleshooting
15	仪器设备	instrument and equipment
16	维修过程中的注意事项	precautions in maintenance
17	传感器接入系统后的检查	check after sensor connecting with system
18	高度指示器接入系统后的检查	check after altitude indicator connecting with system
19	各部件介绍	component description
20	姿态	attitude
21	航向	heading
22	磁航向传感器的校准	magnetic heading sensor calibration
23	航向姿态系统使用注意事项	precautions in ahrs operation
24	系统配套组成及安装位置	component and installation position
25	简要工作原理	simple operating principle
26	检查(内场)	workshop check
27	操作使用	operation
28	涂漆	painting
29	设备拆卸/安装前后的准备和检查	preparation and check before and after equipment removal/installation
30	气象雷达系统简介	radar system introduction
31	雷达的交联接口关系	radar crosslinking interface relationship
32	通电前的检查	check before power-on
33	天线的拆卸/安装	removal/installation of antenna
34	与航向姿态系统的交联检查	check for crosslinking withahrs
35	使用操作程序	operation procedures
36	系统故障分析和处理方式	system failure analysis and solutions
37	惯导系统故障编码及定位	inertial navigation system fault code and positioning
38	状选、控显故障的处理	management of failure in mode selector and control display unit
39	装订及查看安装误差角	binding and viewing installation error angle
40	系统介绍	system description
41	故障隔离	fault isolation
42	主要技术指标(收发机)	main technical data (transceiver)
43	面板介绍	introduction to control panel
44	S模式地址码的设置	S-mode address code setting
45	TCAS 收发机	TCAS transceiver
46	ATC 控制盒	ATC control box
47	指示器	indicator

48	使用说明	operation description
49	罗差测定及调整程序	measuring compass deviation and adjusting procedure
50	结构特点	structure characteristics
51	塔康系统简介	TACAN system introduction
52	塔康的交联接口关系	TACAN crosslinking relationship
53	自检排故	built-in test to correct failure
54	常见故障及排除方法	common failure and troubleshooting
55	面板布置	panel layout
56	需用材料及检测设备	materials and equipment required
57	控制盒面板简介	introduction on the control panel
58	综合显示器面板简介	introduction on the integrated display unit panel
59	基本工作原理	basic operating principle

氧气系统　OXYGEN

1	组成	composition
2	氧气设备的保养	servicing of oxygen equipment
3	拆卸/安装注意事项	precautions during removal/installation
4	常规检查	routine check
5	高压系统气密性检查	air tightness check for the high-pressure system
6	低压系统气密性检查	air tightness check for the low-pressure system
7	环境温度的变化对气密性检查的影响	impact of the ambient temperature change on the air tightness check
8	氧气调节器气密性检查	air tightness check of oxygen regulator
9	氧气调节器高压内腔的气密性检查	air tightness check of the high-pressure inner cavity in oxygen regulator
10	氧气调节器到供氧软管堵盖部分的气密性检查	air tightness check of the section between the oxygen regulator and the oxygen supply hose blanking cover
11	氧气调节器向面罩内供氧情况的检查	check of oxygen supply from the regulator to the mask
12	氧气瓶充氧	charging to oxygen bottle
13	配套使用	component operation
14	特点	features

供水/排污物设备　WATER/WASTE

1	加注	water filling
2	保养	servicing
3	修理	repair

4	污物排出系统	waste drain system
5	抽水马桶的清洗	cleaning of flush toilet
6	污物排出系统	waste drain system

机载辅助动力装置　AIRBORNE AUXILIARY POWER UNIT

1	辅助动力装置系统组成	composition of auxiliary power unit（APU）
2	燃气涡轮发动机的主要数据	main technical data of gas turbine engine
3	油封	preservation
4	转速的调整	speed adjustment
5	燃油消耗量的调整	fuel consumption adjustment
6	进入空转转速时间的调整	adjustment of duration for entering the idling state
7	指示器拆卸/安装	removal/installation of the indicator
8	传感器的拆卸/安装	removal/installation of the sensor
9	排气温度表指示器的拆卸/安装	removal/installation of exhaust gas temperature indicator
10	排气温度表传感器的拆卸/安装	removal/installation of exhaust gas temperature indicator sensor
11	涡轮发电装置温度系统的电阻测量	resistance measurement of turbine generating device temperature system
12	滑油压力的调整	oil pressure adjustment

舱门　DOORS

1	更换登机门的密封胶带	replacement of sealing tape of entry door
2	密封胶带的更换	replacement of sealing rubber tape
3	应急窗门结构	structure of emergency exit hatch
4	应急窗门锁机构	lock mechanism of emergency exit hatch
5	货舱应急窗门密封带的更换	replacement of sealing tape of emergency exit hatch of cargo cabin
6	在试验过程中检查	check during test
7	故障和排除方法	failure analysis and troubleshooting
8	用左、右蓄压瓶供压时检查	check when pressure provided by left and right accumulators
9	液压系统组成及安装位置	hydraulic system composition and installation position
10	功用	function
11	结构与工作	structure and operation
12	结构与主要性能数据	structure and main technical data
13	工作原理	operating principle

14	结构及主要技术性能	structure and main technical performance
15	结构与工作原理	structure and operating principle
16	59 框出入门密封带的更换	replacement of sealing rubber tape for exit hatch at frame 59

机身　FUSELAGE

1	机身结构说明	description of fuselage structure
2	机身气密舱气密性试验	airtightness test for fuselage pressurized cabin
3	机身密封舱蒙皮的检查及维护	check and maintenance for skin of fuselage pressurized cabin
4	橡胶密封带的更换	replacement of sealing rubber tape

窗　WINDOWS

1	座舱玻璃的更换	replacement of canopy glass
2	机头罩玻璃的更换	replacement of nose fairing glass
3	货舱观察窗玻璃的更换	replacement of cargo cabin observation windows

机翼　WINGS

1	更换对接螺栓	butt bolts replacement
2	检查程序	check procedures
3	工作程序	operation procedures
4	襟翼的作用	function of flap
5	滑轮架的保养	pulley mount servicing
6	襟翼的拆卸/安装	removal/installation of flaps
7	副翼的作用	function of aileron
8	副翼调整片的拆卸/安装	removal/installation of aileron trim tabs

螺旋桨　PROPELLERS

1	螺旋桨的拆卸	removal of propeller
2	螺旋桨的安装	propeller installation
3	螺旋桨的检验/检查	inspection/check of propeller
4	元件的安装位置	component installation position

动力装置　POWER PLANT

1	发动机主要性能数据和限制	main performance data and limits of engine
2	动力装置维护的安全要求	safety requirements for power plant maintenance
3	一般技术规定	general technical requirements
4	各种气候条件下发动机的使用	engine operation under various weather conditions
5	发动机防冰加温系统的使用	operation of engine anti-icing heating system

6	发动机灭火与清洗	engine fire-extinguishing and cleaning
7	发动机的清洗	engine cleaning
8	设备、工具和器材	equipment，tools and instrument
9	安装后的检查	check after installation
10	发动机水平测量	engine leveling measurement
11	飞行前检查	preflight check
12	假起动（供油、不点火）	false start（fuel supplied without ignition）
13	冷转发动机（不供油、不点火）	engine cold running（without fuel supply andignition）
14	发动机的起动	engine start
15	发动机的预热和检查工作	engine warming up and check
16	发动机停车的检查	engine shutdown check
17	发动机的冷机与停车	engine cooling and shutdown

点火　IGNITION

1	说明—机身结构	Description—Fuselage Structure
2	功能检查	function check
3	地面起动点火	ground starting ignition
4	空中点火	in-flight start ignition

进气　AIR

| 1 | 防冰压力信号器的校验和调整 | check and adjustment of anti-icing pressure annunciator |

发动机操纵系统　ENGINE CONTROL

1	油门操纵手柄组件检验/检查	inspection/check for throttle control handle assembly
2	短舱连杆机构的检验/检查	inspection/check for nacelle link mechanism
3	钢索的检验/检查	inspection/check for steel cable

发动机指示　ENGINE INDICATING

1	指示器的拆卸/安装	removal/installation of indicator
2	传感器的拆卸/安装	removal/installation of sensor
3	信号转换器的拆卸/安装	removal/installation of rotation speed signal converter
4	振动传感器的拆卸/安装	removal/installation of vibration sensor
5	振动放大器的拆卸/安装	removal/installation of vibration amplifier
6	振动指示器的拆卸/安装	removal/installation of vibration indicator

排气系统　EXHAUST

1	延伸管和外罩的裂纹修理	treatment of cracks on extension pipe and shroud

滑油系统　OIL

1	滑油系统工作介质	operation medium of oil system
2	滑油充填	oil filling
3	指示正确性检查	indication correctness check
4	滑油油量表传感器拆卸/安装	removal/installation of oil quantity sensor
5	滑油油量表指示器拆卸/安装	removal/installation of oil quantity indicator
6	滑油剩油告警灯的检查	check of residual oil warn light
7	滑油油量表指示器的亮度调节	brightness adjustment of oil quantity indicator

起动　STARTING

1	安全规则	safety regulation
2	维护规则	maintenance rule

特种电子设备　SPECIAL ELECTRONIC EQUIPMENT

1	部件功能说明	component function description
2	操作使用	operation
3	常见故障判断与处理	common failures diognosis and troubleshooting

英文索引

adjust control tube / 161

adjust/test / 167

adjustable height / 11

adjustable hook / 184

adjustable inlet / 148

adjustable mechanism / 184

adjustable nozzle / 176

adjustable pull rod / 68

adjustable resistor / 51

adjustable screw / 122

adjustable support arm / 101

adjustable thread rod / 11

adjuster / 161

adjusting bolt / 116

adjusting bonnet / 34

adjusting bushing / 101

adjusting gasket / 141

adjusting joint / 68

adjusting nut / 63

adjusting resistance / 173

adjusting screw / 51,81,101,178

adjusting thread rod / 122

adjusting thread sleeve / 134

adjustment / 24,44,173

adjustment block / 114

adjustment gasket / 167

adjustment procedure / 81

adjustment shim / 77

adjustment/test / 81

advanced ignition / 148

adverse roll / 38

advise established inbound on ILS / 188

aerial delivery / 184

aero (aircraft) oil / 178

aerobatic flight / 188

aerobatics flight test / 188

aerodrome weather minimum / 188

aerodynamic characteristic / 81

aerodynamic configuration / 81,141

aerodynamic control structure / 116

aerodynamic force / 38,116

aerodynamic load / 101,128

aerodynamic performance / 141

aerodynamic vane / 137

aerodynamic variable pitch propeller / 141

aerodynamics contour / 134

aerovan / 184

aero-washing gasoline / 38,101,122

affect / 167

after cargo compartment / 184

after fairing / 118

after fan / 148

after locking device / 184

afterburner / 166

afterburner diffuser / 148

afterburner fuel manifold / 148

afterburner fuel regulator / 162

afterburner fuel-flow regulator / 162

afterburner igniter / 148

afterburner ignition / 148

afterburner jet pipe / 148

afterburner nozzle / 148

afterburner test / 148

aging / 34,122,167

AHRS control box / 89

AHRS power availability / 89

aileron / 1

aileron actuator / 38,24,77,82

aileron actuator deflection angle sensor / 89

aileron balance tab / 24

aileron bay / 137

aileron control system / 82

aileron deflection angle / 89

aileron degree / 38

aileron differential rocker arm / 24

aileron effect / 82

aileron lock / 24

aileron lock mechanism / 68

aileron position feedback sensor / 68

aileron pull rod joint / 24

aileron rocker arm / 24

aileron servo tab / 68

aileron trim tab / 24,68

aiming sight / 89

air adapter / 167

air bag / 135

air bleed overpressure valve / 34

air bleed pressure valve / 81

air bleed switch / 4

air bleed unit / 148

air bleed valve / 24,118,162,167

air bleed valve housing / 167

air bottle / 24

air bottle/pneumatic storage bottle / 34

air cargo / 184

air cargo system / 184

air chamber / 81

air charging device / 60

air charging union / 77

air compressor / 81,116,148,162,178

air compressor blade / 148

air conditioning / 34

air conditioning bay / 128

air conditioning equipment bay / 128

air conditioning pipeline / 4

air conditioning system / 34,81

air damper / 107

airdata / 107

air data computer (ADC) / 89,107

air data measurement control instrument / 107

air data system / 89,107

air delivered by left/right engine / 89

air delivery pipeline / 128

air diffuser duct / 118

air distribution system / 34

air express / 184

air extractor / 24

air filter / 34,77,117

air flow / 167

air freight / 184

air gap / 51

air guider shroud / 77

air horn / 81

air horn system / 117

air horn valve / 34

air identification interrogation signal / 183

air idling / 172

air impingement starting system / 148

air inlet / 167,178

air inlet door / 118

air inlet door actuator / 118

air inlet heating connection signal light / 81

air inlet icing signal switch / 89

air inlet nipple / 114

air inlet shutter / 24

air inlet shutter electric mechanism / 119

air inlet temperature / 162

air inlet/air intake / 81,162

air leakage / 24,128

air leakage quantity / 114

air outlet nipple / 114

air outlet tube / 81

air passage / 148

air pipe / 167

air pipe connector / 81

air pressure / 101

air pressure reducer / 122

air pressure sensor / 107

air pressure signal sensor / 81

air pressurization system / 77

air restricting valve / 81

air scoop / 148

air separator/deaerator / 149,167,178

air source / 34

air speed indicator / 4

air start switch / 4

air starter/air-injection starter / 149

air storage tank / 119

air stream / 149

air supply cock / 90

air supply thermometer / 34,90

air tactics trainer / 188

air temperature / 184

air temperature indicator / 90

air temperature meter / 90

air tightness check/air tightness inspection / 114

air tightness test / 178

air traffic control transponder / 107

air transport / 60

air transportability / 184

air turbine starter / 119,149

air volume flow / 167

air/air distance-measuring / 107

air/air mode / 107

air/ground mode / 107

air/ground relay (A/GR) / 119

air/oil separator / 178,119

air-bleed valve / 167

airborne component / 4,13

airborne crashworthy recording system / 89

airborne electrical equipment / 1

airborne equipment / 1,44,107

airborne marker panel / 107

airborne parachute horn / 60

airborne short-range navigation device / 107

airborne starter/on-board starter / 149

airborne tool / 81,141,166

airborne transponder / 88,183

air-charging nipple / 135

air-charging valve / 90

aircraft aerodynamic contour/configuration / 122

aircraft attitude / 14,38

aircraft axis / 122

aircraft base level line/aircraft level reference datum / 128

aircraft component / 4

aircraft container / 184

aircraft control surface / 38

aircraft control system / 5,68

aircraft deflection / 38

aircraft formation / 44

aircraft horizontal datum line / 90

aircraft inertia / 38

aircraft jacking state / 122

aircraft lateral axis / 38

aircraft motion parameter / 38

aircraft on-board weight / 184

aircraft pallet / 125

aircraft pallet net / 185

aircraft parking state / 122

aircraft power self-calibration circuit / 44

aircraft registered number / 1

aircraft surface / 19

aircraft symbol / 107

aircraft unit load device (ULD) / 185

aircraft wiring manual / 74

aircrew / 24,44,77,81,173

aircrew oxygen system / 24

aircrew seat / 24

air-delivery pipe / 34,81

air-delivery switch / 81

air-delivery valve / 117

air-driven turbine / 149

airdrop / 51,60,117,185

airdrop operator / 122

airdrop parachuting console / 122

airdrop platform / 185

airdrop pull rod / 123

airdrop signal / 90

airdrop system / 185

airdrop weight / 185

altitude / 64,90,141,173,188

altitude alert light / 90

altitude annunciator / 90

altitude code / 107

altitude compensation signal / 38

altitude difference sensor / 38

altitude difference signal / 38

altitude differential pressure gauge / 4

altitude indicator / 90,107

altitude interface signal / 183

altitude limit / 149

altitude loss / 107

altitude setting value / 90

altitude signal / 90

aluminum / 81

aluminum alloy / 34,166

aluminum alloy forging / 123

aluminum alloy plate / 178

aluminum bronze bush / 101

aluminum casting / 17

aluminum foil / 60

aluminum foil membrane / 64

aluminum forging / 119

aluminum liner/aluminum lining / 115

aluminum pipe / 64

aluminum plate / 135

aluminum powder organic silicon heat-resist-ant paint / 178

aluminum sheet / 167

aluminum shell / 64

aluminum sleeve / 101

aluminum wire / 51

aluminum-cased holder / 141

aluminum-color heat-resistant paint / 176

ambient condition / 81

ambient factor / 38

ambient temperature / 24,38,64,81,101,107,141,149,168,173

ammeter / 51,81,141

amperemeter / 90

amplification/enlarge / 90

amplifier / 34,38,162,173

amplifier control / 4

amplifier transformer / 173

amplify / 44,81

amplifying element / 162

amplitude / 74,166

amplitude compatible modulation equivalent / 44

amplitude limiter / 107

amplitude modulation / 44

analog distributor / 90

analog interface / 44

analog quantity / 90

analog sensor / 64

analog signal / 90,107

analog signal interface module / 90

analysis system / 173

anchor nut / 123

aneroid altimeter / 107

aneroid diaphragm capsule assembly / 115

angle / 123,128

angle box / 137

angle fitting / 137

angle of attack (AOA) / 51,107

angle of attack (AOA) sensor / 14

angle of side slip / 107

angle phase difference / 107

angle piece / 117,149

angle protractor / 137

angle sensor / 38

angular box / 168

angular displacement / 38,90

angular displacement sensor / 90

angular rate sensor / 38

angular rate/angular speed / 38

angular sheet / 135

angular speed of turn / 107

autoration angular rate component / 108

auto-regulating system / 178

auto-return-to-zero system/automatic zero return system / 39

auto-transformer / 81,108

autotransformer starter / 149

auxiliary armature / 74

auxiliary control system / 68

auxiliary frame/secondary frame / 128

auxiliary fuel pump / 162

auxiliary igniter / 149

auxiliary inlet / 149

auxiliary instrument / 90

auxiliary material / 101

auxiliary motor / 142

auxiliary oil pump / 24,142,178

auxiliary power unit (APU) / 24,34,149

auxiliary pump / 19

auxiliary ramp / 60

auxiliary thread rod / 11

auxiliary tool / 14

auxiliary travel / 64,101

auxiliary vent valve / 90

auxiliary ventilation valve / 34

auxiliary wheel / 185

available fuel quantity / 74

average airspeed / 188

average reading/arithmatic average value / 128

average transient error / 90

aviation balloon cloth / 137

aviation chemical sanitary agent / 117

aviation cleaning gasoline / 68,78

aviation clock / 5,90

aviation fuel / 74

aviation gasoline / 51

aviation grease / 68

aviation hydraulic fluid / 24,39,78,81,101

aviation kerosene / 74,123,149

aviation lubricating oil / 24

aviation organic glass / 128

aviation rubber plate / 137

aviation sponge rubber plate / 137

aviation standard hose / 78

aviation watch / 5

avoiding / 14

axial acceleration / 90

axial clearance / 172,173,178

axial compensation / 137

axial diffuser / 149

axial force / 135

axial line / 82

axial load / 149,179

axial load (pulling) / 137

axial propeller / 142

axial pulling off force / 60

axial sleeve / 101

axial symmetry inlet / 149

axial-flow compressor / 149

axial-flow compressor passage / 149

axial-flow turbine / 149

axis / 45,68,168

axis of propeller / 142

axle / 1,51

axle load / 185

axle pin / 123

azimuth / 45,90

back power inspection / 45

back pressure / 162

back to the center line / 188

backup oil quantity / 179

backup ring / 119

backward difference / 137

backward valve / 34

baffle / 19,176

baffle-type chamber / 149

bag tank/flexible tank / 24,74

bail / 60

bail-out / 134

bakelite base / 60

bakelized paper / 137

balance / 14,34,68,173

balance arm / 14

balance modulator / 108

balance position / 90

balance spring / 108

balance system / 185

balance tab / 14,24

balance tab control system / 24

balance tab/compensating plate / 128

balance valve / 74

balance weight / 14,117

balanced pulse / 108

balancing circuit / 51

balancing coil / 51

balancing plate / 149

balancing resistor / 51

balancing speed / 142

balancing valve / 5

balk lever / 162

ball bearing / 68

ball compensator / 5

ball helical mechanism / 68

ball holder / 68

ball joint / 34,82,101

ball nipple / 82

ball nut / 68

ball rail / 68

ball socket / 11

ball track / 51

ball transfer panel / 185

ball transfer unit / 185

ball valve / 115

ball-end bracing / 132

ball-head support thread rod / 11

ball-shaped pipe nipple / 179

band changeover selector switch / 90

band suppressor / 90

band switch / 64

band width / 45

bank angle / 39,90

bank angle holding / 39

bank attitude / 39

bank channel / 39

bank coordinated signal / 39

bank drawer / 39

bank measuring axis / 39

banking turn / 39

bank-turn flight/banked turn / 39

barometer altitude / 108

barometric altimeter / 90

barometric solenoid valve / 5

barostat / 162

barrier net / 185

base / 39,82,91,123

base assembly / 39

base leg / 188

base pr ofile / 149

base speed / 188

basic airspeed / 188

basic code / 108

basic definition / 14

basic operating weight / 14

basic performance / 24

basin doorframe / 123

basin type part / 135

basket-tube combustion chamber / 149

batch / 51

battery / 91

battery amperemeter / 91

battery box / 91

battery capacity / 60

battery charge relay / 119

battery compound / 60

battery ignition / 149

battery leakage / 91

blanking cover/blanking cap / 14, 51, 108,149

blanking plug / 101,142

blaster / 64

bleed air (dump) line / 119

bleed air duct / 119

bleed air duct opening / 119

bleed air overpressure indicating light / 119

bleed air over-pressure valve / 117

bleed air pressure indicator / 91

bleed air system / 117

bleed airflow / 149

bleed switch / 119

bleed-air ejector / 119

bleed-air over-pressure switch / 82

bleed-air port / 82

blind nut / 74,123

blink / 45

blink/flash / 168

blinking frequency/glitter frequency / 82

blister / 1

block / 24,34,82,162

block diagram / 39,51,91

block plate / 34

block plug / 82

block speed / 188

blockage / 117,172

blocker door / 149

blocking annunciator / 5

blow away / 24

blowing hose / 168

blowing nipple / 60

blowing pipe / 168

blowout characteristic / 149

boarding / 14

bogie / 14,101

bogie fork ring / 101

boiled water / 117

bolt / 45,135,168,173

bolt brace piece / 101

bond / 137

bonding agent / 115

bonding jumper connector / 82

bonding jumper/bounding wire / 14,45,51, 60,64,149,168,179

bonding layer / 128

bonding limit / 45

bonding resistance / 45,60

bonding resistor / 51

bonding strip / 179

bonding surface / 123

bonding tape / 74

boost charging / 51

booster blades (vane) / 149

booster coil / 149

booster compressor / 149

booster inlet guide vane / 149

booster rotor assemble / 149

booster stator assemble / 149

boosting stage pump / 179

boost-up cavity / 142

bore sight harmonization / 188

bottle body / 24

bottom / 128

bottom beam / 128

bottom chassis / 51

bottom plate / 91

bottom view drawing / 23

boundary / 68

boundary position / 166

boundary-layer bleed inlet / 149

bourdon tube / 64

bow clamp / 74

bow connecting section / 128

bow-shaped clamp / 179

box beam / 128

box body / 128

box bracket / 74

box corrugation plate / 128

box cover / 34,128

box element / 123

box member / 24

box member/box-shaped member / 101

box spanner/box wrench / 68

box-shape framework structure / 123

box-shaped piece / 176

brace / 5

brace joint / 5

brace rod / 11

bracing / 24,68,162

bracing abutment / 132

bracket / 64,101,123

bracket piece / 34

bracket/support / 39

bracking rocker arm / 24

braid shield / 119

braided rubber hose / 117

brake / 68,123

brake accumulator / 23,24,68,78,101

brake assembly / 78

brake balance mechanism / 101

brake clearance / 5

brake distance / 18

brake gear / 60

brake handle / 68

brake lever / 19

brake lock / 68

brake mechanism / 68

brake pin sleeve / 68

brake plate / 5

brake pressure gauge indicator / 91

brake pressure gauge sensor / 101

brake pressure valve / 5

brake pressure-reduction valve / 68

brake pull rod / 68

brake release indicating light / 91

brake release switch / 91

brake rocker / 69

brake shoe / 185

brake valve / 78

braking cable / 5

braking device / 1

braking flange/brake flange / 101

braking mechanism / 172

braking pull rod / 24

braking rocker/brake rocker / 69

braking steel cable / 69

braking system / 24

braking wheel / 101

brass / 179

brass bus / 82

breakdown / 64

breakdown plate / 137

breaking load / 123

break-up corner / 51

bridge arm / 39,74

bridge arm resistance / 179

bridge plate / 128

brightness / 45

brightness adjusting button / 91

brightness adjusting knob / 91

brightness adjusting potentiometer / 91

brightness control knob/brightness adjusting
knob / 45

brightness dimmer / 45

brightness regulator / 45

broadcast station / 108

broken / 108

broken wire / 123

bromide / 69,123

bronze alloy / 24

bronze base / 173

bronze bushing / 179

bruising / 69

brush / 24,34,39,69

brush assembly / 39

brush check access cover / 74

brush handle / 51

brush spring / 51

bubble / 24,135

buckle / 60

buffer / 1,150,162,173,179

buffer accumulator / 24,78

buffer rope / 60

buffer spring / 45

buffer strut / 14

buffer type / 24

buffet / 1

build-in axle / 18

built-in test equipment (BITE) / 91,179

built-in antenna / 60

built-in computer / 91

built-in test button (BIT button) / 108

built-in test normal indicating light / 82

built-in test switch / 82

built-in test/self-test / 91

bulb / 173

bulge / 1,45,51,74,128

bulk cargo / 14,185

bulk cargo door / 185

bulk compartment / 185

bulkhead / 1,82,101,185

bullet nozzle / 176

bullet shaped throttle / 162

bump / 1

bumping test / 188

bundle / 69

buoy / 108

buoyancy / 60

buried ducting / 150

buried engine duct / 150

burn-out / 82

burr / 51,69

burst pressure / 115

bus / 45

bus (bar) / 82,91

bus data / 91

bus signal interface module / 45

bush cover / 101

bushing / 24,34,45,51,74,78,82,173

bushing assembly / 82

bushing clamp / 166

bushing dimension / 101

bushing/bearing bush / 101

bushing-type capacitor / 51

butt / 82

butt attachment frame / 5

butt bolt / 168

butt comb part / 5

butt connector/butt joint / 82

butt cover plate / 137

butt flange edge / 176

butt joint / 51,128

butt joint section / 82

butt joint strip plate / 19

butt landing gear mounting trolley / 101

butt nut / 142

butt position / 5

butt profile / 137

butt rib / 137

butt seam / 137

butt section / 5

butt socket / 5

butt step difference / 176

butt strut / 137

butterfly bolt / 91,39

butterfly plate / 117

butterfly sheet / 115

butterfly spring / 101

butterfly throttle / 162

butterfly valve / 82

butting mode / 137

butt-joint frame / 128

butt-joint plate / 128

butt-joint seam / 128

button / 51,102

button assembly / 39

button light / 82

button lock catch / 5

button seat / 45,69

button set / 45

button-head bolt / 34

button-head rivet/mushroom-head rivet / 128

buzz / 60

by-pass door / 150

bypass valve / 117,168,179

cabin / 24,34,135,168

cabin altitude / 188

cabin altitude and pressure differential gauge / 91,108

cabin altitude annunciator / 91

cabin altitude differential pressure gauge / 34

cabin altitude pressure difference gauge / 5

cabin audio recording system / 91

cabin crew training equipment / 188

cabin door / 19

cabin door signal / 91

cabin emergency pressure release switch / 91

cabin movable door / 34

cabin oxygen supply / 117

cabin pressure / 24

cabin pressure release switch / 91

cabin service trainer / 188

cable / 45,102

cable bundle / 51

cable control / 150

cable cover / 45

cable drum / 19

cable drum wheel / 102

cable end / 51

cable interconnection / 51

cable laying/cabling / 51

cable pipe / 134

cable plug / 45,64,82,102,168,173

cable plug connector / 74

cable quadrant / 168

cable shield / 119

cable shroud / 137

cable socket / 64

cable through / 51

cable-insulating layer / 168

caddice ends / 60

cadmium-plating / 137

caging error / 108

calculating cam / 162

calculating element / 162

calculating lever / 162

calculation / 60

calculation formula / 14

calibrate / 14

calibrated airspeed (CAS) / 188

calibration / 1,45,91,108,162,173

call on down wind / 188

call procedure turn completed / 188

call starting procedure turn / 188

calling address / 45

calling setting state / 45

calling status / 45

cam / 51,60,82,142,168,179

cam mechanism / 82

cam/bulging / 142

cambered blade pr ofile / 150

cancel / 14

cannular combustion chamber / 150

canopy / 60,82,135

canopy skeleton / 1

cantilever high-wing / 137

cantilever moment / 150

can-type combustion chamber / 150

canvas / 24,74

canvas jacket / 142

cascade vane / 177

cascade vane segment / 177

case / 150

case type assembly / 166

casing / 150,162

casing assembly / 69,117

casing treatment / 150

casing/case / 24,82

casing/housing / 39

cast / 34

cast aluminum / 75

caster assembly / 185

caster tray / 185

casting / 123

casting aluminum alloy / 128

casting control system / 60

casting magnesium alloy / 128,166

castle nut / 102

castor / 18

catalytic igniter / 150

catch pin / 69

cathedral angle / 5

cathode-ray tube / 108

cause / 82,102,168,173

caution / 14,23,168

caution obstacle / 108

caution terrain / 108

caution wake turbulence / 188

caution zone / 128

cavitation / 162

cavitation corrosion/cavitation erosion / 75

cavity / 34,82,135

C-clamp / 60

ceiling / 39,52,68,82,128

ceiling altitude/peak altitude / 188

ceiling light / 117

ceiling light switch / 91

ceiling panel / 185

cell / 52

cement ground / 137

center console / 150

center distance / 123

center electrode / 166

center guide/restraint / 185

center hole / 150

center instrument panel / 52, 45, 60, 69, 82,179

center instrument panel / 162,168,173

center of gravity / 1,14,91,185

center power distribution device / 52

center damper / 185

center vertical gyro / 39

center wing / 1,34,39,64,128

center wing fairing / 128

centering mechanism / 102

centering muff / 142

centerline / 128

center-return spring / 39

central rocker / 123

central warning system / 91

central-body inlet / 150

central-body supersonic inlet / 150

centralize / 45

centralizing spring / 39

centrifugal acceleration / 64,102

centrifugal balancing weight / 119,142

centrifugal compressor / 150

centrifugal compressor diffuser / 150

centrifugal force / 34,39,142

centrifugal hydraulic tachogenerator / 162

centrifugal mechanical tachogenetator / 162

centrifugal mechanism / 142,150

centrifugal nozzle / 150

centrifugal oil-mist separator / 179

centrifugal pitch / 142

centrifugal regulator / 119

centrifugal speed governor / 108

centrifugal speed regulating mechanism / 142

coil resistance / 173

coil winding / 39,52,173

coincidence / 123

coincidence amplifier / 108

cold air / 150

cold air safety valve / 78

cold drawing steel-wire / 45

cold end / 64

cold front / 189

cold junction compensation (CJC) / 119

cold run/crank / 52

cold start/dry motoring / 150

cold state / 52

cold stream / 150

cold-weather start / 150

cold-work hardening / 123

collar / 25

collection / 173

collector / 91

collector ring / 177

collide / 20

color alphanumeric picture / 108

color strip / 23

colorized flare / 128

colorized signal flare / 129

column / 14

comb plate / 172

combat flight / 25

combat simulator / 189

combi / 185

combination cooling blade / 150

combination key / 45

combination valve / 5

combination valve/assembly valve / 5

combined bracket / 69

combined control / 69,92

combined flat nose pliers / 69

combined harness-shoulder harness / 60

combined lock mechanism / 25

combined receiving device / 45

combines compressor / 150

combustion casing / 179

combustion chamber / 142,150

combustion chamber casing / 168

combustion chamber/combustor / 64,82,168

combustion efficiency / 150

combustion in parallel layes / 150

combustion intensity / 150

combustion mixture-ratio regulator / 162

combustion product / 150

combustion theory / 150

combustor casing / 64,150

combustor cooling / 150

combustor outer casing / 150

combustor wall temperature / 150

commander / 18

common brush / 129

common cable / 52

common connecting point / 52

common dome light / 5

common rib / 135

common scale / 14

communicating pipe / 78

communication / 18

communication agreement / 92

communication audio terminal / 45

communication control / 92

communication oil line / 142

communication protocol / 45

communication radio set / 52

communication system / 25,45

communication valve / 78,150

communicator / 14,35,45

communicator AC instrument panel / 92

communicator circuit breaker board / 45,52,
69,82,92,142,179

communicator console / 45

communicator control panel / 45,92

communicator DC instrument panel / 52,92

communicator PDB / 52

communicator radio instrument panel / 92

communicator seat / 25

communicator working desk / 45

commutate / 60

commutating segment/commutator segment / 52

commutating spark / 52

commutator / 52

compact picture / 162

compass / 1

compass deviation / 108

compass fast slave button / 92

compass lighting switch / 92

compass sensor / 137

compensate rod / 162

compensating wire / 173

compensating wire bundle / 173

compensation / 39,35,162,168

compensation coil / 52

compensation winding / 52

compensation wire / 119

compensator / 35

complementary charging / 52

complex hydrocarbon aviation lubricating oil / 179

complex shape / 166

complex weather condition / 82

component / 1,14,142

component position/element position / 46

component vibration damping / 52

component/element / 52,82

composite / 115

composite error / 92

composite material / 1

composite power supply PDB / 52

composite trainer / 189

composited circuit / 82

composition / 82

compositor / 150

compound bristle brush / 129

compound motor/compound-wound motor / 52

compound power / 82

compress / 14,168

compressed air / 25,52,108,168

compressed washer / 115

compressed-air starter / 150

compressibility correction / 189

compressing / 35

compression amount at parking / 102

compression bolt / 135

compression bush / 11

compression disk / 75

compression force / 25

compression nut / 64,150

compression of tyre / 18

compression pad / 150

compression quantity / 102

compression spring / 168

compression state / 123

compression system / 35

compressor / 35

compressor adjustment / 150

compressor air flow / 150

compressor air inlet duct / 119

compressor blade / 142

compressor blade root damper / 150

compressor casing/compressor case / 25,150

compressor characteristic / 150

compressor control / 168

compressor efficiency / 150

compressor element stage / 150

compressor fan cowling / 150

compressor front roller bearing / 150

compressor impeller / 35

compressor rotor / 151

control coil / 92

control column / 18

control component / 39

control condition / 39

control device / 1,142

control display / 6

control display unit (CDU) / 92,108

control distribution accessory / 78

control element / 162

control flux / 39

control force / 69

control hand wheel / 69

control handle / 39,69,102

control knob / 39

control lever / 142

control linkage / 69

control mechanism / 14,25,39,69,108

control mode / 46

control panel / 64

control pull rod / 25,123

control pulley mechanism / 25

control signal / 64

control stand / 172

control steel cable / 1,25

control steel cable assembly / 117

control stick / 39,102

control stick/control column / 25

control surface / 1,39,92

control surface angle protractor / 39

control surface deflecting angle / 39

control surface lock / 1,39,172

control surface lock control handle / 92

control surface lock control system / 25

control surface lock system / 25,69

control surface moment / 39

control switch / 69,78,92,142

control switching circuit / 39

control system / 142,151

control velocity / 40

control wheel / 69

control wheel framework / 69

control winding / 40

control wiring / 1

controlled object / 172

convection cooling blade / 151

conventional frame / 123,135

conventional rib / 138

convergent nozzle / 177

convergent-divergent inlet / 151

convergent-divergent nozzle / 177

converter / 64,92,173

converter / 64,92,108

convertible cable / 40

convex block / 60,102

convex lug / 53

cool / 35,168

cool path / 35

cooled combustion chamber / 151

cooler / 179

cooling accessory / 117

cooling air / 53

cooling air collector / 119

cooling air crossover / 119

cooling air fan / 119

cooling air shroud / 119,151

cooling air shut off valve / 119

cooling air valve / 119

cooling assembly / 118

cooling effectiveness test of turbine blade / 151

cooling fan / 53

cooling fan shut off valve limit switch / 119

cooling oil / 79

cooling path outlet header / 35

cooling pipe valve component / 35

cooperation surface / 75

coordinate / 14

coordinated resolver / 108

cowling plate / 142

cowling/fairing / 35

cowling/fillet / 133

crab angle / 92

crack / 25,35,60,64,123,168,173

crack-arrest hole / 177

crane / 14,123,142,151

crane hook / 143

crane sliding rail / 123

crank / 1,60,151

crank arm / 172

crank linkage/link mechanism / 40

crank shaft / 102

crank/cold run / 119,151

cranking / 151

crash / 83

crash axe / 60

crash condition / 185

crash survivability / 46

crashworthiness techniques / 92

craze / 135

craze crack / 53

crazing / 69

crease / 1

crescent groove / 123

crest factor / 53

crew member / 14,185

crew seat / 20

critical angle of attack / 92,108

critical angle of attack annunciator / 92

critical angle of attack power / 92

critical angle of attack sensor / 92,108

critical angle of attack signal system / 92

critical angle of attack voltage / 92

critical angle of attack warning light / 92

critical AOA warning / 46

critical engine / 189

critical speed / 189

critical temperature / 64

cross beam / 123

cross component force / 60

cross joint / 69

cross section / 123,166

cross section area / 129

cross track distance / 108

cross valve / 35,92

cross wind / 189

cross wind landing / 189

cross wind leg / 189

crossbeam / 123,129

crossfeed manifold pressure switch / 162

crossfeed switch / 92

cross-generator starting / 151

crosshead pin joint / 60

crosslink / 60

crosslink component / 40

crosslinking relationship / 183

crossover valve / 35

cross section / 18,123,129,166

cross-section area / 129

crosswind / 172

crosswind landing / 189

cruise airspeed / 189

cruise status / 189

cruise-in altitude / 189

cruise-out altitude / 189

cruising altitude / 189

crystal filter / 46

cup / 75

cup-shape cylinder / 179

curling / 177

current collector / 151

current consumption / 173

current cross / 143

current divider / 53

current divider contactor / 119

current frequency / 108

current intensity / 35

dead-center ignition / 151

deaerating ability / 168

deaerator / 20

deceleration / 174

deceleration ratio / 174

decelerator fairing/speed reducer fairing / 133

decision altitude/decision height / 189

deck angle / 185

decline / 38

decoding / 183

decolorization / 1

decompression orifice plug / 40

decompression panel / 185

decoration / 174

decoration cover / 117

decoration panel / 20,129

decorative inner skin / 123

decreasing / 168

dedicated protractor / 92

deep scrape / 179

defect / 136,138

defect/flaw / 69

defective part / 53

deffector rail / 185

deflecting nozzle / 177

deflection / 138

deflection angle / 5,40

deflection angle limiting mechanism / 69

deflection angle sensor / 6

deflection indicator / 69

deflection limiter / 25

deflection measuring device / 102

deflection mechanism / 69

deflection of inclined strut / 14

deflection position / 14

deflector / 151

deform / 46

deformation / 53,133,136,138,174

defuel/drain / 133

defueling access cover / 138

defueling port / 138

degree- of-freedom / 109

degumming/unbonding / 138

dehydrated carbon dioxide / 25

dehydrated kerosene / 138

deicer boot / 143

deicing / 25

de-icing / 83

deicing fluid / 25

deicing switch / 143

deicing timer / 143

delaminate / 53,136

delamination / 1

delay / 83

delay circuit / 83

delay error / 174

delay lamination limiter / 151

delay time / 143

delay valve / 6

delay/lag / 69

delayed ignition / 151

delayed start / 151

delayer / 109

deleterious gas / 25

deliver / 14

delivery flight test / 189

delivery order / 185

delivery period / 75

delivery receipt / 185

delivery time / 185

demagnetize / 53

demand mechanism flow / 115

demand oxygen supply system / 115

demand regulator / 115

demand valve / 115

demodulate / 46

density / 53

differential-pressure sensor / 35

diffuser / 75

diffuser case / 151

diffuser efficiency / 151

diffusion cover plate / 35

digital process / 46

digital scale / 179

digital signal / 109

digital volume potentiometer / 46

digital wheel window / 109

digital-analog hybrid computer / 172

digitron / 46

dihedral angle / 6,138

diluent stator combustor / 151

dilution zone / 151

dim light / 189

dimension / 40

dimmer tab / 102

dimming potentiometer / 6,46

diode / 53,83

dip stick / 20,25,119,179

diphenyl ethers lacquer / 53

dipping / 124

dipstick / 20

direct current (DC) / 174

direct current power / 83

direct current voltage / 64

direct ignition / 151

direct proportion / 115,174

direct-current (DC) electronic system / 119

direct-flow combustion chamber / 151

direction / 174

direction scanning rate / 109

directional objective signal / 119

directional sensitivity / 109

direct-to flight plan leg / 109

dirigible fabric/balloon fabric / 138

dirt / 18

dirty fuel test / 172

disassemble / 1,11

disassembly / 11

disassembly check / 53

disbond / 168

disc part / 102

disc valve / 179

discard / 78

disc-drum rotors / 151

discharge / 83

discharge coefficient / 151

discharge needle / 46

discharge tube / 25

discharger / 46

discharging pulse / 46

disconnecting / 168

disconnecting mechanism / 25

discrete magnitude signal / 109

discrete signal / 109

discriminatory criterion / 1

disc-type annular combustor / 151

disc-type rotor / 151

disengagement / 138,174

disengaging lock / 61

disk / 75

dispart / 14

displacement / 83,174

displacement signal / 40

display / 92

display capability / 109

display process unit / 6

display radar / 109

display range / 109

display window / 46,109

disposal / 20

disruptive field intensity / 46

dissolving / 25

distance / 14

distance advisory / 109

distance between stringers / 129

down wind leg / 189

download / 46

downloading duration / 92

downlock / 2,20

downlock actuator / 78

downlock cable rocker arm / 124

downwind landing / 189

downwind speed / 189

drag / 152

drag coefficient / 152

drag link / 177

drain cock / 75

drain cup / 172

drain fitting / 119

drain fuel pipe / 152

drain funnel / 75

drain hose / 25

drain line / 119

drain outlet / 78

drain pipe / 143

drain pipe connector / 75

drain plug / 119

drain plug/oil draining / 179

drain stand pipe / 119

drain threaded plug / 25

drain tube / 179

drain valve / 6,11,26,129

drain valve cover / 117

drainage / 20

draining mechanism / 129

draining port / 14

draw back / 26

drawing No. / 14

drift / 109

drift angle / 61

drift angle error / 109

drift out to the right / 189

drifting amount / 102

drifting deflection mechanism / 61

drinking water equipment/potable water equipment / 117

drip pan / 119

drip-proof test / 189

drive mechanism / 53

drive shaft / 79

drive shaft/transmission shaft / 53

driven gear / 26,143

driver stage / 46

driving / 40

driving assembly / 69

driving box / 179

driving circuit / 46

driving component / 64

driving disk / 83

driving element / 143

driving end / 53

driving gear / 53,83

driving lever / 102

driving lever pin / 40

driving mechanism / 6

driving part / 53

driving pipe / 69

driving piston/servo piston rod / 102

driving pull rod / 179

driving rod / 6,26,69,78

driving rod bracket / 26

driving rod/driving lever / 78

driving shaft / 83,152

driving shaft/transmitting shaft / 102

drop control box / 6

droplet combustion / 152

dropping personnel / 46

dropping pipe / 26

drop-press forming part / 124

drum / 69

drum wheel / 2,26,61

drum-type rotor / 152

dry / 185

ejection pipe / 133

ejection radiating switch / 83

ejection radiation airflow / 117

ejection radiation valve / 179

ejector / 83

ejector heat-radiation bleed airflow / 152

ejector nozzle / 177

ejector radiating tube / 83

ejector radiation electric mechanism / 26

ejector radiation switch / 93

ejector switch / 83

ejector type nozzle / 152

eject-radiating air inlet window / 119

elapsed time indicator access / 119

elastic contact / 93

elastic deformation / 70

elastic element / 162

elastic washer / 83

elasticity / 83,124

elasticity deformation / 53

elbow / 83,177

elbow assembly / 83

elbow bracket / 133

elbow nipple / 26

elbow outlet / 179

elbow pipe / 35

elbow union / 26

elbow/elbow bend pipe / 143

electric accelerometer / 93

electric arc / 53

electric bell / 93

electric bonding seat / 75

electric brake state / 40

electric bridge / 75,83,143

electric bridge principle / 93

electric bridge theory / 109

electric brush / 75

electric cable / 6

electric control circuit / 83

electric deicer / 133

electric dryer / 35

electric energy / 166

electric erosion ignition plug / 120

electric erosion layer / 166

electric field intensity / 46

electric floating-type quantity indicator / 78

electric gyro / 109

electric heater / 109

electric heating anti-icing system / 83

electric heating glass / 83,136

electric heating mechanism / 6

electric heating pipe / 118

electric heating water tank / 26

electric ignition / 152

electric iron / 53

electric load / 102

electric mechanic instrument panel / 93

electric mechanism / 26,83

electric mechanism extension rod / 26

electric mechanism pull rod / 26

electric mechanism valve / 168

electric motor / 143

electric oven / 118

electric percussion device / 61

electric plug / 65,83,143

electric power station / 20

electric principle / 83

electric propeller / 143

electric pump / 78

electric signal / 40

electric signal gun / 61

electric socket / 75

electric thermometer / 109

electric winch / 26,53,61

electric zero alignment mark / 93

electric zero datum point / 93

electrical bridge / 40

electrical butterfly valve / 35

electro-thermal ice protection system / 83

element damage / 180

element panel / 109

elevation motor / 109

elevator / 2,6,15,20,26,40

elevator actuator / 40

elevator control system / 70

elevator deflection angle / 93

elevator driving shaft / 20

elevator illumination switch / 93

elevator lock / 26

elevator rocker arm / 26

elevator rocker tube / 26

elevator suspension lug socket / 26

elevator trim tab / 70

elevator trim tab control system / 26

elevator trim tab deflection mechanism / 6,26

elevator trim tab hand wheel / 93

elevator-type loader / 185

eliminate / 136

elimination / 20,70,84

ellipse / 136

ellipse degree / 70

elongation / 2,174

ELT control box / 93

embedded training / 189

emergency access / 6

emergency access cover / 6

emergency air inlet / 152

emergency airdrop indicating light / 93

emergency airdrop PDB / 54

emergency airdrop switch / 93

emergency brake / 26

emergency brake handle / 124

emergency brake valve / 78

emergency bus bar/emergency bus / 54, 84,93

emergency cargo airdrop / 93

emergency control / 124

emergency control circuit / 93

emergency control hydraulic fluid line / 124

emergency control system / 124

emergency cut off button / 40

emergency cut- off dropping fuse / 54

emergency door / 20,129,136

emergency door indicating light / 93

emergency door opening / 93

emergency dropping fuse / 54

emergency exit / 2,124

emergency exit hatch / 129

emergency extension system / 70

emergency feathering / 143

emergency feathering shutdown / 143

emergency feathering switch / 143

emergency forced landing / 65

emergency fuel regulator / 162

emergency handle / 115

emergency horizon / 93,109

emergency horizon brightness adjusting knob / 93

emergency horizon power switch / 93

emergency horizon/standby horizon/standby gyro horizon / 54

emergency hydraulic feathering / 143,152

emergency hydraulic fluid line / 124

emergency ignition / 152

emergency landing / 189

emergency light / 93

emergency locator / 61,93

emergency locator transmitter (ELT) / 93

emergency maintenance / 46

emergency network / 54

emergency opening / 124

emergency oxygen supply / 26,115

emergency oxygen supply mechanism / 115

emergency power supply / 54

emergency power supply box / 6,54

emergency power supply control switch / 93

emergency power supply switch / 54,93

emergency power- off button / 70

emergency pressure release switch / 6,129

emergency pressure release unit / 35

emergency reducing valve / 162

emergency retracting/extending changeover switch / 70

emergency shutdown / 172

emergency spin recovery devices / 189

emergency start cut off button / 93

emergency steel cable / 129

emergency unlocking rocker / 102

emergency up/down access cover / 129

emergency window / 2,20

emergent descent / 26

emitter follower / 46

empennage / 18

empty (weight) aircraft / 15

empty bottle / 26

empty bottle weight / 26

empty cavity / 124

empty tank / 75

empty weight / 15

enclosure / 124

encoder / 6

encoding / 93

end brush box bracket / 143

end brush holder / 143

end bulkhead / 138

end cover / 143

end cover eyelet / 124

end face / 65

end face slotted tooth / 152

end face teeth / 143

end nut / 70

end play / 70

end point / 40

end rib / 70

end spline / 143

end stop / 185

end stop assembly / 185

end surface / 40

end thread / 138

endurance on board / 189

endurance test / 189

energizing contact / 168

energy consumption brake / 61

enforcing belt / 61

engage / 143

engaged voltage / 143

engagement / 40,70

engine / 15,20

engine accessory casing / 84

engine air inlet fairing / 65

engine air inlet heating indicating light / 93

engine air inlet icing indicating light / 93

engine air intake anti-icing valve / 26

engine air intake system / 168

engine airborne kit / 180

engine airborne tool / 168

engine altitude chamber / 152

engine anti-icing system / 169

engine appearance / 152

engine automatic shutdown device / 6

engine auto-shutdown indicating light / 93

engine auxiliary mounting bracket / 180

engine axis / 152

engine bracing / 180

engine close-loop control / 172

engine combustion chamber/engine combustion combustor / 162

engine compressor casing / 166

engine continuous operation duration / 152

engine control / 26

engine control rob / 166

engine coupling / 54

engine curve test / 84

engine efficiency / 152

engine electronic control / 172

engine emergency feathering shutdown / 76

engine emergency feathering shutdown knob / 93

engine emergency shutdown system / 172

engine exhaust gas temperature indicator / 93

engine exhaust gas thermometer / 174

engine exhaust pipe / 133

engine fault indicating light / 93

engine fault signal relay / 143

engine feathering / 93

engine fire-extinguish device / 93

engine first running / 174

engine fixing joint / 26

engine gear box / 78

engine ground running-up / 54

engine high-temperature area / 169

engine I / 23

engine I capacitor / 180

engine I fuse / 180

engine ignition contactor / 166

engine indication and crew alert system (EI-CAS) / 163

engine inlet / 84

engine inlet guide heating switch / 93

engine inlet hot air / 169

engine inlet icing signal system / 169

engine inner cavity / 65,152

engine inner cavity fire-extinguishing button / 93

engine in-operation / 143

engine instantaneous vibration value / 152

engine instrument / 93

engine internal fire extinguishing system / 26

engine life / 152

engine lifting service ladder / 180

engine lighting switch / 93

engine link / 152

engine lower cover / 180

engine lubrication / 180

engine main/auxiliary mounting joint / 152

engine model / 152

engine mount / 152

engine mounting / 2,6,133,138

engine nacelle / 2,6,20,26,54,84,143,152, 169,180

engine nacelle pulley bracket / 26

engine nacelle theoretical configuration / 169

engine nacelle-wing fairing / 152

engine oil / 180

engine oil line / 143

engine oil tank/pressure relief door / 152

engine open-loop control / 172

engine operation parameter / 152

engine out- of-operation / 143

engine over-vibration indicating light / 93

engine over-vibration indicator / 93

engine parameter / 143

engine power control system / 172

engine power output shaft / 143

engine propeller / 166

engine pylon/strut / 152

engine revolution indicator / 93

engine rotating speed / 13

engine running-up / 180

engine service ladder / 180

engine shuf off valve / 163

engine shutdown / 143,152,163,180

engine shutdown relay / 143

engine shutdown solenoid valve / 163

engine shutdown switch / 6,93

engine speed reducer / 143,180

engine start / 152,163

engine start button / 93

engine start fuel pipe / 163

engine start fuel reducing solenoid valve

exceeding / 15,20

excess air coefficient / 153

excessive braking / 2

excessive control / 2

excessive fuel / 153

excessive fuel supply / 153

excessive fuel system / 153

excessive fuel tank/residual fuel tank / 75

excessive pressure / 35

excessive pressure limiter / 35

excessive pressure limiting mechanism / 35

excessive pressure valve / 35

excessive pressure/residual pressure / 102

excessive water box / 118

excitation coil / 40

excitation flux / 40

excitation interface / 109

excitation return circuit / 84

excitation voltage / 40

excitation winding / 54

exciting circuit/excitation circuit / 54

exciting coil / 54

exciting current / 54

exciting manner / 61

excrement pail / 118

exhaust / 133

exhaust and ventilation cooling system / 153

exhaust case / 177

exhaust device / 180

exhaust duct / 120,180

exhaust duct shroud / 120

exhaust emission / 153

exhaust exit of fire bottle / 65

exhaust gas / 84

exhaust gas system / 177

exhaust gas temperature（EGT）/ 153,
163,169

exhaust gas temperature probe / 120

exhaust gas temperature sensor / 153,177

exhaust gas/waste gas / 138

exhaust hole / 84,163

exhaust jet test of simulation external stream
/ 177

exhaust manifold / 177

exhaust muffler / 120

exhaust nozzle test / 177

exhaust pipe / 120,153,169,177,180

exhaust pipe adapting ring / 120

exhaust pipe blanking cover/blanking cap
/ 153

exhaust port / 138

exhaust tailpipe / 20

exhaust temperature indicator / 94

exhaust valve / 6

exhaust-gas turbine / 153

exit / 94

exit feathering relay / 143

exit hatch / 6

exit point / 2

exit temperature distribution / 153

exit temperature distribution coefficient
/ 153

expanding space / 26

expanding temperature transmitter / 163

expansion / 26

expansion chamber / 109

expect approach clearance / 190

expect ILS approach / 190

expect long final approach / 190

expect straight-in approach / 190

expectation value / 109

experimental and research airplane / 190

expiration / 61

explode / 75

explosion cap / 94

explosion protection / 65

explosion starting device / 65

explosion-pro of light / 75

fairing part / 133

fairing support plate / 153,169

false feathering / 144

false signal / 65

false start / 153

false start/wet motoring / 153

false weld / 84

fan / 153

fan blade spacer / 153

fan blades / 153

fan disc / 153

fan exhaust / 177

fan outlet guide vane / 153

fan pressure ratio control / 172

fan shaft / 153

fan switch / 6,94

fan thrust reverse / 177

fan trim balance screw / 153

fan turbine / 153

faring cap / 144

fast slaving speed / 109

fastener / 2,46,54,70

fastening belt/fixing strap / 153

fastening bolt / 136,180

fastening condition / 18

fastening nut / 46

fastening part / 35

fastening pulley / 61

fastening screw / 124,136,180

fastening washer / 54

fatigue-resistant performance / 138

fault / 109,133

fault code / 94,109

fault description / 109

fault flag / 109

fault indicating light of engine / 144

fault isolation / 109

fault level / 109

fault light / 6,94

fault part / 180

fault relay / 120

faulty component / 94

faulty indicating light/failure indicating light / 54

faulty signal light / 54

faulty warning circuit / 84

faying surface / 54,75

faying surface sealing / 124,129,138

feather / 84

feathering angle / 144

feathering button / 6

feathering check cut off switch / 94

feathering circuit / 144

feathering cock / 78

feathering control panel / 94,144

feathering cycle / 144

feathering hub / 144

feathering indicating light / 94

feathering pipeline / 144

feathering position / 153

feathering procedure / 144

feathering propeller / 144

feathering pump / 6,20,26,144,180

feathering pump oil supply pipe joint / 180

feathering ready / 144

feathering ready circuit / 144

feathering ready indication light / 6

feathering ready relay / 144

feathering relay box / 6

feathering signal light / 94

feathering standby oil quantity / 26

feathering stop / 144

feathering system / 54,94,144,153

feathering time / 144

feathering timing mechanism / 144

feature / 84

feed diagram / 61

feed pipe / 180

fine felt / 70

fine filter / 6,163

fine filter blocking annunciator / 75,163

fine filter blocking indicating light / 94

fine filter blocking signal switch / 94

fine filter element / 40

fine pitch line / 6

fine plain / 138

fine rope / 70

fine sand paper / 70,129,133

fine white cloth / 129

fine-pitch oil chamber / 144

fine-pitch oil line / 144

fine-pitch oil line pressure annunciator / 144

fine-pitch pipe line pressure annunciator / 180

fine-pitch stop / 144

finished plug / 94

finished product / 15

fire / 26

fire bottle / 6,20,26,65

fire bottle detonator indicating light / 65

fire bottle inner cavity / 26

fire circuit / 65

fire detection / 65

fire detection system / 65

fire extinguishing / 65,84

fire extinguishing agent / 65

fire extinguishing button / 65,94

fire extinguishing check box / 65

fire extinguishing check switch / 65,94

fire extinguishing circuit / 65

fire extinguishing control box / 65

fire extinguishing device / 65

fire extinguishing equipment / 26,75

fire extinguishing joint casing / 65

fire extinguishing pipeline / 65

fire extinguishing selection switch / 65

fire extinguishing selector switch / 94

fire extinguishing signal / 65

fire extinguishing system / 26,153

fire extinguishing test button / 65

fire extinguishing test changeover switch / 65

fire pipe / 6

fire protection system / 65

fire relay box / 65

fire repeat relay (FRR) / 120

fire shutdown relay / 120

fire signal / 65

fire source / 26,65

fire switch / 6

fire truck / 120

fire wall / 20,54,120,133

fire warning button indicating light / 65

fire warning sensor / 6,65

fire warning system / 54

fire/explosion / 26

fire-extinguishing / 84

fire-extinguishing control / 7

fire-extinguishing microswitch / 7

fire-extinguishing relay box / 7

fire-proof electromagnetic valve / 75

fire-proof performance / 26

fire-proof switch / 75

fire-proof wall / 153

fire-protection control / 54

fireseals / 153

firewall / 20,54,120,133

fire-warning light / 169

fire-warning signal light / 65

firing pin / 26,65

first calibration / 169

firtree inline with scallop / 153

fit / 70

fitting / 7,133

fitting clearance / 153

fitting precision / 70

flat pipe connector / 115

flat wrench / 70

flat-head axle / 180

flat-head pin / 61

flat-head rivet / 129

flat-head shaft / 124

flexible cable control / 70

flexible chock / 75

flexible connection / 75

flexible connection part / 78

flexible connector / 163

flexible mechanical control system / 70

flexible pin / 61

flexible shaft / 54,163

flight accident recorder / 70

flight airspeed / 190

flight altitude / 40,84,190

flight archives / 94

flight attitude / 40

flight code / 109

flight condition / 144

flight control box / 40

flight control system / 27,70

flight data field check processor / 46

flight data recorder / 7,46,94

flight data recording system / 46,94

flight data recording system field tester / 46

flight data system / 46

flight data workshop check processor / 47

flight date / 7

flight day / 144

flight determination of buffet boundary / 190

flight direction / 153

flight distance / 109

flight document / 2

flight engineer / 15,47,61,65

flight engineer fan switch / 94

flight engineer seat / 27

flight envelop / 190

flight envelope / 153

flight environment data system / 109

flight flutter test / 190

flight idling / 190

flight instrument / 94

flight landing signal light box / 94

flight load measurement / 190

flight parameter / 94

flight parameter field check processor / 94

flight path / 109

flight platform / 109

flight record / 20

flight route/flight course / 109

flight safety / 84

flight simulator / 190

flight speed / 84

flight statistical data / 94

flight stress measurement / 190

flight test / 15

flight test mission sheet / 190

flight test modifications / 190

flight test of stability and control / 190

flight test program / 190

flight test/trial-flight / 190

flight time / 94

flight time driving / 94

flight time handle / 94

flight training / 94

float / 61,115

float assembly / 75

float valve / 7,27,75

floating anchor / 61

floating anchor nut / 138

floating shaft / 163

float-type fuel lever transducer / 163

float-type oil indicator / 180

floor / 174

floor access cover / 174

floor crossbeam / 129

folding power drive unit / 186

folding side restraint / 186

follow-up device / 7

follow-up mechanism / 27

follow-up piston/slave piston / 79

follow-up pull rod / 7,70

follow-up system / 103

follow-up unit / 84

food cabinet / 118

foot button / 7

foot console / 70,94

foot heating switch / 7

foot pedal / 27,40,61,70,103,129

foot voltage / 84

force component pad / 61

force transfer member / 129

force-bearing frame / 153

forced landing / 7,190

forced landing fire microswitch / 65

forced landing fire relay / 65

force-distributing chock / 61

forced-landing fire extinguishing / 94

fore locking device / 186

foreign force / 124

foreign object / 11,15,115

forging aluminum alloy / 129

fork / 18,124

fork fixed head / 47

fork joint / 18,40,70

fork lug / 103,124

fork rocker arm / 27

fork rod / 61

fork shoulder shaft / 70

fork sleeve / 70

forked lug / 103

forked member / 47,70

forklift capability / 186

forklift pocket / 186

format / 15

formation airdropping and parachuting / 61

formation light / 7,94,138

formation signal / 54

forming / 129

forward cabin floor door / 27

forward cabin floor door lock actuator / 27

forward cabin oxygen supply check button / 94

forward cabin oxygen supply signal / 94

forward difference / 138

forward fluid resistance / 36

forward inboard cowling / 153

forward inclination / 70

forward outboard cowling / 153

forward power / 47

forward thrust lever / 163

forward/backward travel / 103

four-blade propeller / 144

four-hole socket / 7

four-pole DC reversible automatic vent motor / 61

fraction / 20

fracture / 18,36,47,138

fracture surface / 124

fragile element / 174

frame / 2,11,15,124,133,138

frame door / 124

frame edge / 136

frame edge/flange / 124

frame location / 7

frame plane / 124,129

frame plate / 124

frame web / 133

framework / 84,124,136

free end / 65

free travel / 40

free turbine / 153

free water content of inlet air / 36

free-stream tube area / 153

gas turbine starter / 154

gas welding / 118

gaseous oxygen / 115

gas-generator turbine / 154

gas-injecting zone / 138

gasket / 54,61,84,115

gasket/spacer / 180

gasoline / 27,54,61,136

gas-turbine air starter / 154

gas-turbine compressor / 154

gate / 65

gauze / 36,69,84,103

GCA approach / 190

gear / 36

gear box / 54,120

gear fuel pump / 120

gear peak / 79

gear plate joint/toothed plate joint / 103

gear pump / 79,144

gear reducer / 76

gear rotating angle / 40

gear selector / 164

gear shaft / 54,70

gear vale / 144

gear valley / 79

gears pair / 70

gear-type high-pressure fuel pump / 164

general calling address / 47

general cargo / 186

general check / 110

general current / 144

general distribution box / 7,180

general information / 23

general list / 186

general power distribution board / 54

general purpose freight / 186

general purpose trainer / 190

general trainer / 190

general wiring diagram / 40

general-purpose tool / 84,110

generator / 2,7,36,95,120

generator control panel / 2

generator control unit / 120

generator cooling air overboard port / 154

generator current divider / 7

generator excitation winding / 120

generator power-on switch / 7

generator voltmeter / 95

geographic longitude/latitude / 110

geometric measurement / 15

geometric twist / 138

geometrical altitude / 190

geometrical angle / 76

germanium rectifier / 174

G-force / 186

gimbal axis / 40

given speed / 144

gland / 76

glass / 20

glass assembly / 124,136

glass cloth / 65,129

glass cloth laminate / 138

glass fiber / 65

glass fiber reinforced plastic / 2

glass fiber reinforced plastic oxygen bottle / 27,115

glass fiber reinforced plastics honeycomb / 129

glass heating / 95

glass heating circuit fuse / 84

glass heating pipe / 7

glass heating switch / 95

glass heating transformer / 84

glass insulating cloth / 84

glass of navigator's compartment / 129

glass tape / 54

glass wadding gasket / 118

glide / 40

ground hydraulic test cart / 79

ground hydraulic testing system / 70

ground idling / 190

ground navigation station / 47

ground object / 110

ground oil truck / 27

ground oxygen charging cart / 27

ground power / 103

ground power cart / 103

ground power supply / 7,54,144

ground power unit (GPU) / 54,76,169

ground power/ground electrical power source / 84

ground pressure source / 79

ground proximity warning system (GPWS) / 47,95,110

ground pump truck / 103

ground reception station / 61

ground refueling connector / 27

ground run test / 180

ground simulation test of primary control system / 190

ground speed / 95,110,190

ground speed variance / 110

ground start / 154

ground starter/ground-powered starter / 154

ground starting envelope / 154

ground starting ignition / 166

ground station / 95,110

ground target / 110

ground taxiing / 103

ground temperature / 20

ground wire / 40

ground-controlled landing / 190

grounding brush / 18

grounding cone / 7,47

grounding strip / 110

grounding wire/earthing wire / 2,18,27,47, 54,175

group zero tank / 15

group zero tank without refueling / 15

guaranteed period / 27

guard valve / 79

guide blade / 36

guide bushing / 84

guide clip plate / 103

guide header / 36

guide plate / 71,76,125,129,138

guide post / 36

guide pulley / 71

guide pulley bracket / 71

guide rail / 103,129

guide rail (assy) / 186

guide rod / 71

guide seat / 71,79

guide sleeve / 79,125

guide sliding rail / 125

guide vane / 154

guided landing / 190

guider / 7,71,76,129

guiding groove / 84

guiding information / 110

guiding mechanism / 71

gun oil / 138

gust / 40,190

gyratory angular velocity / 40

gyro / 15,110

gyro azimuth / 110

gyro control task manager failure / 110

gyro drift quantity / 110

gyro group / 7

gyro magnetic compass / 110

gyro measuring axis / 41

gyro output gradient / 41

gyro platform / 41

hail / 2

hairspring / 41,65,120

hairspring resistance / 174

half compass drift error / 110

half size container / 186

half-ball valve / 79

half-ring lock / 79

half-round head rivet / 169

half-round head screw / 55

hammer / 138

hand and foot control mechanism / 20

hand clamp / 55

hand pump / 7,11,79

hand transmission device / 61

hand vacuum pump / 41

hand wheel / 18,27,84,115

hand wheel assembly / 84

hand wheel control mechanism / 71

hand wheel disk / 172

handbag / 61

hand-electric pump system / 27,71

handle / 11,20,27,136,164

handle control speed / 41

handle with care / 186

handling fitting / 138

handling pulley / 61

hand-powered winch / 103

hang/suspend / 61

hangar / 18,71

hanging start / 154

hard alert area / 110

hard aluminum plate/duraluminsheet / 138

hard anodizing treatment / 129

hard center / 41

hard chromium / 138

hard core / 36

hard inorganic glass / 138

hard landing / 2

hard paper plate / 65

hard-anodizing / 129

hardness / 15,95

harmful gas / 115

harmonic content / 55

harmonic distortion / 47

harness / 95

haul / 186

have the ground in sight / 190

hazardous cargo / 186

head airflow / 144

head assembly / 144

head case / 27

head wind / 154,190

header body / 36

heading / 15,71,84,110

heading angle / 95

heading angle setting / 110

heading attitude system / 110

heading attitude system indicator / 110

heading channel / 41

heading deviation / 41,95

heading deviation warning signal / 95

heading indicator / 41

heading integrated signal / 41

heading linkage box / 41

heading maintaining flight status / 41

heading preset signal / 41

heading setting knob / 41

heading signal / 41

heading stability signal / 41

head-on airflow / 84

headset / 47

headset/earphone / 47

headwind / 154

headwind landing/upwind landing / 190

heat / 36

heat collector / 180

heat conduction direction / 55

heat energy / 79,84

heat exchange / 84

heat exchanger / 180

heat exchanger core / 36

heat exchanger core assembly / 36

heat insulation layer / 36

heat knife / 84

heat loss / 154

heat path / 36

heat preservation measure / 55

heat screen / 27

heat shield / 84,120

heat shroud / 120

heat source / 27,138

heat transfer efficiency / 36

heated container / 186

heater / 11,27

heater sleeve pipe / 154

heating / 27

heating and air supply electric mechanism / 169

heating annunciator / 84

heating area / 65

heating check button / 84

heating check indicating light / 95

heating component / 169

heating ejector system / 36

heating element / 84,110

heating muff / 65

heating path inlet header / 36

heating path outlet header / 36

heating pipe valve component / 36

heating power / 110

heating power supply / 84

heating resistance wire / 169

heating resistance/heating resistor / 84

heating signal / 84

heating signal control box / 84

heating switch / 7,84

heating time / 27

heating type / 85

heating unit / 27

heating valve / 36,144

heating warning box / 85

heating warning control / 7

heating window access cover / 133

heating wire / 85

heating wire plug / 154

heat-insulating material / 85

heatpro of asbestos ring / 120

heat-resistance shield / 133

heat-resistant alloy steel plate / 120

heat-resistant enamel paint / 166

heat-resistant non-alkali strap / 166

heat-resistant stainless steel / 133

heat-shrinkable tube / 55

heavy vehicle axle load / 129

height / 27,125

height difference / 15

height overall / 186

helical actuator / 71

helical disk / 71

helical electromagnet / 120

helical mechanism / 71

hemispherical combustion chamber / 154

heterodyne radio / 110

hexagon flat nut / 115

hexagon head bolt / 71

hexagonal head bolt / 55,71

hexagonal head screw / 144

hexagonal nut / 71

hexagonal screw / 166

hexagonal socket nut / 125

HF radio set control box / 95

high altitude oxygen annunciator / 7

high altitude oxygen-using annunciator / 85

high energy (spark) igniter / 154

high energy ignition exciter / 154

high frequency (HF) radio set / 27,47

high frequency (HF) signal / 47

high frequency (HF) transmission signal box / 47

hi-lock nut / 139

hinge / 2,61,125,130,136,139

hinge bracket / 36

hinge hole / 125

hinge pin / 139

hinged cover plate / 71

hinged cowling / 154

hinging bolt / 135

hinging joint / 28

hiosting gear / 186

hoist beam / 130

hoist bucket / 15

hoist fitting / 154

hoisting weight / 144

holding bracket / 47

holding bracket/mounting bracket / 55

holding stand / 164

hollow bolt / 180

hollow shaft / 55,139

hollow support plate cover / 169

hollow worm / 71

homogeneous combustion / 154

honeycomb / 20

honeycomb tube / 180

hook / 18,47,103,125,133

hoop / 177

horizon / 41

horizon gyro platform / 10

horizon indication / 41

horizon selection switch / 95

horizontal axis / 169

horizontal axis line / 177

horizontal condition / 11

horizontal correction system/principle / 41

horizontal deflection angle / 61

horizontal distance threshold / 110

horizontal flow direction / 139

horizontal position / 144

horizontal position indicator / 110

horizontal situation indicator (HIS) / 95

horizontal sliding rail / 61

horizontal stabilizer / 2,85

horizontal tail / 7,20,47,85

horizontal wing tip / 2

horn joint/flared join / 85

horn mouth / 65

horn release button / 95

hose / 2,28,79,103,110,144,180

hose connector / 28

hose union / 95

hot air / 85,139

hot air duct / 169

hot air stream / 133

hot bath / 65

hot end / 65

hot key / 47

hot oil / 144,169

hot shower / 28

hot state / 55

hot stream / 154

hot-air pipe / 154

hour hand / 95

housing / 47,110

housing screw/compression screw / 28

housing/casing / 55

hub / 144

hub assembly / 144

hub case / 144

hub spinner / 15,55

hub spinner/hub fairing / 2, 7, 20, 85, 133,155

huck rivet / 139

hull bottom / 61

humid heat test / 55

humidity / 136

hydraulic accessory / 20

hydraulic accumulator / 71,120

hydraulic actuator / 71,125

ice and rain protection / 85

ice and rain protection system / 28

ice block / 12

ice crystal / 76

ice-detector main warning light / 85

ice-throwing / 130

icing / 20,36,85

icing annunciator / 7,85,95

icing annunciator relay / 169

icing area / 85

icing detecting system / 85

icing detection / 169

icing indicating light / 95,169

icing region / 169

icing signal control box / 169

icing signal switch / 95

ideal combustion / 155

identification / 61

identification area / 7

identification code / 110

identification code pulse / 110

identification color circle / 71

identification mark / 55

identification of friend or foe system (IFF) / 184

identification of friend-or-foe (IFF) control box / 95

identification signal / 110

identification signal tone / 110

identifying number / 7

identity identification / 184

idle speed governor / 164

idle valve / 164

idler rocker / 71

idling / 155,164,169,180

idling pressure / 79

idling speed / 144

igloo / 186

igniter / 20,55,66,120,155

igniter boss / 120

igniter cam / 121

igniter lead / 121

igniter lead connector / 121

igniter plug / 121

igniter/igniting/ignition chamber / 155

ignition adjusting lever / 155

ignition boss / 155

ignition cable / 155

ignition cam / 155

ignition characteristic / 155

ignition circuit breaker / 166

ignition coil / 7,155

ignition delay / 155

ignition distributor / 155

ignition energy / 155

ignition exciter / 121

ignition harness / 155

ignition lead / 155

ignition lead connector / 155

ignition magnetos / 155

ignition plug / 7

ignition power supply system / 166

ignition relay / 155

ignition switch / 95,121,155

ignition test / 190

ignition time / 155

ignition time lever / 155

ignition transformer / 155

illuminate/blink/glitter / 85

illuminating / 169

illuminating light / 55

illumination / 85

illustration placard / 130

ILS approach / 190

imaging / 61

immediate relight / 155

immerse / 28,85

immersion thermocouple / 121

inflammable / 186

inflate / 61

inflation equipment / 103

in-flight engine start switch / 95

in-flight ignition / 155

in-flight simulator / 190

in-flight starting envelope / 155

in-flight starting ignition / 166

information subunit / 110

inhalation resistance / 115

initial angle / 41

initial approach / 190

initial breakdown / 8

initial elasticity / 110

initial ignition / 155

initial parameter setting / 110

initial position / 41,95

initial position setting / 110

initial power / 71

initial pressure / 103

initial value / 15

initialization / 47

initiating combustion / 155

injection interface / 47

injection power / 169

injection switch / 8

injector / 55

inlet / 74

inlet additive drag / 155

inlet buzz / 155

inlet casing/inlet case / 155

inlet cowling / 155

inlet dynamic characteristic / 155

inlet gas vane / 21,85

inlet guide heating / 95

inlet guide icing signal / 95

inlet guide valve / 121

inlet guide vane / 85

inlet guide vane heating pipe / 85

inlet guide vanes (IGV) / 155

inlet hole / 66

inlet lip / 155

inlet mass-flow coefficient / 155

inlet moisture content / 36

inlet oil pressure / 41

inlet oil temperature / 145

inlet operating condition / 155

inlet pipe / 8,28

inlet port / 8

inlet pressure / 66

inlet pressure/intake pressure / 164

inlet starting / 155

inlet surg / 155

inlet test / 155

inlet throat / 155

inlet total pressure recovery coefficient / 155

inlet union / 66

inlet/outlet pressure / 155

inlet/return line / 103

inner cavity / 12,55

inner cavity (of airtight joint) / 28

inner chamber / 115

inner cone / 177

inner cylinder / 18,28

inner cylinder of jack / 12

inner edge/flange / 130

inner labyrinth / 169

inner marker (IM) / 110

inner oil vent pipe / 180

inner ring / 71

inner ring/inner gimbal / 41

inner section / 85

inner skin / 71,85,125

inner structure / 2

inner surface / 125

inner/outer bushing / 103

inner/outer frame edge strip / 133

inner/outer point / 155

insulation tape / 55

insulator / 47,180

insulator surface / 167

intake / 2,164

intake port / 169

integral amplifier / 110,121

integral rib / 139

integral tank / 76,139

integral thrust reverser / 177

integrated auto throttle servo mechanism / 172

integrated circuit / 41,96

integrated test bed / 36

integrating circuit / 66

intelligibility rate / 47

inter modal / 186

intercept the localizer / 190

interchangeability / 47,66,76,180,186

intercom / 8,96,147

intercom button / 71

intercom control / 47

intercom control box / 96

intercom expanding equipment / 47

interconnect / 36,41,71

interconnected bus / 96

interface / 36,110

interface board / 96,110

interface circuit / 47

interface sub-unit / 110

interface unit failure / 110

interfacing relationship / 184

interference / 71

interference input method / 47

interference source / 47,164

interior / 23

interlining cloth / 61

interlock / 41

interlock device / 145

interlock relay / 55,66,85

interlock unit / 103

interlocking cable / 71

interlocking contactor / 55

intermediate and image frequency rejection ratio / 110

intermediate frequency (IF) signal / 47

intermediate frequency/video circuit / 110

intermediate level / 47

intermediate rocker arm / 8

intermediate zone / 155

intermediate-frequency / 110

intermediate-pressure (IP) compressor / 155

intermediate-pressure turbine (IPT) / 155

intermittent sound / 61

internal bus / 47

internal compression inlet / 155

internal handle / 125

internal point/inner point / 15

internal resistance / 55

internal stream characteristic test of exhaust nozzle / 177

internal tiedown / 186

interrogation/response / 110

intersection point / 130

in-trim landing / 190

intruder / 111

inverse heading view / 76

inverse proportion / 174

inversion output / 55

inversion unit / 55

inverter / 8,55

inverter bridge / 55

inverter power box / 76

iron bird / 190

iron core / 41

iron core sleeve / 55

iron stick / 61

iron yoke / 55

iron-nickel alloy / 174

landing gear changeover switch / 96

landing gear compressed state / 8

landing gear control / 96

landing gear extension / 96

landing gear free state / 8

landing gear signal / 96

landing gear structure / 2

landing gear(LG) system / 28

landing gear up/down switch / 172

landing gear well / 8,21,47,85,115

landing gear well door / 79

landing lane / 191

landing light / 2,125

landing light control / 96

landing run / 191

landing signal light box / 103

landing taxiing distance / 145

landing weight / 15

lap joint / 130

lapping paste / 96

large storage module / 96

large-scale integrated circuit / 96

large-storage data memory array / 96

laser inertial/satellite integrated navigation
system / 96

laser strapdown inertia/satellite integrated
navigation system / 96,111

laser strapdown inertial / 41

lashing ring / 186

latch / 121,133

late entry (insertion) function / 47

late ignition / 156

lateral acceleration / 96

lateral balance / 139

lateral centrifugal force / 145

lateral component force / 41,145

lateral component moment / 145

lateral control / 71

lateral controllability / 71

lateral displacement / 139

lateral guide / 186

lateral leveling / 15

lateral member / 130

lateral section / 130

lateral side / 41

lateral surface / 169

laterial element / 85

lath / 133

latitude / 96

latitude potentiometer / 111

latitude setting / 111

Laval tube / 111

lavatory / 21

layout / 76,96

LCD screen / 111

lead brass / 139

lead filter / 156

lead out / 71

lead screw / 169

lead seal / 28

lead sealing / 36,41,76,103,136,156

lead sealing block / 180

lead sealing piece / 76

lead signal / 41

lead wire / 85

lead wire port / 55

leading edge / 2,8,21,36,85,139

leading edge ejector / 85

leading edge fairing / 8

leading edge skin / 133

lead-out end / 175

lead-out wire / 71

leaf spring/spring leaf / 111

leak / 28,36,85

leak pipe / 28

leakage / 2,21,103

leakage quantity / 130

leaking liquid / 156

longitudinal static stability coefficient / 191

long-range flight / 41

loop ceramic insulator / 167

loop resistance / 121

loose / 2

loose weld / 180

loosen / 169

looseness / 21,139

loosening / 36

lose time / 191

louver / 104

low air pressure tolerance / 175

low altitude vent electric valve / 8

low calorific value / 156

low cycle fatigue test / 191

low frequency / 48

low fuel pressure system / 164

low oil pressure (LOP)driver / 121

low oil pressure light / 180

low oil pressure shut down relay / 121

low oil pressure switch / 121

low oil pressure timer (LOPT) / 121

low oil pressure warning / 180

low oil quantity switch / 121

low power / 96

low pressure / 191

low pressure fuel filter / 121

low pressure fuel line / 121

low pressure gauge / 28

low pressure rotor / 156

low pressure start / 156

low pressure stator / 156

low temperature grease / 55

low temperature lubricating grease / 28,139

low temperature test / 191

low velocity airdrop / 186

low velocity platform airdrop / 186

low-air-pressure test / 191

low-altitude vent valve / 28

low-altitude ventilation system / 36

low-carbon alloy structure steel / 130

low-drag and noise-pro of / 48

low-drag inlet / 156

low-energy ignition plug / 167

lower access cover / 8

lower arm / 71

lower bracing / 21

lower braking pull rod / 28

lower cargo compartment / 186

lower cover / 36,164,181

lower cover plate / 139

lower deck / 186

lower deck container / 186

lower edge strip / 130,139

lower emergency door / 8

lower front panel / 130

lower fuselage / 8

lower joint / 125

lower left bracing rod / 145

lower limit position / 71

lower longeron / 130

lower panel / 71,130,133,139

lower pressure cases / 156

lower rear panel / 130

lower section / 2,85

lower semi-frame / 130

lower shroud / 121

lower sideband (LSB) / 48

lower spring pull rod / 28

lower surface / 133

lower torque arm / 28

low-frequency socket / 111

low-fuel pressure indicator / 96

low-level ventilation valve / 36

low-lever warning sensor / 164

low-noise amplifier / 111

low-noise antenna / 111

low-noise receiver / 111

magnetic sensor / 145

magnetic system of high-tension magnetic / 156

magnetic testing / 71

magnetic variation pointer / 111

magnetism / 175

magnetize / 2,55

magneto (impulse) starter / 156

magneto ignition / 156

magnetron / 111

magnifier / 55,76,104,125

magnifying glass / 3

main air channel / 117

main air-bleed pipe / 8

main cable / 55

main cargo compartment / 186

main component / 130

main computer module/host computer module / 48

main control panel / 96

main control processing module / 184

main control system / 71

main controller unit / 48

main deck / 186

main display format / 48

main driving rod / 71

main driving wheel system / 96

main engine control / 172

main flap / 139

main fuel pump / 21,76,164

main icing warning light / 96

main jacket band / 62

main joint / 104

main landing gear / 3,12,104,125

main landing gear door / 29,125

main landing gear shock strut / 29

main landing gear brace / 104

main landing gear inclined brace / 104

main landing gear suspending ring / 29

main landing gear well / 55,130

main load-bearing member / 125

main oil pump / 181

main oil pump boosting stage / 181

main performance / 156

main piston rod / 12

main pressure relief valve / 42

main pulley / 29,62

main recorder / 96

main rod / 18

main shaft / 8,115,175

main shock strut / 29

main tank boost pump pressure switch / 162

main technical data / 12,18,85,156

main technical performance / 48

main technical specification / 36,48

main vibration shock absorber / 156

main warning light / 169

main wheel / 104

main wheel brake / 71,104

main wheel braking mechanism / 71

main wheel tread / 8

main/auxiliary oil pump / 156

main/auxiliary fuel pump / 156

main/upper/middle/lower link / 156

mainfold / 85

mainframe / 48

maintainability / 186

maintenance / 3,21,36,139

maintenance access / 8

maintenance access cover / 8,104

maintenance bulkhead / 8

maintenance check / 3

maintenance description / 23

maintenance door / 8

maintenance practice / 12,18,85,133

maintenance simulator / 191

maintenance technician / 55

major area / 8

make a left short circle / 191

make a long approach / 191

make a normal approach / 191

make a short approach / 191

make a short traffic pattern / 191

make a trial on ILS approach / 191

mal-function / 85

malfunction training / 191

mandatory life limit / 3

mandrel / 18,79

mandril / 36

maneuver load / 191

maneuverability / 96

maneuvering flight / 42

maneuvering performance / 104

manganin / 36

manual / 16

manual button / 104

manual calling / 48

manual control / 8

manual control fire bottle / 66

manual feathering / 145,156

manual feathering button / 145

manual fire extinguishing / 66

manual fuel consumption / 29,96

manual fuel consumption pump switch / 96

manual loading / 186

manual nut / 12

manual oar / 62

manual pressure refueling / 29,76

manual refueling quantity / 29

manual regulating system / 181

manual shutdown / 172

manual unfeathering / 145,156

manual unloading / 186

manual valve / 125

manufacture technology / 55

manufacturer / 55

mark / 104,111

mark line / 96,104,139

mark sleeve / 55

marker beacon / 111

marking / 16,186

marking line / 125

mask union / 115

mass / 29,85,156

mass balance test / 139

mass block / 96

master /standby changeover switch / 8

master control switch (MCS) / 121

master display / 111

mat / 121

match / 145

match of inlet and engine / 156

matching of exhaust nozzle and aircraft / 177

matching of exhaust nozzle and engine / 177

material / 16,85,139

matrix / 23

max/min fuel pressure limiter / 164

maximum capacity / 186

maximum cross section / 130

maximum cruise speed / 191

maximum deflection angle / 104

maximum diameter / 12

maximum flow / 42

maximum gross weight / 186

maximum height / 12

maximum indicating position / 96

maximum oil consumption / 85

maximum oil return pressure / 85

maximum opening / 125

maximum operating altitude / 191

maximum operating condition / 85

maximum operating limit speed / 191

maximum operating pressure / 66

maximum output power / 48

maximum overload / 3

maximum overload value / 96

maximum payload / 186

maximum permissible operating speed / 191

maximum scale value / 66

maximum speed governor / 164

maximum take off weight / 16

maximum threshold speed / 191

maximum torque / 156

mayday signal / 62

mean aerodynamic chord / 8,16,96

mean ear enclosure sound insulation factor / 48

mean value / 16

mean-time-to-repair / 48

measure / 29,175

measured angle / 42

measured time driving / 96

measured time handle / 96

measurement / 29

measurement data / 16

measurement of flight speed / 191

measurement of level flight performance / 191

measurement of Mach number / 191

measurement of position error in airspeed system / 191

measurement of powered descent performance / 191

measurement of pressure altitude / 191

measurement of primary control system characteristics / 191

measurement of vertical climbing performance / 191

measuring / 145

measuring accuracy / 164

measuring cup / 79

measuring instrument / 55,104

measuring item / 16

measuring personnel / 71

measuring point / 16,42,71

measuring range / 111

measuring tool / 16

measuring tool/measuring equipment / 139

mechanic adjusting value / 175

mechanic energy / 86,164

mechanic impurity / 29,79,164,181

mechanic indicating rod / 29,125

mechanic vibration frequency / 86

mechanic weighing scale / 16

mechanic's footboard / 62

mechanical amplification / 42

mechanical angular displacement / 96

mechanical control / 172

mechanical damage / 21, 56, 115, 125, 145,156

mechanical driving system / 71

mechanical energy / 72

mechanical impurity / 76

mechanical indication rod / 29

mechanical lock / 104

mechanical pitch stop / 145

mechanical pitch-control mechanism / 145

mechanical pitting / 56

mechanical propeller / 145

mechanical property / 175

mechanical rotation signal / 164

mechanical self-lock system / 56

mechanical system / 130

mechanical zero position / 42

mechanical-resistance action / 104

mechanism / 36

medical oxygen standard / 29

medical staff / 48,62

medium and long range communication / 48

medium marker (MM) / 111

medium zone / 8

megger / 76

melt / 169

member / 12,139

minimum height / 12

minimum night altitude / 191

minimum operative voltage / 66

minimum terrain clearance (MTC) / 111

minimum threshold speed / 191

minute hand / 96

mirror / 118

mirror light / 118

misalignment angle / 42,111

misalignment signal / 96

misplace / 3

missed-approach altitude / 191

mission control center (MCC) / 62

mission trainer / 191

mixed combustion chamber / 156

mixed compression inlet / 156

mixed electronic-hydromechanical control system / 172

mixed oxygen / 115

mixed-flow compressor / 156

mixed-flow turbine / 156

mixer / 111,156

mixing chamber / 36,86

mixture / 29,86

mixture (fuel) / 156

mixture (oil) / 156

mode button / 111

mode selection knob / 184

mode selector failure / 111

mode selector knob / 111

model/designation / 76

modem module / 48

modification quality / 16

modular structure / 164

modulate / 48

modulated frequency / 111

modulated wave / 111

modulating stage / 56

modulation / 62

modulator / 111

moisture / 12,86,175

moisture collector / 121

mold / 135

molding / 125

moment / 16,76,104,136

moment motor excitation winding / 42

monitor / 56,175

monitor procedure / 36

monitor reset circuit / 96

monitoring capability / 111

monitoring range / 111

monostable circuit / 175

mooring / 21

mooring equipment / 62

mooring fitting / 139

mooring fitting joint / 62

mooring net / 62

mooring steel cable / 62

mooring strap / 62

Morse code / 111

mosquito repellent tissue / 62

mother board / 48

motion mechanism / 3

motion system / 191

motor / 36,42,56,86,96

motor ventilating hood / 169

motor-tachometer / 42

mould fungus / 156

mould part / 104

moulding / 133,139

moulding fitting / 139

moulding flange / 139

moulding part / 130

moulding rubber / 133

moulding rubber ring / 130

mount/dismount / 16

mountain / 111

mountain peak / 111

natural icing flight test / 191

natural vibration frequency / 72

navigation / 18

navigation cabin / 3,16,97,130,136

navigation compartment / 8

navigation compartment/navigation cabin / 56

navigation computer / 42

navigation deviation / 111

navigation indicator / 111

navigation instrument / 97

navigation light / 3,97,139

navigation parameter / 111

navigation radar bay / 130

navigation satellite / 111

navigation selector indication / 42

navigation signal / 56

navigation source selection / 111

navigation station / 111

navigation system / 3,29

navigation system relay box / 42

navigator / 8,16,36,42,48,86,136

navigator cabin / 62

navigator circuit breaker box / 145

navigator connection button / 42

navigator console / 97

navigator control knob / 42

navigator controller / 8

navigator fluorescent light / 97

navigator indicator / 97

navigator instrument panel / 42,97

navigator oxygen annunciator / 97

navigator seat / 29

navigator side cover plate / 97

navigator turn knob / 42

NDB approach / 191

near-earth satellite / 62

needle / 79

needle bearing / 3,139

needle file/handle file / 56

needle flow restrictor valve / 79

needle plunger valve / 86

needle valve / 36

negative（not approved）for straight-in approach / 191

negative control / 86

negative feedback brush / 97

negative feedback potentiometer / 97

negative pole / 56,145,170

negative thrust / 8,145

negative thrust auto-feathering / 145,157,164

negative thrust auto-feathering annunciator / 181

negative thrust auto-feathering circuit / 145

negative thrust check device / 145

negative thrust feathering check button / 97

negative thrust feathering check switch / 145

negative thrust feathering sensor / 145

negative travel / 145

negative wire / 56,86,181

negative/positive pressure / 76

net calling address / 48

net mesh / 186

net weight / 76,145,181,186

neutral gas / 56

neutral gas bottle / 29,66

neutral gas explosion cap / 97

neutral gas fire extinguishing button / 97

neutral gas heating / 97

neutral gas indicating light / 97

neutral gas interface / 76

neutral gas selection switch / 97

neutral gas system / 129

neutral position / 16,29,56,72,86,104,125,181

neutral slot / 42

neutral soap water / 86,115

neutral soapsuds / 29,177

nicked teeth / 62

nickel / 86

nickel cadmium battery / 56

nickel enamel wire / 181

nickel wire / 66

nickel-cadmium alkaline battery / 56

nickel-cadmium battery / 121

nickel-chrome resistance wire / 111

nickel-plating / 115

nipple / 66

nipple/union / 104

nitrogen / 9,62,104

nitrogen bottle / 79,104

nitrogen cavity / 29

nitrogen charging pressure / 29,72,104

nitrogen charging unit / 29

nitrogen gas / 136

nitrogen pressure / 29

nixie tube / 184

noise / 21,48

noise reduction function / 48

noise suppressor / 177

noiseless chain / 72

no-load start / 157

no-load voltage/zero load voltage / 56

nomenclature / 56,104,170

nominal dimension / 186

nominal pressure / 12

nominal section area / 56

non-adjustable pull rod / 72

non-circular cross section / 130

nondestructive testing manual / 16

noninflammable varnished cloth / 118

non-memory channel/non-stored channel / 48

non-metal part / 170

non-metallic scraper / 139

non-operating condition / 12

non-operating end / 66

non-pressurized cabin / 130

non-pressurized panel / 130

non-shielding wire / 56

nonsteady combustion / 157

nonsustaining combustion / 157

non-twisting wire / 56

non-volatile memory / 111

norm / 186

normal acceleration / 97

normal atmospheric pressure / 157

normal brake / 79

normal cargo airdrop / 97

normal closed contact / 42,86

normal combustion / 157

normal control circuit / 97

normal control line / 125

normal list / 111

normal opening / 125

normal operation / 29,86,170

normal parking compression value / 29

normal pressure / 29

normal retraction/extension system / 72

normal slaving speed / 112

normal start / 157

normal temperature / 86,112

normalized maintenance / 48

normally closed contact / 56

normally open contact / 56,86

normally opened contact / 56

normal-shock inlet / 157

northing acceleration / 97

northing speed / 97

nose / 3,12,16

nose cap/dome / 145

nose dome/nose fairing / 130

nose fairing / 8

nose fairing glass / 136

nose jack / 12,29

nose landing gear (NLG) / 3,12,18,21

nose landing gear door / 125

nose landing gear extension / 97

nose landing gear suspension ring / 29

nose landing gear well / 16,130

nose landing gear wheel bearing / 29

nose landing gear wheel shaft / 29

nose landing light / 97

nose pliers / 125

nose radome/nose fairing / 86

nose ring cowling / 157

nose shock strut / 29

nose wheel / 8,16,21,29

nose wheel control mechanism / 79

nose wheel fork / 12

nose wheel landing gear / 72

nose wheel steering / 79,97,104

nose wheel steering actuator / 29

nose wheel steering bracket / 18

nose wheel steering clamp / 29

nose wheel steering control cable / 29

nose wheel steering follow-up mechanism / 29

nose wheel steering guide device / 18

nose wheel steering mechanism / 3

nose wheel support / 12

nose/main landing gear (NLG/MLG) / 16,72

nose-down pitching (steering) moment / 42

nose-up signal / 42

notch / 62,112

note / 16

nozzle / 136,175

nozzle base drag / 177

nozzle cowling / 157

nozzle efficiency / 177

nozzle expansion ratio / 177

nozzle guide vane (NGV) / 170

nozzle pressure / 157

nozzle ring / 36

nozzle thrust coefficient / 177

nozzle thrust ratio / 177

null error/zero position error / 42

null position error / 112

nut / 12,42,48,86,139,157

nut gear / 170

nut washer / 79

nut-cover / 18

nylon / 72

nylon cushion block / 139

observation / 12,86

observation access cover / 104,125

observation value / 112

observation window / 125,130,136

observe / 16

observer / 18

observing angle / 62

obstacle / 18

obstacle ahead / 112

obstruction / 12

occupant oxygen system / 29

odd number / 8

off-axis inlet / 157

off-course distance / 97

offset / 16

off-set pliers / 56

offset propeller hub / 145

ohmmeter / 66,170

oil / 3,16,21,29

oil accumulation / 181

oil assembly / 29

oil bag / 181

oil barrel / 157,181

oil cavity / 145

oil chamber / 66,104

oil consumption / 157,181

oil container / 104

oil control system / 181

oil cooler / 121

oil cooler inlet line / 121

oil cooler outlet line / 121

oil cylinder / 145

oil deflector/oil retainer / 181

oil delivery nipple / 42

oil department / 56

oil distributing piston / 42

oil distributing valve (rod) / 86

oil distribution bush / 145

oil distribution system / 181

oil distribution valve / 145

oil drain cock / 121

oil drain hose / 181

oil drain pipe / 145

oil drain port / 8

oil drain screw plug / 181

oil drain sleeve / 145

oil drain valve/oil drain cock / 157

oil drip tray / 181

oil ejection radiation system / 181

oil ejector radiation switch / 97

oil ejector radiation valve / 117

oil eject-radiating electric valve / 170

oil filler access / 157

oil filter / 29,42,121,145

oil filter bypass light / 181

oil filter bypass warning / 181

oil filter differential switch / 181

oil filtering accuracy / 157

oil flow / 157

oil gauge sensor / 181

oil groove / 145

oil heat dissipation capacity / 157

oil hole / 145

oil indication / 181

oil inlet / 42,145

oil inlet cavity / 145

oil inlet hole / 181

oil inlet temperature / 170

oil leakage / 181

oil level sensor / 121

oil lever switch / 121

oil line / 145,170

oil lubrication system / 170

oil mist / 181

oil mist separator / 170

oil nozzle / 29

oil passage orifice / 181

oil passing blind nut / 42

oil pipe unit / 145

oil plug / 36

oil plumbing / 181

oil pressure / 146,175,181

oil pressure annunciator / 8

oil pressure gauge / 181

oil pressure indicator / 181

oil pressure pump / 121,181

oil pressure sensor / 157

oil pressure transmitter / 181

oil pro of device / 56

oil pump / 8,21,29,121,146

oil pump assembly / 146,181

oil quality / 181

oil quantity / 181

oil quantity gauge / 29,181

oil quantity gauge indicator / 97

oil quantity gauge switch / 97

oil quantity indicator / 29

oil quantity indicator sensor / 157

oil quantity transmitter / 181

oil radiating changeover switch / 97

oil radiating efficiency / 181

oil radiating ejector / 181

oil radiator / 21,29,86,133,170,181

oil radiator shutter / 29

oil radiator shutter indicator switch / 97

oil remaining warning light / 97

oil return / 42

oil return line / 86

oil return nipple / 86

oil return pipe / 79

oil return position / 80

oil return valve / 12

oil sampling / 121

oil scavenge cavity / 181

oil scavenge filter / 29

oil scavenge pump / 121,157

oil scavenge stage / 29

oil scavenge tube / 181

oil sealing device / 170

oil seepage / 181

oil separator / 181

oil stain / 56

oil suction pipe / 181

oil sump / 121,181

oil supply / 181

oil supply duct / 181

oil supply pump pipe / 181

oil system / 29,157,181

oil system indicating / 181

oil tank / 8,16,21,29,121,126,146,157

oil tank antisiphon device / 181

oil tank filer port / 121

oil tank inner casing / 181

oil tank sediment / 181

oil tank/oil storage / 181

oil temperature auto-regulating circuit / 181

oil temperature control box / 181

oil temperature indicator / 181

oil temperature regulating system / 181

oil temperature regulator / 182

oil temperature sensor / 157

oil tray / 72,136

oil trunk / 182

oil turn pipe adapter / 182

oil vapor separator / 177

oil vent pipe / 21

oil vent tank / 182

oil viscosity / 182

oil/fuel filter / 21

oil/fuel vent tank / 21

oil-air chamber / 104

oil-air separator / 29

oiler / 29

oil-in cavity / 182

oil-mist separator / 8,21,157,182

oil-penetrating valve / 42

oil-return passage / 80

oil-return valve / 121

oil-to-fuel heat exchange / 121

ointment / 104

oleo-nitrogen damper strut / 104

oleo-nitrogen type / 104

omni-directional caster / 186

on condition check / 3

on start / 157

on-aircraft inspection / 72

on-board battery / 121

onboard circuit / 56

onboard power source / 66

one-side-entry combustor / 157

one-way buffer / 126

one-way delay valve / 126

one-way restricting function / 126

on-line programming / 97

on-site programming / 97

open air-cycle system / 36

open bearing / 56

open circuit / 56,182

open end spanner / 16

open fire source / 29

open shelter / 29

open spanner/open wrench / 72

open wooden lookum / 29

open wrench / 80

operation surface / 56

operation temperature / 30

operation voltage range / 86

operation weather limits / 191

operation/maintenance / 86

operational flight test / 192

operational flight trainer / 192

operational fuel / 157

operational oil / 157

operator / 30

optical system / 62

optimum oil temperature / 157

orange yellow enamel paint / 18

organic glass / 136

oriented organic glass / 126,136

orifice / 182

original position / 86,97,126,170

original position check / 139

original pr ofile / 157

ornament plate / 8

orthogonal excision network / 42

orthogonal excision stage / 42

oscillating divergent curve / 42

oscillation signal / 97

oscillatory combustion / 157

out board guide/restraint rail / 186

outboard aileron / 139

outboard engine / 16,139

outboard wing / 3,8,21,48,96,139

outdoor / 16

outer casing nut / 30,164

outer chamber / 116

outer cylinder / 12,21,104,126

outer cylinder joint / 104

outer cylinder lug / 104

outer cylinder pipe / 104

outer edge / 130

outer electrode / 167

outer fringe/outer rim / 72

outer marker / 112

outer of nut / 12

outer ring / 72

outer section / 86

outer shroud / 80

outer sidewall / 133

outer skin / 86,126,133

outer surface / 126

outlet / 133

outlet guide vanes / 177

outlet port / 8

outlet pressure / 66,182

outlet relay / 66

outlet tube / 170

outlet union / 66

outline dimension / 56

out- of- control / 86

out- of-band interference / 48

out- of-tolerance / 72

output / 146,175

output cable / 42

output circuit / 56

output end / 66

output force / 42

output frequency / 56

output interface format / 112

output parameter accuracy / 112

output power / 56

output range / 112

output shaft / 56,170

output signal / 86

output signal channel / 86

output swing angle / 86

output switch / 56

output synchronizer / 42

output voltage / 42,56,97

outseam / 167

outside diameter / 186

outside dimension / 48,186

oxygen flow indicator / 30

oxygen flowmeter / 62

oxygen high-pressure pipe / 116

oxygen hose / 30

oxygen instrument panel / 97

oxygen mask / 66

oxygen mask microphone / 48

oxygen part / 30

oxygen percentage mechanism / 116

oxygen pipe / 72

oxygen pressure gauge / 8

oxygen pressure reducer / 30

oxygen receiving valve / 116

oxygen regulator / 8,30

oxygen relay box / 97

oxygen signal light / 97

oxygen source gauge / 9,30

oxygen source pressure gauge / 116

oxygen special wrench / 116

oxygen supply annunciator / 97

oxygen supply check button / 9,97

oxygen supply hose / 116

oxygen supply signal check button / 97

oxygen supply signal light / 97

oxygen supply signal release button / 97

oxygen supply valve / 116

oxygen switch / 9

oxygen system / 30

oxygen valve / 30

oxygen warning system / 97

oxygen-resistance / 116

packing / 121

packing box / 30,146,157

packing box bracket / 157

packing list / 187

packing ring / 104

pad / 30,86,133

page down / 48

paint / 130,133

paint coat / 76,133

paint layer / 126

paint remover / 72

paint stripping position / 30

painting color / 23

paired centripetal ball / 42

pallet / 187

pallet attachment fitting / 187

pallet lock / 187

pallet net / 187

panel / 3,9,16,37,86,130,136

paper roll / 118

parachute / 62

parachute cord / 56

parachute cord recovery circuit / 97

parachute holder / 116

parachute jumps flight / 191

parachute opener / 62

parachute steel cable / 62

parachute support / 62

parachute-cord box / 62

parachute-opening shock / 62

parachuting / 117

parachuting airborne component / 16

parachuting signal / 56

parachuting signal device / 62

parachuting windshield plate / 62

paraffin paper / 21,121

parallel altitude code / 112

parallel arrangement / 130

parallel connection / 86,97

parallel rating factor / 56

parallel resistance / 182

parallelism / 16

parameter / 3,37,48

parameter adjustment potentiometer / 42

parametric down-converter / 48

parametric up-converter / 48

paratrooper/parachutist / 62

pilot valve / 117

pilot/copilot / 86

pilot/copilot console / 98

pilot/copilot instrument panel / 98

pilot's stand / 130

piloted air-blast fuel nozzle / 157

pin / 12,18,30,42,48,56,66,105,112,146, 157,170,175

pin connector / 56

pin groove / 133

pin hole / 105

pin number / 42

pin rod of the depression throttling valve / 42

pin rod slide valve / 42

pin shaft / 105

pin sleeve / 9

pin socket / 62

pin support / 62

pin valve / 146

pinion / 66

pipe / 21,30,157,170

pipe clip / 118

pipe connector / 9,86

pipe end / 86

pipe head / 105

pipe holder / 66

pipe joint / 42

pipe joint/coupling / 37

pipe joint/union / 170

pipe line / 133,182

pipe port / 117

pipe shaft / 30

pipe temperature sensor / 37

pipe wall / 72

pipeline / 12

pipeline heater / 66

pipeline high-temperature limiting sensor / 37

pipeline low-temperature limiting sensor / 37

piston / 12,21,30,86,170

piston aero-engine ignition system / 157

piston aero-engine starting ignition device / 157

piston cavity / 146

piston chamber / 66

piston compressor / 157

piston cylinder / 12

piston full travel / 126

piston push rod / 80

piston ring / 76,170,182

piston rod / 12,21,30,86,105,126

piston rod end / 42

piston rod sleeve / 126

piston travel / 126

piston unit / 146

pit / 3,18,126

pitch angle / 42,98

pitch change actuator / 146

pitch channel / 42

pitch contact / 42

pitch drawer / 42

pitch lock mechanism / 146

pitch potentiometer / 43

pitch scale / 43

pitch variation mechanism / 146

pitch variation mode / 146

pitch variation range / 146

pitching test / 192

pitch-reversing mechanism / 146

pitch-setting mechanism / 146

pitot pressure / 112

pitot pressure adapter nozzle / 98

pitot pressure hose / 98

pitot pressure input limit / 112

pitot pressure union / 98

pitot tube / 98,112

pitot tube heating indicating light / 98

pitot tube system tester / 98

polyurethane adhesive fluid / 126,139

polyvinyl butyral rubber sheet / 136

polyvinyl chloride (PVC) / 57

polyvinyl chloride foam plastics / 139

pop rivet/hollow rivet / 72,126

porous gas turbine blade / 157

porous section / 48

port of destination / 187

port of embarkation / 187

port throttle / 164

portable carbon dioxide fire bottle / 30

portable explosion-proof light / 30

portable fire bottle / 21,66

portable light / 9

portable oxygen system / 30

portable signal light socket / 98

position / 21

position error / 192

position feedback sensor / 43

position feedback sensor rod / 43

position indicator / 98,112,182

position longitude-latitude / 112

position measuring instrument/position level instrument / 16

position/location / 86

positioning / 12

positioning gasket / 72

positive and negative sintering polar plate / 57

positive control / 86

positive displacement hydraulic motor / 121

positive displacement pneumatic motor / 121

positive electricity / 86,182

positive pole / 66,170

positive pulse signal / 175

positive travel / 146

positive wire / 57

positive/negative acceleration / 98,105

positive/negative angular acceleration / 105

positive/negative error / 175

positive/negative pole / 175

positive/negative travel / 175

post-flight inspection / 146

potable water / 157

potential difference / 175

potential strength / 49

potentiometer / 43,57,76,86,98,112,182

potentiometer shaft / 42

potentiometer winding / 98

powder chamber / 66

powder metallurgy braking disc / 105

power / 170,175

power amplification unit / 49

power amplifier / 66,182

power amplifier module / 49

power capacity / 49

power circuit breaker / 49

power consumption / 37,49,87,182

power consumption equipment / 57

power control / 172

power distribution board / 30,57

power distribution box / 146

power distribution cabinet / 118

power distribution equipment / 57

power distribution panel / 57

power distribution protection device / 57

power distribution system / 3,37

power drive unit / 187

power feeder / 3

power filter / 112

power filtering box / 184

power indicating light / 49

power level / 49

power lever / 59,74,100,172

power lever angle (PLA) / 157,164

power limiter / 9,164

power mechanism / 80

power module / 49

propeller anti-icing / 98

propeller anti-icing heating / 9

propeller blade / 21,31,146

propeller blade icing / 146

propeller blade root / 146

propeller blade root cowling / 146

propeller brake / 87

propeller brake mechanism / 146

propeller braking distribution box / 9

propeller casing / 146

propeller casing unit / 146

propeller digital synchronize / 146

propeller feathering / 182

propeller governor / 146

propeller heating indicating light / 98

propeller heating switch / 98

propeller hub / 16,158

propeller hub spinner / 140

propeller over-speed / 146

propeller pitch-control mechanism / 146

propeller ready indicating light / 98

propeller root fairing / 134

propeller root fillet / 146

propeller rotating speed / 146

propeller rotating zone / 72

propeller shaft / 146,158

propeller stop relay / 57

propeller stop release indicating light / 98

propeller stop release mid-pitch valve / 9

propeller stop release switch / 98,146

propeller synchronizer / 146

propeller tip / 3

property / 87

proportional amplifier / 121

proportional control valve / 121

proportioner/quantilizer / 105

protection / 134

protection circuit / 57

protection cloth / 21

protection panel / 105,131

protection performance / 98

protection ring / 87

protective belt / 57

protective coating / 37

protective cover / 9,16,57,98,182

protective diode / 87

protective layer / 126

protective layer/coating / 131

protective lid/cover / 37

protective skin / 131

protective sleeve / 126

proximate traffic (PT) / 112

pull back the throttle / 192

pull force / 18

pull ring / 18

pull rod / 9,12,16,21,72,80,87,126

pull rod linkage / 72

pull up / 112

pull up and go around / 192

pulley / 3,18,31,49,126

pulley bracket / 3,126

pulley groove / 126

pulley mount / 9

pulley rim / 126

pulley shaft mechanism / 72

pulley support / 140

pull-in bandwidth / 112

pulling force / 57

pulling propeller / 146

pully bearing / 31

pulsating button / 57

pulse / 49,112

pulse amplifier / 87

pulse circuit / 98

pulse current / 62,67

pulse drive / 182

pulse generator / 87

pulse peak / 112

radiation switch / 87

radiator / 9,37,80,182

radiator shutter / 158

radio / 21

radio altimeter (RALT) / 98,112

radio altimeter brightness adjusting knob / 98

radio altimeter indicator / 98

radio and radar facilities / 192

radio communication / 3

radio communication equipment / 3

radio communication keying / 98

radio communication system / 49

radio compass / 9,49

radio equipment / 3

radio frequency (RF) signal / 184

radio frequency amplifier unit / 49

radio frequency assembly / 112

radio frequency switch module / 184

radio interference / 57

radio magnetic indicator / 98,112

radio navigation system / 112

radio set / 9

radio set button / 72

radio set protector / 49

radio talk button / 98

radioactive requirement / 87

radio-frequency antenna / 112

radome / 3,9,21

raid of hail / 3

rail groove / 105

railing for parachute discharging operator / 62

rain protection / 175

rainfall rate / 112

raise/fall / 16

ram air / 37,76

ram compressor / 158

ram scoop / 121

ram-air turbine / 158

ramie rope / 72

ramie thread / 58,98

ramjet inlet / 158

ramp / 187

ramp door / 9,16

ramp door locked indicating light / 98

ramp door open indicating light / 98

ramp lock actuator / 80

randomly-checking flight test / 192

range / 16

range ring / 112

range-to-go / 98

rapid repeat start / 158

raster circuit / 175

raster voltage / 175

ratchet / 158

ratchet clutch / 158

ratchet mechanism / 62,98

rate control valve / 121

rate gyro / 43

rate gyroscope casing/rate gyro box / 43

rated DC voltage / 146

rated indicating light / 98

rated lifting height / 12

rated load / 67,182

rated operation rotating speed / 146

rated output power / 57

rated power / 146

rated state / 158

rated status / 170

rated switch / 98

rated value / 31,57,76

rated voltage / 37,87,175

rated weight / 31,67

raw rubber tape / 116

RC time delay circuit / 57

reaction force / 31

reading / 87

redundancy design / 116

redundant/absent equipment / 17

reed / 76

reel / 62

reference code / 112

reference phase / 112

reference point / 57

reference surface / 17

refilling port/charging port / 57

reflecting screen / 31

reflective mirror / 43

reflective sticker / 99

refrigerated container / 187

refrigeration system / 37

refuel / 31

refueling / 17,21

refueling adapter / 76

refueling altitude / 192

refueling automatic control system / 31

refueling bonding jumper / 77

refueling cart / 80

refueling command / 31

refueling connector / 31

refueling controller / 9

refueling distribution board / 57

refueling distribution box / 31

refueling gun / 31

refueling personnel / 31

refueling pipe / 31

refueling port / 31

refueling power distribution board / 9,77

refueling power distribution box / 9

refueling procedure / 31

refueling pump / 9

refueling quantity / 31

refueling selector switch / 31

refueling switch / 31

refueling system / 9

refueling truck / 27,77

refueling valve / 9

register / 112

regular compass alignment / 112

regulardistribution / 146

regulated parameter / 172

regulating / 72

regulating bearing / 182

regulating cover / 37

regulating nut / 12

regulating screw / 146

regulating shim / 170

regulating sliding rail / 31

regulating system / 158

regulating turnbuckle / 21

regulation / 21

regulator / 9,158

reheat combustion chamber/after-burner / 158

reheat ignition/after burning ignition / 158

reheat regulator / 165

reigniter chamber / 158

reinforce / 37

reinforced access frame / 140

reinforced edge / 134

reinforced edge strip / 126

reinforced frame / 3,17,131

reinforced groove / 17,126,134,140

reinforced pad / 135,173

reinforced partition / 135

reinforced plate / 131,140

reinforced rib / 140

reinforced ring / 134

reinforced rod / 182

reinforced splice / 77

reinforced stringer / 131

reinforced strip / 131

reinforced strut / 140

reinforcement / 126

reinforcing access frame / 147

residual oil quantity signal / 182

residual oscillation / 192

residual pressure / 126,131

residual snow / 31

residual travel / 105,126

residual water / 31

resin / 131

resistance / 87

resistance box / 9,182

resistance compensation / 57

resistance lad current / 105

resistance load / 37,57

resistance signal / 87

resistance temperature / 113

resistance temperature transmitter / 165

resistance value / 67

resistance wire / 67,87

resistive current / 67

resistor / 37,43,182

resistor disc / 87

resistor/resistance / 147

resolution advisory / 113

resolution alert / 113

resonance operating frequency / 175

response light / 184

response mode switchover knob / 184

response probability / 184

response signal / 184

restore / 17

restraining chock / 62

restraint / 126

restraint system / 187

restricting diameter / 67

restricting grid / 80

restricting heater / 67

restricting machine / 87

restricting nipple / 67

restricting nozzle / 67

restricting orifice / 31,67,126

restricting orifice/restricted orifice / 43

restricting venturi tube / 117

restrictor / 147,173,182

restrictor valve / 80,87

retainer / 62,72,126

retard / 37

retard the throttle to idle / 165

retarded combustion / 158

retract / 72

retractable actuator / 3

retractable end restraint / 187

retractable guide restraint / 187

retractable guide roller / 187

retractable light / 9

retraction / 126

retraction circuit / 105

retraction travel / 72

retraction/extension / 17,80

retraction/extension actuator / 22,31,105

retraction/extension mechanism / 12,31

retraction/extension position / 105

retraction/extension space / 126

retraction/extension test / 3

return / 170

return filter / 80

return line / 173,103

return pipe / 170

return spring / 43,105

return-to-zero / 99

return-to-zero limit / 43

reversal installation / 126

reverse current / 57

reverse flow / 67

reverse force / 57

reverse heading / 43

reverse magnetic field / 57

reverse pressure / 67,105

reverse process / 49

reverse sequence / 126

reverse thrust control link / 165

reverse thrust lever / 165,177

reverse thrust lever interlock actuator / 165

reverse voltage / 43

reversed procedure / 105

reverse-flow check valve / 121

reverse-flow combustion chamber / 158

reverse-pitch propeller / 147

reverser / 37,113

reverser actuator / 177

reverser control rod / 165

reverser deploy / 177

reverser throttle interlock / 177

reverser unlocked switch / 177

reverser/commutator/converter / 57

reverse-thrust nozzle / 177

reversible hydraulic motor / 72

revolutions per minute (r. p. m) / 175

rheostat / 9,57

rheostat/variable resistor / 57

rib / 3,9,37,49,87,126

rib location / 9

rib plate / 140

rib-sealing cavity / 182

rich running / 192

rigger load / 187

rigging pin hole / 165

right rotary knob / 113

right system accumulator / 31

right system cock / 31

right system pressurization connector / 31

right traffic pattern (circuit) / 192

right view drawing / 23

right-hand rotation propeller / 147

rigid connection / 99

rigid control / 72

rigid control linkage / 72

rigid driving part / 72

rigid pull rod / 72

rigid support / 126

rigidity / 140

rigidity/stiffness / 134

ring amplifier / 113

ring spring/spiral spring / 105

ripple / 131

rivet / 3,22,37,87,105,126,134,136

riveted structure / 126

riveting placard / 87

rocker arm / 3,9,12,18,31,37,43,87,126,
136,140

rocker arm assembly / 9

rocker arm bracket / 9,134

rocker bracket / 72

rocker group / 72

rocker lever / 77

rocker lug / 31

rocker switch / 158

rocker valve / 116

rod / 22

rod body / 72

roger / 192

roll angle / 99

rolled weld / 177

roller / 31,72,105,131

roller bearing / 80,165

roller device / 131

roller gear plate/roller toothed plate / 105

roller track / 178

roller tray assembly / 187

roller with bush / 31

rolling / 140

rolling rate oscillation / 192

rolling sleeve / 173

rolling take-off method / 192

rollout stop / 187

rollway / 62

root / 72

root mean square current / 57

sandy soil / 105

sanitation system / 118

satellite integrated navigation system / 43

sawtooth generator / 182

scale / 99,176

scale beginning point / 182

scale/dial / 57,67

scallop / 158

scan / 49

scanner / 113

scanning channel group / 49

scanning driver / 113

scanning driver base / 113

scatter / 65

scavenge filter / 22

scavenge filter/oil return filter / 182

scavenge pump / 165,182

scavenge stage / 182

scavenge stage of main oil pump / 170

scenario training / 192

scheduled and unscheduled servicing / 32

scheduled flight altitude / 43

scheduled heading / 43

scheduled lubrication / 32

scheduled maintenance / 134

scheduled maintenance check / 3

schematic diagram / 13,43,49,87

scissor / 43,140

scissors difference / 73,140

scoop channel / 67

scrap / 87

scrape / 63

scratch / 13,22,57,126,136,140

screen / 22

screen intensity / 113

screw / 13,22,32,87,134,140

screw-in elbow joint / 116

screw-lifting mechanism / 13

scrub / 77

scrunch / 73

sea level / 184

seal drain port / 122

seal element / 131

sealant / 131,136,140

sealant stem / 140

sealed bearing / 58

sealed double-pole changeover switch / 58

sealed type induction coil / 167

sealing / 13,32,127,171

sealing adhesive putty / 73

sealing cap / 136

sealing cavity / 171,177

sealing compound / 147

sealing cover / 58

sealing cup ring / 165

sealing device / 105

sealing device/unit / 87

sealing element / 131

sealing gasket / 182

sealing grease / 140

sealing guide sleeve / 105

sealing material / 113

sealing pad / 182

sealing partition / 58

sealing place / 58

sealing plate / 131,140

sealing plug/airtight-plug / 140

sealing protection / 58

sealing putty / 58,105,131,136

sealing rib / 140

sealing ring / 13,32,43,87,105,127,165,171,183

sealing ring mounting slot / 43

sealing rubber / 22

sealing rubber pad / 22

sealing rubber pipe / 22

sealing rubber ring / 105,140

sealing rubber tube / 127

sensor model / 67

sensor rotor / 122

sensor stator / 122

separation / 134

separation connector / 80

separation disk / 73

separation net / 187

separation point / 127

separator joint / 22

sequence start / 158

sequencing oil temperature switch / 122

sequential action / 127

sequential engaging / 147

serial bus transmitting/receiving circuit / 99

serial connection / 87

serial data / 99

serial data bus / 99

serial number (S/N) / 87,171

serial port / 184

series coil / 58

series connection / 176

series number (S/N) / 17,176

series resistance / 183

series start coil / 63

series-wound motor / 58

serious accident / 32

serious rust / 32

serrate spring washer / 73

serrated gasket / 140

serrated pad / 131,140

serrated plate / 131

service / 87

service condition / 67

service ladder / 13,17,37,43,49,58,67,87,
127,131,140

service ladder/working ladder / 105

service life / 32,80,99,105,183

servicing / 19,32,150

servo (driver) / 43

servo (driver) system/follow-up system / 43

servo circuit / 99

servo motor / 77

servo plate / 113

set altitude / 99

setter / 99

setting / 113

seven-position distribution cock / 73,80

seven-position distribution valve / 32

severe towing / 3

sew / 73

shaft / 22,32

shaft assembly / 80

shaft base/axle base / 37

shaft coupling / 147

shaft end long groove / 171

shaft end section/shaft end surface / 171

shaft gear / 43

shaft head / 32

shaft horse power (SHP) / 165

shaft pin / 63

shaft power / 147

shaft rod / 19

shaft sleeve / 73,88,105

shake / 171

shaking failure / 113

sharp edge / 140

sharp nose pliers / 63

sharp turning / 19

sharp-lip inlet / 158

shave / 136

shear / 147

shear force / 19

sheath / 73

shed / 19,134

sheet code / 9

sheet drawing / 10

sheet-metal part / 127

shell / 67

side-lobe pulse / 184

sideslip angle / 43

sideslip indicator / 113

sidewall / 58

sight oil separator / 122

sighting angle / 63

sighting system / 63

sign / 131

signal / 88

signal box / 99

signal circuit / 171

signal control box / 99

signal converter fault light / 99

signal crosslinking box / 184

signal damper / 113

signal detection amplifier / 165

signal flare / 10,17,99

signal format / 99

signal gun / 10,43

signal gun chamber / 63

signal gun control box / 99

signal gun control box cover plate / 99

signal indicator / 99

signal liaison / 63

signal light / 32,67,73,105,171176

signal lose detector / 122

signal monitor / 122

signal operating point / 113

signal pipe / 118

signal point error / 113

signal processing module / 184

signal processing unit/signal processor / 113

signal processor / 165

signal processor unit / 49

signal safety device / 67

signal sampling / 37

signal sensor / 8,22

signal source / 49,88,99

signal switch assembly / 165

signal system / 105

signal union / 32,67

signal volume / 165

signal weight / 88

signal window / 99

signal-to-noise ratio / 49

signature / 32

silent chain / 32

silica gel / 122

silicate glass / 136

silicon pressure sensor / 165

silicon rectifier / 122

silicon rubber / 58,131

silicon rubber sleeve / 176

silicone rubber piece / 80

silicone rubber section / 134

silk cloth / 32,58

sill lock / 187

sill roller / 187

simple toilet / 131

simplified control box / 49

simplify start procedure / 167

simulation icing flight test / 192

simulation test / 58

simulator / 113

simultaneous ignition / 159

single actuator / 127

single arm access cover lock / 10

single gear / 43

single line power supply / 58

single line system / 167

single metal braking disc / 105

single oil line / 147

single pendulum / 99

single set address / 49

single shot recoil / 63

single sideband (SSB) / 49

single stage gear oil pump / 183

single stage high pressure gear pump / 183

socket / 49,58,67,73,88,140

socket wrench/socket spanner / 77

socket/receptacle / 171

soft alert area / 113

soft brush / 32

soft cloth / 32

soft key / 113

soft tank/flexible tank/bag tank / 73

softening coefficient / 159

solder / 58

solder/braze / 58

soldering point / 58

solenoid clutch/electromagnetic clutch / 43

solenoid valve / 10,67,73,80,106,147

solenoid valve of starting fuel supply / 10

solid anodization / 63

solid grain / 80

solidification / 140

solid-state storage / 99

solubility / 106

solvent / 131

sonic inlet / 159

sonic speed / 192

sound pickup / 49

sound/noise / 131

sound-suppression nozzle / 178

space movement / 73

spacer / 134

spacer block / 135

spacer bushing / 159

spacer/washer / 58,159

span / 140

span length / 140

spanner/wrench / 88,136

spar / 10,58,134,140

spare cap / 58

spare part / 171

spare wire / 58

spark / 58

spark duration / 159

spark energy / 159

spark igniter / 159

spark plug / 159

spark quenching circuit / 88

spark-plug ignition / 159

spatial condensation / 50

speaking status / 50

special anti-magnetic screwdriver / 113

special blanking cover/blanking cap / 37

special bracket / 32

special cargo / 187

special condition / 67

special container / 116

special cotton uniform / 67

special hand wheel / 32

special identification button / 184

special note / 23

special position replay / 184

special rotation table / 43

special state / 17

special tool / 58

special washer / 80,134

special-shaped sleeve / 106

specific location / 183

specification / 50,187

specification of airplane flying qualities / 192

specification of the flight test mission / 192

specified level / 58

specified value / 32,58,106,176

specify / 17

specimen / 17

speed bias torque motor / 147

speed control / 173

speed governor / 22,113,165,183

speed handle / 43

speed on go-around / 192

speed pressure/dynamic pressure / 171

speed reducer / 58,171,183

spring tab / 171

spring tension / 80

spring trim tab / 10

spring vibration absorber / 99

spring washer / 58,73,88,127,171,183

sprinkle / 88

square column / 73

square flange / 37

squelch function / 50

squib / 32

stabilivolt / 88

stabilized voltage source / 88

stabilizing shock absorber / 106

stable airflow / 43

stable buffer / 32

stable buffer outer cylinder / 32

stable climb / 73

stable shock strut / 32

stable state/stable condition / 43

stacking / 187

stage ignition / 159

staggered arrangement / 131

staggering / 63

stagnate / 88

stagnation / 3,22,58

stain spot / 73

stainless steel / 67

stainless steel ball handle lock / 127

stainless steel plate / 140

stainless steel rivet / 140

stainless steel strip / 88

stainless steel wire / 80

stall / 88

stall flight test / 192

stall warning buffeting / 192

stall warning system / 73

stalling / 63

stalling speed / 192

stamp/press/punch / 80,140

stamping / 135

stamping part/pressing part / 136

stand / 73

standard / 187

standard atmosphere condition / 159

standard atmosphere pressure / 73

standard code / 80

standard part / 73,140

standard pressure altitude / 113

standard temperature / 73

standby control switch / 73

standby interface / 184

standby power supply fuse / 58

standby static changeover switch / 113

standing-wave ratio / 50

starboard throttle / 165

start / 22,171

start bleed valve / 159

start contactor / 122

start control drum / 173

start control push-pull cable / 173

start control system / 58

start fault light / 99

start fuel spray / 167

start indicating light / 99

start interlock relay / 122

start latch relay / 122

start lever / 165

start lever assembly / 165

start motor / 122

start relay / 122

start relay box / 167

start system / 165

start time / 159

starter / 122

starter anti-fire switch / 99

starter button / 159

starter changeover switch / 99

starter clutch / 122

steady pull-up method / 193

steady straight flight method / 193

steady turn method / 193

steady-flow combustion / 159

steel ball / 73

steel ball lock / 106

steel ball lock spring / 77

steel ball seat / 80

steel ball tightener / 73

steel bolt / 131

steel bottle / 32

steel bushing / 88,127

steel cable / 10,32,50,73,106

steel cable clamp / 58

steel cable control / 17

steel cable control mechanism / 32

steel cable pad / 63

steel cable tension / 127

steel cylinder / 73

steel drop-forging rod head / 106

steel fork joint / 50

steel liner/steel lining / 116

steel pipe / 73

steel section / 131

steel structure part / 3

steel tape / 17

steel thread plug / 32

steel tube / 134

steel wire / 134

steel wire braided hose / 80

steel wire pin / 38

steel wire retainer / 43

steer to the left / 193

steering actuator bracket / 32

steering force / 44

steering mechanism / 32

steering radius / 10

steering wheel / 10

step / 127

step connection / 140

step difference / 140

step motor / 100

step signal / 44

step-ailerons method / 193

step-rudder method / 193

stick-core / 183

stick-fixed maneuver margin / 193

stick-free maneuver margin / 193

stick-free static stability margin / 193

stiffened edge / 178

stiffener hole/strengthened pit / 131

stiffening / 135

stiffening strut / 135

stiffness / 131

still pot / 32

stipulation / 3

stop block / 19,127

stop bushing / 178

stop cam / 38

stop clamping plate / 73

stop gear / 58

stop joint / 131

stop limit position / 19

stop nut / 13,106

stop pad / 171

stop pin / 13,22,32,73,127

stop plate / 73

stop refueling command / 32

stop releasing indicating light / 147

stop releasing switch / 147

stop ring / 106,183

stop ring / 183,127

stop screw / 80,88,127

stop shaft / 13

stop travel / 73

stop washer / 58

stop watch / 73,88,106

stop way / 193

terrain model / 193

tertiary-air door / 178

test / 50,88,114

test bed / 73

test bench/test stand / 59

test condition / 187

test date / 187

test equipment / 132,171

test loading / 187

test result / 187

test room / 132

test run / 160

test socket / 114

test value / 114

tester / 88

testing similar condition / 160

tethered hovering method / 193

textolite / 73

textolite gasket / 67

theoretical clearance / 178

theoretical configuration / 141

thermal barrier / 187

thermal bimetal / 165

thermal container / 187

thermal decomposing / 33

thermal expansion valve / 77

thermal load / 160

thermal part / 160

thermal plug / 4

thermal pressure relief / 122

thermal radiation / 33

thermal relay / 88

thermal resistance / 183

thermal safety valve / 127

thermal sensitive element / 67

thermal shielding/heat shield / 122

thermal switch / 88

thermal water bottle / 33

thermal water tank / 33

thermistor / 88,176

thermo couple assembly / 67

thermo electromotive force / 67

thermo potential / 67

thermo potential value / 67

thermocouple / 67,165,176

thermoelectric instrument / 122

thermo-joint/hot junction / 176

thermometer / 59,100

thermos bottle / 118

thermos bottle seat / 118

thermostat / 122

thickness / 127,136

thinner lead mat / 165

thiokol sealant / 167

thread / 50,73,88

thread broken / 127

thread die / 67

thread groove / 33

thread hole / 88

thread mechanism / 73

thread neck / 67

thread peeling- off / 63

thread rod / 13

threaded connector / 73

threaded plug / 33

threaded rod / 183

three-arm rocker / 73

three-axis accelerometer / 100

three-blade propeller / 147

three-digit number / 10

three-dimensional flow / 160

three-dimensional inlet / 160

three-hole straight plug / 80

three-phase AC generator / 176

three-phase AC power circuit / 59

three-phase converter / 10

three-point socket / 80

three-position distribution cock / 73

time limit / 4

time measurement / 100

time procedure annunciation / 167

time threshold / 114

timed acceleration fuel control unit / 122

time-lag fuse / 59

timer adjuster / 165

timer assembly / 118

time-to-go / 100,114

time-to-station / 114

timing circuit / 100

timing device / 10

timing mechanism / 59,88,160

timing precision / 100

tin crimping / 59

tin soldering / 59

tire footprint loading / 187

toggle switch / 184

toilet basin / 118

toilet bottom hose / 118

toilet device / 118

toilet tub / 118

tolerance / 17,33,100

tolerance of acceleration / 176

tolerance of vibration / 176

tolerance on fit / 74

tone key / 50

tongue piece / 141

tool / 106,171

tooling equipment / 81

tooth strip / 141

top cap / 13

top cover / 77

top emergency exit / 127

top emergency exit hatch of cockpit / 81

top view / 10

top view drawing / 23

torch igniter / 160

torque / 160

torque arm / 4,22,106

torque auto feathering circuit / 165

torque auto-feathering sensor / 10

torque automatic feathering sensor / 171

torque feathering check switch / 147

torque feathering interlock throttle angle / 148

torque measuring mechanism / 148

torque measuring pump / 183

torque measuring sensor / 148

torque meter / 10,59,100,176

torque meter switch / 100

torque moment / 116

torque moment measuring gauge / 148

torque motor / 44

torque oil pressure / 148

torque pressure / 148

torque pressure gauge sensor / 160

torque sensor / 22

torque spanner / 106

torque spring / 132

torque wrench / 59,77,136,141,160

torquer / 100

torsion difference / 17

torsion rod / 44

torsional spring / 127

total air temperature / 100,114

total bleed airflow / 160

total capacity / 81,584

total cargo load / 132

total flow sampling / 50

total fuel indicator / 165

total load bearing / 141,584

total resistance / 183

total sampling rate / 100

total temperature / 165

total temperature sensor / 114

total temperature tube / 160

total time / 148,584

transonic propeller / 148

transonic speed / 193

transonic stage / 160

transonic turbine / 160

transparency / 88

transparent mask / 116

transparent shroud / 114

transpiration cooling blade / 160

transponder / 114

transport / 33,187

transport aircraft / 4

transportable / 187

transportation and storage / 50

transportation trolley / 160

transporter / 187

transverse beam / 4

transverse butt-joint seam / 132

transverse load-carrying member / 135

transverse partition / 127

transverse positioning / 127

trapezium shape / 141

trapezoid / 132

trapezoidal screw thread / 33

trapped fuel / 17

travel / 33,88,177

tread / 23

treadway / 187

trend / 74

trial speed / 193

triangle box beam / 132

triangle inclined bracing / 22

triangle inclined strut / 106

triangle partition / 132

triangle truss / 106

triangular frame / 13

triangular mark / 176

triangular rocker / 74

triangular spar / 127

tricycle retractable type / 106

trigger connector / 33

trigger joint / 68

trim / 74

trim tab / 4,10,17,100

trim tab control rocker arm / 33

trim tab method / 193

trim tab pull rod / 33

trimmer / 114,135

trimmer/brightness regulator / 114

triode / 59,88,106

trolley / 33

trouble shooting / 13,114,171,176

trouble/failure/fault/malfunction / 17

troubleshooting / 13,38,89,114,171,176,184

true airspeed / 100

true cruise airspeed / 193

true heading / 100

true value / 176

true/magnetic heading / 114

truss head bolt/mushroom-head bolt / 141

TSI switch / 100

tube passing / 141

tube root heater / 171

tube shaft / 74

tube-annular combustion chamber / 160

tubular beam / 135

tune / 50

tuning precision / 50

tuning time / 50

turbine blade / 160,171

turbine blowing cover / 10,160

turbine blowing cover cooling air inlet / 171

turbine blowing shroud / 178

turbine casing / 171

turbine cooler / 22

turbine cover / 160

turbine disk / 171

turbine efficiency / 160

ultraviolet ray illuminant / 74

unable to approve straight-in approach / 193

unbalanced engine thrust / 44

uncertificated ULD / 188

underslung inlet / 161

under-voltage protection / 59

under-voltage surge / 59

underwater locator beacon / 50,100

unevenness/heterogeneity / 114

unfeathering circuit / 148

unfeathering indicating light / 100

unfeathering pressure annunciator / 148

unfeathering solenoid valve / 10

unfeathering time / 148

unglued / 74

uniform airtight load/mean airtight load / 132

uniformity / 176

union / 13

union thread / 33

union/transition nozzle / 77

unipolar-converted / 68

unit / 17

unit/device / 89

unit/module / 38

universal compressed air source / 117

universal coupling bearing / 33

universal joint / 33

universal partition / 135

universal protractor / 74

unload / 17

unload valve/unloading valve / 81

unlocking actuator / 33

unlocking retraction extension actuator / 106

unnecessary signal / 50

unpacked state / 33

unpaved-runway landing / 193

unpowered wheeled device / 63

unpressurized area / 63

unreasonable bending / 127

unscheduled maintenance check / 4

unscheduled servicing / 33

unscrew / 33

unshrouded propeller / 148

unthreaded section / 74

up wind leg / 193

up/down vibration energy / 106

uplock / 4,33

uplock hook / 33

uplock/downlock / 106,132

uplock/downlock mechanism / 106

uplock/downlock rocker arm / 106

upper access cover / 10

upper airfoil / 141

upper and lower torque arms / 33

upper arm / 74

upper braking pull rod / 33

upper butt fitting / 141

upper cavity / 148

upper cover / 38

upper cover assembly / 33

upper edge strip / 132

upper fairing / 134

upper front panel / 133

upper fuselage / 10

upper joint / 127

upper longeron / 132

upper panel / 132

upper rear panel / 132

upper section / 89

upper semi-frame / 132

upper shroud / 122

upper shroud mount bracket / 122

upper sideband (USB) / 50

upper sideband report / 50

upper skin / 141

upper torque arm / 33

upper/lower access cover / 134

vent bolt / 13

vent connecting disk / 77

vent handle / 22

vent hole / 33

vent pipe / 33,59,183

vent pipe fairing/vent pipe shroud / 141

vent plug / 77,148

vent port / 22

vent pressure / 183

vent screw / 13

vent system / 10,77

vent tank / 33

vent tube / 183

vent valve / 10

vent valve/exhaust valve / 38

ventilation / 171

ventilation condition / 89

ventilation drain valve / 10

ventilation hole / 132

ventilation outlet / 134

ventilation pipe / 171

ventilation pressurizing joint / 127

ventilation system / 171

ventilation valve / 38

ventilation window / 132,136

ventilation window sliding device / 33

ventilator / 59

venting system / 183

verify / 17

vernier / 74

vernier caliper / 59

vernier potentiometer / 13

vertical acceleration / 100

vertical accelerometer / 10

vertical axis / 100

vertical axis line / 178

vertical beam / 141

vertical channel / 44

vertical climbing method / 193

vertical condition / 13

vertical control / 100

vertical deflection / 17

vertical fin / 22

vertical gyro / 44

vertical landing speed / 63

vertical partition / 132

vertical polarization / 50

vertical restraint device / 188

vertical shaft / 33

vertical sliding bracket / 63

vertical sliding rail / 33

vertical speed / 100,114

vertical speed indicator (VSI) / 10,100,114

vertical speed pointer / 114

vertical speed scale / 114

vertical stabilizer / 4

vertical stabilizer area / 10

vertical tail / 50

vertical tail span / 10

vertical tail tip / 63

vertical tail/fin / 10

vertical tracking speed / 44

very high frequency (VHF) / 59

very high frequency (VHF) radio set / 50

very-low altitude / 50

VHF Omni-direction range (VOR) / 59

VHF/UHF radio set / 100

vibration / 13,132,171

vibration absorber / 127

vibration acceleration / 89,106

vibration amplifier / 176

vibration disc /shock absorber disc / 161

vibration frequency / 106

vibration indicator / 59

vibration isolator / 122

vibration mechanism / 114

vibration meter / 11

vibration overload coefficiency / 161

warm-up time / 114

warning / 171,176

warning altitude knob / 114

warning bell / 100

warning bell control mechanism / 74

warning bell ringing button / 100

warning board / 106

warning cancelling inertia / 68

warning cursor / 114

warning electromotive force / 68

warning flag / 114

warning inertia / 81

warning light / 100,114,176

warning placard / 50,81

warning placard/warning board / 50

warning signal / 50

warning signal light / 114

warning signal light built-in test / 114

warning signal siren / 173

warning signal siren placard / 100

warning temperature / 68

warp / 178

warranty period / 63

washbasin / 118

washer / 50,59,127,176

washer/gasket/spacer / 89

washing agent / 132

washing gasoline / 106

washing hole / 13

washing water / 118

washroom oxygen supply signal light / 100

washroom oxygen supply warning bell / 100

waste / 22

waste disposal cart / 118

waste water / 118

watch out the altitude / 193

water bag / 63

water based cleaning agent / 33

water boiler / 63

water bubble / 100

water charging cover / 118

water charging nipple / 118

water drain hole / 101,135,141

water drain pipe / 118

water drainer / 38

water entering pipe / 118

water faucet/water tape / 118

water injection / 161

water jar / 118

water jet / 38

water leakage pipe/water-leaking pipe / 132

water purifying system / 33

water separator / 38

water steam / 89

water supply/waste disposal equipment / 33

water-cooling / 4

waterline / 188

waterpro of cloth / 59

water-soluble acid and soda / 183

watertight sealing / 132

wave beam / 114

wave crest / 141

wave filter / 59

wave trough / 141

wave-carrier / 50

waveform / 59

waveguide / 11,114

waveguide nipple / 114

waveguide tube / 4

waxed paper / 77

waxed ramie string / 59

waypoint / 114

waypoint number setting / 114

weak DC voltage signal / 166

weak running / 193

weapon tactics trainer / 193

weather map/weather chart / 193

weather radar / 114,193

参 考 文 献

[1] 华人杰.英汉航空航天新词典[M].上海:上海科学普及出版社,1999.

[2] 英汉航空词典编写组.英汉航空词典[M].4版.北京:商务印书馆,1998.

[3] 李福崇.英汉航空技术专用缩略语手册[M].成都:成都飞机设计研究所,1997.

[4] 杨伟.英汉航空航天缩略语词典[M].北京:航空工业出版社,2009.

[5] 航空工业部第一技术情报网,英汉航空缩略语词典编辑组.英汉航空缩略语词典[M].西安:陕西人民出版社,1986.

[6] 李公昭.英汉军事大词典[M].上海:上海外语教育出版社,2006.

[7] 清华大学.英汉技术词典[M].北京:国防工业出版社,1986.

[8] 世界国防科技工业概览编委会.世界国防科技工业概览[M].北京:航空工业出版社,2004.

[9] 潘家祯.科技文献检索手册[M].北京:化学工业出版社,2001.

[10] 中国航空工业总公司.货运飞机专用技术设计指南[M].北京:航空工业出版社,1997.

[11] 中国人民解放军总装备部批准.GJB 3968-2000 军用飞机用户技术资料通用要求,General requirement for military aircraft user's technical data[S].2000,04.

[12] ATA Specification 100-Manufacturer's Technical Data, Revision 29, 1990.

[13] Reithmaier, Larry. The Aviation/Space Dictionary[M]. Blue Ridge Summit: TAB Books Inc., 1989.

[14] Paul Garrison. The Illustrated Encyclopedia of Genderal Aviation[M]. Blue Ridge Summit: TAB Books Inc., 1990.

[15] Ocran E B. Dictionary of Air Transport and Traffic Control[M]. London: Granada Technical Books, 1984.

[16] Gunston Bill. Jane's Aerospace Dictionary[M]. London: Jane's Information Group Ltd., 1989.

[17] Military Acronyms. http://www.dtmedia.co.uk/no.htm